The Bereitschaftspotential

Movement-Related
Cortical Potentials

The Bereitschaftspotential

Movement-Related Cortical Potentials

Edited by

Marjan Jahanshahi

Sobell Department of Motor Neuroscience & Movement Disorders
Institute of Neurology
University College London
London, UK

and

Mark Hallett

Human Motor Control Section
National Institute of Neurological Disorders & Stroke
National Institutes of Health
Bethesda, Maryland

Kluwer Academic/Plenum Publishers
New York, Boston, Dordrecht, London, Moscow

Library of Congress Cataloging-in-Publication Data

The Bereitschaftspotential: movement-related cortical potentials/edited by Marjan
Jahanshahi, Mark Hallett.
     p.   cm.
    Includes bibliographical references and index.
    ISBN 0-306-47407-7
     1. Motor cortex.   2. Electroencephalography.   3. Movement disorders.   I. Jahanshahi,
Marjan. II. Hallett, Mark.

QP383.15 .B475 2003
612.8'252—dc21

2002040791

ISBN 0-306-47407-7

©2003 Kluwer Academic/Plenum Publishers, New York
233 Spring Street, New York, New York 10013

http://www.wkap.nl/

10 9 8 7 6 5 4 3 2 1

A C.I.P. record for this book is available from the Library of Congress

Printed in the United States of America

PREFACE

The Bereitschaftspotential (BP), also known as the readiness potential, was first described by Kornhuber and Deecke in 1964. It is a negative wave that was originally recorded from the surface of the scalp beginning 1 to 1.5 s prior to self-paced movements. The BP has become a well-established tool in the motor physiology laboratory, and from recent publications in scientific journals it is evident that the BP is also very much alive and well as a research tool. Its amplitude or latency have been shown to be impaired in neurological disorders such as Parkinson's disease and in patients with focal lesions of the frontal or parietal cortices. There has been a surge of interest in the BP, with new methods for its measurement such as MEG, intracranial recordings, and concurrent EEG and PET/fMRI. The measurement of the lateralised readiness potential has proven to be an interesting and useful refinement.

However, when reviewing the relevant literature, there are still a number of fundamental questions relating to the BP that remain unanswered, such as its likely generators, whether it has several components, and if so whether these reflect the concurrent or sequential activity of different brain regions. The particular factors that affect its amplitude, slope and latency, the neurochemical and pharmacological influences on the BP, the precise nature of the processes that contribute to the BP, and the function, significance or value of the BP remain to be clarified.

The aim of this book is to bring together some of the most important findings from the literature in a single volume, to highlight and address the pertinent outstanding questions, and to identify the key topics for future investigation in this area. This book was put together in honour of Hans Kornhuber and Lüder Deecke. We hope that its contents gives a flavour of the work done on the BP during the last 38 years and is proof of the longevity and importance of Kornhuber and Deecke's discovery.

Marjan Jahanshahi
Mark Hallett
London & Bethesda, 2002

CONTENTS

INTRODUCTION

FROM SURFACE TO DEPTH ELECTRODES

DIPOLE SOURCE MODELING AND THE GENERATORS OF THE BEREITSCHAFTSPOTENTIAL

THE BEREITSCHAFTSPOTENTIAL: WHAT DOES IT MEASURE AND WHERE DOES IT COME FROM?

Marjan Jahanshahi[1] and Mark Hallett[2]

[1]Sobell Department of Motor Neuroscience and Movement Disorders,
Institute of Neurology
The National Hospital for Neurology & Neurosurgery
Queen Square, London WC1N 3BG, UK

[2]Human Motor Control Section, NINDS
NIH Building 10 Room 5N226
10 Center Drive MSC 1428
Bethesda Maryland 20892-1428
USA

INTRODUCTION

The Bereitschaftspotential (BP) is a negative cortical potential which develops beginning 1.5 to 1 s prior to the onset of a self-paced movement (see Figure 1). The BP was first described by Kornhuber & Deecke in 1964, which makes it about 38 years old. Before the days that citation indices and impact factors came into vogue, one criterion for the significance of a research finding or paper was whether it led to other research studies. By this criterion, the BP has been incredibly influential. The BP has become a well-established tool in the motor physiology laboratory (Marsden et al, 1986), and from recent publications in scientific journals it is evident that the BP is also very much alive and well as a research tool. Its amplitude, slope, and/or latency have been shown to be impaired in neurological disorders such as Parkinson's disease (Dick et al, 1989; Jahanshahi et al, 1995; Cunnington et al, 1995; Praamstra et al, 1996a), Huntington's disease (Johnson et al, 2001), dystonia (Van der Kemp et al, 1995; Deuschl et al, 1995), cerebellar disease (Shibasaki et al, 1978; Verleger et al, 1999; Wessel et al, 1994), and psychiatric disorders such as schizophrenia (eg Singh et al, 1992; Fuller et al, 1999; Northoff et al, 2000) and depression (Khanna et al, 1989; Haag et al, 1994) and in patients with focal lesions of the thalamus (Shibasaki, 1975; Green et al, 1999; Feve et al, 1996), basal ganglia (Feve et al, 1994; Kitamura et al, 1996) cerebellum (Shibasaki et al, 1978; Ikeda et al, 1994; Gerloff et al, 1996), prefrontal (Shibasaki, 1975; Singh & Knight, 1990; Honda et al, 1997) or parietal (Knight et al, 1989; Singh & Knight, 1993) cortices. In the last decade, there has been a surge of interest in the BP with the demonstration that besides simple movement parameters such as force (Kutas & Donchin, 1980) and rate (Mackinnon et al, 1996) higher order motor processes such as movement complexity (Benecke et al, 1985; Simonetta et al, 1991) and mode of movement selection (Jahanshahi et al, 1995; Touge et al, 1995; Praamstra et al, 1996a; Dirnberger et al, 1998) affect its amplitude or slope. In the last decade or so, methods other than scalp electroencephalography (EEG) have been used to quantify the BP or its equivalents measured with magnetoencephalography (MEG, Cheyne et al, 1989; Kristeva et al, 1991; Erdler et al, 2000), intracranial EEG recordings (Neshige et al, 1988; Ikeda et al, 1994; Rektor et al, 1994; Jahanshahi et al, 2001), combined EEG and positron emission tomography (PET, Jahanshahi et al, 1995; MacKinnon et al, 1996), combined EEG and functional magnetic resonance imaging (fMRI, Ball et al, 1999), combined EEG and MEG (Nagamine et al, 1994) or combined MEG and PET (Pederson et al, 1998). The discovery of the lateralised readiness potential (LRP) in the late 1980s (Coles & Gratton, 1986; Smid et al, 1987), has proven to be a useful refinement not only in the study of motor preparatory processes but also in covert cognitive processing.

However, when reviewing the relevant literature, there are still a number of fundamental questions relating to the BP that remain unanswered, such as its likely generators, how to dissect its different components, and whether these reflect the concurrent or sequential activation of different brain regions. Furthermore, the precise nature of the processes that contribute to the BP also remains to be clarified.

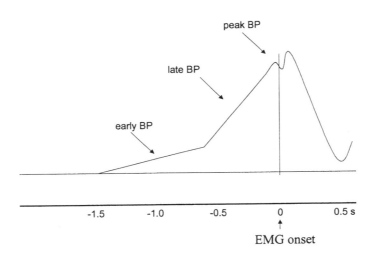

Figure 1:   A schematic representation of the time course and components of the Bereitschaftspontential (BP) prior to movement onset.

AIMS OF THE BOOK

In editing this book we had several aims. First, to bring together some of the most important findings from the literature on the BP and other movement-related cortical potentials in a single volume. However, the coverage is not intended to be comprehensive. For example, the cellular mechanisms of generation of slow potentials, the biochemical and pharmacological influences on BPs, and models of BP generation are not extensively covered. Developmental studies of the BP (Chiarenza et al 1995) are not included. Also, with exceptions (see chapters by Ikeda & Shibasaki, Pfurtscheller & Neuper), the main focus is on recording of the BP prior to hand movements, and BPs prior to movements of other effectors such as the feet and the eyes are not extensively covered. Similarly, because of space limitations, we have been unable to ask all investigators who have been prominent in the field to contribute to the book, but we hope that their contribution is reflected in the content of the chapters.

Our second aim was to include some information about motor preparatory and anticipatory processes from other key sources such as animal studies and investigations using functional imaging with PET or fMRI or MEG, as well as other EEG measures that have been employed to assess motor preparatory processes such as recordings of the contingent negative variation (CNV), event-related desynchronization (ERD) and the Lateralized Readiness Potential (LRP). Again, because of space limitations, other methods such as reaction time or computational approaches to the study of motor preparation have not been included. Single cell recordings in animals, PET/fMRI, MEG and EEG approaches to investigating motor preparatory processes each have their advantages and disadvantages.

The spatial and temporal precision provided by single cell recordings in primates is perhaps unsurpassed by any of the other techniques. However, as outlined in the chapter by Wise, interpretation of the functional significance of single-cell data can be complex due to conceptual problems such as representation of functions by parallel distributed networks rather than by single brain areas, the difficulty of distinguishing sensory from motor signals, and the large number of potentially confounding spatial variables. With these caveats in mind, the chapter by Wise considers (i) the contribution of ramp-to-threshold activity in response selection (ii) the build up of anticipatory activity resulting in motor readiness, a non-specific increased tendency to act, or as a form of attentional filter suppressing inputs that would trigger inappropriate responses and (iii) how elegant experimental designs can demonstrate that delay-period or set-related activity in the premotor cortex or posterior parietal cortex reflect motor preparation rather than alternative processes such as spatial working memory or spatial attention. Animal lesion studies have also been important in providing information relevant to the BP. For example, in the monkey, such lesion studies have established that the surface negative, depth-positive field potentials in the motor cortex that precede self-paced movements are abolished or reduced following unilateral cerebellar hemispherectomy (Sasaki et al, 1979) and that the same effect is produced by temporary cooling of the dentate nucleus (Tsujimoto et al, 1993).

Compared to the EEG, the MEG has the advantage of adding good spatial resolution to temporal resolution in the ms range, thus allowing better localization. The main drawback is, of course, that radial sources are not reflected in MEG activity which primarily reflects intrasulcal activation (Lopes da Silva, 1996). MEG evidence relevant to the BP are discussed in several chapters particularly those by Mackinnon, Toma & Hallett, Kristeva-Feige, and Deecke & Kornhuber. The advantage of PET and fMRI is that they provide simultaneous information about the functional anatomy of the whole brain both cortical and subcortical structures. This together with the high spatial resolution means that patterns of functional and effective connectivity between components of distributed networks can be examined. The advent of event-related fMRI, in more recent years has added the benefit of better temporal resolution in the seconds range. The contributions of PET, fMRI and event-related fMRI to investigation of motor preparatory processes is the topic of the chapter by Toni and Passingham and also covered by MacKinnon, Kristeva-Feige and Deecke & Kornhuber in their chapters.

Other EEG-derived measures of movement-related activity such as the CNV, ERD and LRP have been as influential and crucial in clarifying movement-related activity as the BP. The CNV first described by Walter et al (1964), is a slow negative wave that develops in the interval between a warning (S1) and a "go" (S2) stimulus. As outlined in the chapter by Brunia, the CNV is considered to reflect anticipation for a forthcoming signal and preparation for execution of a response, as well other processes such as time estimation and uncertainty. In addition to the comprehensive chapter by Brunia, information about the CNV is provided in a number of other chapters. In their chapters, Ikeda & Shibasaki and Rektor provide useful information about the CNV compared to the BP from intracranial recordings. The results of some of the studies which have recorded the CNV in patients with movement disorders are considered in the chapter by Praamstra, Jahanshahi & Rothwell. In the final chapter by Deecke & Kornhuber, the effect of subcortical surgery in patients with Parkinson's disease on the CNV is reported.

The term ERD was introduced by Pfurtscheller and Aranibar (1977) who described techniques for measuring it. Alpha activity is considered to reflect relaxation and prevalence of alpha in the EEG is therefore equated with reduced activation of underlying brain areas. Conversely, increased activation of underlying brain areas is assumed to block or decrease the amplitude of alpha activity. Thus, while desynchronisation of EEG alpha frequencies (ERD) is considered to reflect cortical activation (Niedermeyer, 1997), synchronisation or the increase of alpha activity (ERS) characterises cortical areas at rest or in an idle state (Pfurtscheller, 1992). In their chapter Pfurstcheller & Neuper consider the effect of movement-related factors such as movement of different effectors, active voluntary movement versus passive somatosensory stimulation, and higher order processes such as internal vs external pacing of movements and motor imagery on ERD, and the conclusions about oscillatory activity in the motor system based on this evidence. The co-occurrence of focal ERD with ERS in neighbouring areas (focal ERD/surround ERS) and their possible functional significance as a mechanism for attentional focusing on a motor subsystem, as well as the issue of cross-talk' between sensorimotor areas are also covered. ERD assessed during intracranial recording is also considered in the chapters by Rektor, Ikeda & Shibasaki who note interesting dissociations between measures of ERD and the BP.

The LRP, first described in the late 1980s (Coles & Gratton, 1986; Smid et al, 1987), is a measure of the lateralised section of the BP and is considered to reflect preferential motor preparation of one hand prior to movement initiation and execution. In their lucid chapter, Eimer & Coles demonstrate why the LRP has become an important measure of covert cognitive processing and illustrate this with its applications to investigate subliminal perception, implicit learning, stimulus-response compatibility, the Simon effect, serial versus parallel models of information processing and to investigate the processes of withholding/inhibiting a response after its release by a 'go' stimulus. The chapter by Praamstra, Jahanshahi & Rothwell demonstrates how measurement of the LRP in patients with Parkinson's disease has allowed further exploration of motor preparatory processes in this disorder beyond that possible with the BP.

Our third aim in this book was to collate relevant information about the generators of the BP and processes that contribute to it from a variety of sources. Thus the chapters reflect several different but complementary approaches to addressing the issue of the likely sources or generators of the BP. First, source density analysis to identify the generators is discussed in some detail in the chapters by Toma & Hallett, MacKinnon, and Kristeva-Feige. Second, while the topographical distribution of the BP in scalp recordings provides less reliable information about the generators of the BP because of the distortions possible with volume conduction, intracranial recordings constitute more reliable information about this. The chapters by Ikeda & Shibasaki and Rektor describe intracranial recordings of the BP and consider possible cortical and subcortical sources. Finally, some albeit, less direct information about the possible sources of the BP can be obtained from the study of patients with focal lesions of the brain (see chapter by Gerloff) or investigations of the BP in clinical populations such as patients with movement disorders where the pathophysiology is relatively well mapped out (see chapter by Praamstra, Jahanshahi & Rothwell) or in psychiatric populations such as schizophrenia where evidence has highlighted the importance of specific configuration of symptoms and medication on the BP (see chapter by Westphal).

Fourth, we hoped that the chapters would highlight some of the outstanding issues and pertinent questions that may light the way for future research on movement-related cortical potentials. To this end, a number of chapters end with consideration of some of the questions that need to be addressed in future studies. Below, we have highlighted the main questions that remain unanswered and have provided a summary of the key issues for future work raised in the various chapters.

Finally, while it is customary to honour prominent members of the scientific community on retirement (or even after death!), with Lüder Deecke and Hans Kornhuber, we considered it timely to do so while they are still scientifically active and engaged in research. The fact that interest in the BP continues to be as strong today as when it was first reported by Kornhuber & Deecke (1964, 1965), almost forty years ago, is proof of the longevity and importance of their discovery. We hope that the contents of this book convey a flavour of the way in which research on the BP has progressed in the intervening years.

There is a degree of overlap in the content of some of the chapters in the terms of the pertinent issues discussed or the literature reviewed. Where it exists, we have not removed such overlap, because the same empirical results are often considered from a different perspective by the various contributors.

WHAT IS THE BP A MEASURE OF?

What physiological and psychological processes does the BP measure? The BP is an event-related potential. That is, the onset of the BP is time-locked to an event. The event which 'elicits' the BP is an actual, intended or imagined movement. This means that the BP is a movement-related potential. Given that the initiation of the BP *precedes* the onset of movement by 1.5 to 1 s, in the original formulation by Kornhuber & Deecke (1965), the BP was portrayed as a 'readiness potential', that is as an index of motor preparation. On the basis of evidence of the types of tasks or experimental manipulations that affect the amplitude or slope of the BP, subsequent investigators have considered the BP to reflect a host of other processes. If the argument that the late CNV wave and the BP are identical is accepted (see chapter by Brunia), then the set of cognitive, motivational and motor processes such as anticipation and expectancy (Gaillard, 1977), attention, preparatory set, time estimation, information processing, and motivation which are considered to be reflected by the CNV (Rockstroh et al, 1982) could also be pertinent to the BP. Others have suggested that the BP reflects an intention to act which remains unconscious for a proportion of its time course (Libet et al, 1983), an index of resource mobilisation (McCallum, 1993), effort (Wessel et al, 1994, Kristeva, 1991), and timing of movements (Deecke et al, 1985). For example, the evidence of cortical negativity associated with imagery (Uhl et al, 1990), together with the sensitivity of the BP to inertial load (Cui et al, 1996) and greater activation of the SMA and larger readiness fields with MEG observed for movements of the impaired than the unimpaired hand in a patient with hemi-parkinsonism (Cunnington et al, 2000) has led to the proposal that the BP reflects the degree of effort associated with movement (see chapter by Deecke & Kornhuber). On the basis of evidence showing that when task presentation is made contingent on increased surface negativity in key cortical areas, subsequent performance is facilitated (eg Bauer & Nirnberger, 1981), in his chapter, Lang suggests that the decision 'when to act' may be made contingent on spontaneous increases of cortical excitability in relevant motor areas.

It is possible that different frontal and prefrontal areas differentially contribute to these various cognitive, motivational, and motor processes involved in the BP (Jahanshahi et al, 2001). For example, while prefrontal areas may be engaged by the decision-making necessary for response selection, timing or initiation/suppression of action; the pre-SMA/cingulate motor area and lateral premotor cortex may be involved in preparatory processes; in contrast to the activation of the SMA proper and motor cortex which may be more directly associated with the actual initiation and execution of the movement. An important first step in identifying the generators of the BP may be clarification of the specific motor, cognitive and motivational processes that contribute to its genesis. This would require development and use of tasks that would allow dissociation of the different processes associated with recording of movement-related potentials. For a discussion of the processes that contribute to the BP see the chapters by Lang, Rektor, Ikeda & Shibasaki, MacKinnon, Libet, Deecke & Kornhuber, as well as those by Brunia and Eimer & Coles.

ONE OR MORE BP COMPONENTS?

In their original and later publications on the BP Kornhuber & Deecke (1964, 1965) made a distinction between a slowly rising phase of the waveform and a change in the steepness of the slope of the BP which occurred from around 400 to 500 ms prior to movement onset. It was proposed that the early slowly rising negativity and the later change of slope respectively reflected activation of the SMA and M1 (Deecke et al, 1976; Deecke, 1987). In the literature different terminology have been used to refer to these early and later phases of the BP. The early slowly rising negativity has been variably referred to as the BP, early BP, BP1, and negative slope 1 (NS1). The late BP, BP2, NS', NS2, and motor cortex potential are the terms applied to the second phase of negativity. The early and late BP components can be discerned in subdural recordings as well (eg Neshige et al, 1988; Jahanshahi et al, 2001). In a later publication, Deecke, Scheid & Kornhuber (1969) distinguished a third negative component occurring about 50 to 60 ms prior to movement onset, the Motor Potential (MP), which is the point of maximum negativity over the hand area contralateral to the moving hand, also referred to as the peak BP or peak NS'. Different features of the MP have also been described, such as the parietal (ppMP) and the frontal peak (fpMP) (Tarkka & Hallett, 1991). In addition, the pre-motion positivity (PMP), is also discernable as a sudden positive deflection in the EEG mainly over parietal electrodes about 80 to 90 ms prior to movement onset (Deecke, Scheid & Kornhuber, 1969). Movement onset is followed by a positive shift referred to as the reafferent potential. Instead of imposing uniformity of terminology across the chapters, we have allowed the contributors to retain the terms commonly used by their group, since this mirrors use of different terminology in the literature.

Use of diverse terminology (see the chapters by Lang and also Ikeda & Shibasaki for a discussion of this) is accompanied by differences in opinion about whether one or two or more components can in fact be discerned and whether the components reflect different processes and/or contribution of diverse generators (see below). Some (Botzel and Schulze, 1996) have even suggested that the distinction between the early, late and peak BP is arbitrary. As noted by Lang in his chapter, the various BP components are usually clear in group averages but not always in the averaged BP waveforms of individual subjects. However, the original distinction between different BP components was not solely based on observation of a change of slope of the waveforms. The early, late and peak BP differ in terms of their distribution over the scalp. While the early BP is bilaterally symmetrical, the late and peak BP are asymmetrically distributed and maximal over the contralateral precentral areas (eg Deecke, Scheid & Kornhuber, 1969; Shibasaki et al, 1980). These issues are discussed in the chapters by Lang, Deecke & Kornhuber, MacKinnon, Ikeda & Shibasaki. Furthermore, there is experimental evidence showing dissociation of the different BP components, thus supporting their distinction and the possibility that they relate to different physiological processes. We will summarize some of the evidence which supports such a distinction.

- The early and late BP components are differentially sensitive to pharmacological manipulation. In normal subjects, acute administration of dopaminergic medication increases the amplitude of the 'early' but not the 'late' BP (Dick et al, 1987). A similar effect is obtained in patients with Parkinson's disease (Dick et al, 1987).
- Movement disorders such as Parkinson's disease and cerebellar disease differentially impair the BP components, suggesting that the striatal-frontal and cerebello-frontal circuits primarily contribute to specific BP components. When patients with Parkinson's disease are assessed off medication, the amplitude of the early but not the late BP is reduced (Dick et al, 1989; Jahanshahi et al, 1995). In contrast, in patients with cerebellar degeneration or atrophy, it is the late BP which is absent or reduced (Tarkka et al, 1993; Wessel et al, 1994).
- The early and late BP components are also differentially affected by various experimental manipulations. For example, the nature of the movement, whether it is self-initiated or externally triggered by a stimulus affects the late but not the early BP component both prior to finger movements (Jahanshahi et al, 1995) and saccades (Jahanshahi et al, 2000). In contrast, the regularity of the movement, which affects the extent of motor preparation possible influences the early BP recorded prior to finger movements (Jahanshahi et al, 1995) or saccades (Jahanshahi et al, 2000).

WHERE DOES THE BP COME FROM?

Increase of extracellular K^+ concentration and decrease of Ca^{++} concentration activate brain regions and result in negative slow potentials (Caspers, 1980). Thus, surface negative potentials such as the BP and CNV are considered to represent neuronal activation, whereas surface positive potentials such as the P300 reflect neuronal inhibition (Skinner & Yingling, 1977). There is evidence that slow negative cortical potentials such as the BP represent excitatory post-synaptic potentials (EPSP) (Caspers & Speckman, 1974). While the BP is considered to primarily arise from neuronal activity, it has been proposed that glial processes also make a major contribution to the genesis of slow potentials such as the CNV and the BP (Bauer et al, 1993; Laming, 1989; 1993; Roitbak et al, 1987).

In terms of the likely sources of the BP, two main questions have preoccupied investigators in this field. First, what are the most likely generators of the BP? Second, what is the time course of activation of these generators? We will briefly consider each of these questions.

(i) What are the most likely generators of the BP?

With the rapid technical developments that have occurred in the last two decades, the search for the generators of the BP has become gradually more sophisticated. From reliance on topographical analysis of a limited number of scalp electrodes, investigators have moved to high density EEG, commonly with 64 electrodes (eg Cui et al, 1999). With the advent of MRI, co-registration of EEG and MRI data (eg Ball et al, 1999) has provided the possibility of more accurate and precise localization. This expansion and improvement of recording techniques has been coupled with better approaches and methods for source analysis, the latest of which, distributed source modeling has the advantage that it does not require *a priori* assumptions about the number or location of sources (see chapter by Kristeva-Feige).

In the literature, three main hypotheses about the likely generators of the BP can be identified (see MacKinnon chapter). These are:

1.   The early BP is generated by the SMA, whereas the late BP is generated by later activation of the contralateral motor cortex.

2.   Both the early and late BP are generated by bilateral activation of the motor cortex with no or minimal contribution from the SMA.

3.   Both the early and late BP are generated by bilateral activity in the motor cortex as well as the SMA.

In relation to these hypotheses, the term SMA should be considered to include both pre-SMA and SMA proper as well as the cingulate motor area (CMA) (see chapters by Kristeva-Feige, Ikeda & Shibasaki, Deecke & Kornhuber). The EEG and MEG evidence relevant to these hypotheses are discussed in the chapters by MacKinnon, Toma & Hallett, Kristeva-Feige, Lang, Ikeda & Shibasaki, Deecke & Kornhuber. From the information covered in these chapters it is evident that the number and precise location of the dipoles identified crucially depends on the methodology used. First, while the majority of studies (eg Ball et al, 1999; Cui et al, 1999; Toro et al, 1993) have focused on dipole source modeling of slower movements separated by at least 5 s or longer which are most commonly used to measure the BP, others have investigated the dipole sources of fast repetitive movements (eg Gerloff et al, 1998a,b; MacKinnon et al, 1996). Second, the particular BP components modelled differ across studies. For example, the dipole source modelling study of Toro et al (1993) was concerned with the interval from 200 ms before to 200 ms after movement onset and as a result focused on the peak BP and the post EMG motor potential, and therefore has no implications for the likely generators of the earlier BP components. In contrast, others (eg Boetzel et al, 1993; Praamstra et al, 1996b; Cui et al, 1999) have used longer time intervals from 1500 ms before movement onset. Third, the specific nature of the movement used differs across studies which may influence the degree of involvement of the SMA which is particularly engaged by more complex and bimanual movements. While the majority of studies have used simple movements such as brisk movements of the right index or middle finger (eg Boetzel et al, 1993; Ball et al, 1999; Cui et al, 1999; Pedersen et al, 1998; Toro et al, 1993), Praamstra et al (1996b) applied dipole source modelling to data collected in a 'mode of movement selection' paradigm comparing free selection of one of 4 possible movements with stereotyped performance of a single movement. Fourth, the various studies differ in terms of crucial aspects of the dipole source modelling process, such as the head model, the source model and the source space models employed, the *a priori* assumptions about the number, location or direction of sources and hence the constraints imposed.

The BP has also been recorded from intracranial electrodes positioned in the vicinity of subcortical structures such as the basal ganglia and the thalamus (see Chapter by Rektor). This work shows that the BP can be recorded in the caudate, putamen and pallidum and some areas of the thalamus. It is possible that the BP recorded on the scalp in surface recordings actually represents the summed activity of dipoles generated in subcortical as well as cortical structures (see chapter by Rektor). This proposal has empirical support not only from studies of the BP in patients with movement disorders such as Parkinson's disease and cerebellar disease (See chapters by Praamstra, Jahanshahi & Rothwell, Ikeda & Shibasaki, Deecke and Kornhuber), but also from animal studies. Single unit recordings in the globus pallidus of cats (Neafsey et al, 1978) and the caudate nucleus or putamen of monkeys (Shultz & Romo, 1992) have shown movement-related activity prior to self-initiated movements in these subcortical structures. As noted above, cerebellar hemispherectomy in monkeys abolishes the surface negative, depth positive depth potentials recorded in the motor cortex prior to self-paced movements (Sasaki et al, 1979).

(ii) What is the time course of activation of the SMA and M1?

Okano & Tanji's (1987) recordings of neuronal activity in the SMA, premotor cortex, and M1 prior to self-paced and visually triggered movements in monkeys is of particular relevance to the question of the time course of the contribution of these areas to the BP. They observed two types of neuronal activity in these areas: long-lead activity which started 1 to 2 s prior to movement onset and short-lead activity which occurred within 480 ms prior to movement onset. The long-lead neuronal activity which was particularly prevalent among SMA neurons was mainly observed prior to self-paced movements and less frequently before triggered movements. In contrast, short-lead activity was observed in neuronal recordings in the SMA, premotor cortex, and M1 and prior to both self-paced and triggered movements. Subsequently, Shima et al (1991) reported such long-lead neuronal activity occurring 500 ms to 2 s prior to self-paced movement in single cell recordings in the anterior cingulate as well.

With regard to the time course of activation of the SMA and M1 prior to self-paced movements used to record the BP in man, two main hypotheses have dominated the field. According to the first hypothesis, the BP reflects serial activation of the SMA and M1, with SMA activation preceding the M1 activation. The second hypothesis proposes that both the SMA and M1 are active in parallel and that the SMA activation is sustained until shortly before or after onset of EMG activity. The relevant issues and pertinent evidence concerning these hypotheses are discussed in the chapters by Lang, MacKinnon, Kristeva-Feige, Toma & Hallett, Ikeda & Shibasaki, and Deecke & Kornhuber. As reviewed in these chapters, the current source density analysis by Cui et al (1999) and Ball et al (1999), the concurrent MEG and PET data of Pedersen et al (1998), and the event-related fMRI studies by Cunnington et al (1999, 2002) are some of the evidence that provide support for SMA activation preceding M1 activation during the BP recorded prior to self-paced movements.

WHAT MAY THE FUTURE HOLD?

We hope that the content of the book will convey some of the important work on the BP and motor preparation that has been completed in the last 38 years since the BP was first described. Below are some of the issues that have been flagged up by the different contributors that need to be addressed in future.

- ### *The BP paradigm*

The classical BP paradigm requiring self-paced movements of a finger separated by 5 s or more has a limitation. The nature of the movement is pre-determined and the rate is pre-defined. This can lead to relative 'automatic' performance of the movements across trials. Real life movements are faster, more spontaneous and more complex. The demonstration that the BP can be recorded prior to spontaneous movements (Keller & Heckhausen, 1990), prior to sequential button pressing (Cunnington et al, 1995), or when one movement among a set of possible responses has to be selected (Praamstra et al, 1996a; Touge et al, 1995; Dirnberger et al, 1998) demonstrates that the use of less constrained movements for recording of the BP is feasible. The use of more object or goal-directed movements in future studies is raised by Lang and Kristeva-Feige. One of the limitations of research on response selection and motor preparation using reaction time (RT) paradigms is that while these indices of overt behaviour measure the interval between stimulus presentation and response initiation, the 'covert' processing in the 'black box' intervening between the stimulus and response remains untapped. The lateralized readiness potential which has provided a tool for charting the time course of such covert processing is likely to be employed for revealing the inner workings of the 'black box' while engaged in cognitive and motor tasks (see chapter by Eimer & Coles).

- ## *Technological and methodological developments*

The improved spatial resolution afforded by high density EEG combined with better approaches and methods for source analysis has been instrumental in identifying the likely generators of the BP. With the improved temporal resolution of event-related fMRI, Cunnington et al (1999) have demonstrated that it is possible to record a BP equivalent in the hemodynamic response and to chart the time course of such a 'Bereitschafts-BOLD' response. Continuous EEG concurrent with event-related fMRI seems to be a reality now (Lemieux et al, 2001), a development which can pave the way for more precise localization and more accurate information based on topographical and source analysis. Brain areas do not operate in isolation and even the simplest movements are performed through the coordinated activation of distributed neuronal assemblies. With such concurrent EEG and fMRI it would be possible to examine patterns of functional and effective connectivity in cortical and subcortical networks involved in the generation of the BP. For example, using fMRI it has been shown that the pre-SMA, CMA, SMA proper as well as the lentiform nucleus are activated with a self-initiated sequence of movements (Cunnington et al, 2002) and it should be possible to map such subcortical-cortical connectivity in normals and alterations of it in patient groups in future studies. Future research on the BP is likely to benefit from a multimodal approach, combining EEG/MEG with PET/fMRI (see chapters by Toma & Hallett, Kristeva-Feige, MacKinnon).

- ## *Other generators of the BP*

The concentration of cortical motor areas, M1, SMA, CMA and lateral premotor cortex in the frontal cortex means that this part of the brain has been the main focus of BP studies. The contribution of the parietal cortex to the BP is suggested on the basis of electrophysiological and functional imaging evidence (see chapters by MacKinnon, Rektor, Gerloff, Toni & Passingham). The contributions of cortical regions such as lateral intraparietal area or subcortical areas such as the striatum, thalamus and the cerebellum to the BP have been less widely studied in man than frontal areas and will probably be more extensively investigated in future. More anterior areas of the prefrontal cortex such as the dorsolateral prefrontal cortex may also contribute to the BP, possibly depending on the nature of the movement particularly when decision-making or selection about the nature or timing of the movement is involved. This prefrontal involvement in the BP also requires further evidence (see chapters by Ikeda & Shibasaki and Rektor). The contribution of the ipsilateral M1 to the BP also clearly necessitates further investigation (see chapters by Mackinnon, Toma & Hallett, Rektor). As noted above, to date, the search has been for distinct generators of the BP, the technological and methodological advances of recent years point the way towards mapping the interactions of the distributed networks involved in BP generation.

- ## *Clarification of the processes that contribute to the BP*

Design of appropriate experimental paradigms has allowed the dissociation of anticipatory attentional processes from motor preparation. For example, as described in the chapter by Brunia, the 'Stimulus Preceding Negativity' (SPN) which can be recorded prior to stimuli providing 'knowledge of results' (Brunia & Damen, 1989) as well as before instruction cues with predictable onset times (see Bocker & Brunia, 1997 for review) reflects anticipatory attention unconfounded by motor preparation. In more recent years, similar attempts at dissociating the processes that contribute to the BP have been made. For instance, Dirnberger et al (2000) examined BPs prior to regular (same movement repeated across trials), alternating (alternated between right or left index button presses) and random (randomly selected to press buttons with left or right index or middle fingers) tasks. The alternating and random tasks which involved the same degree of switching between movements across trials and hence considered similar in terms of motor preparation were associated with BPs with higher amplitudes than the regular task in which the same movement was performed across trials. The additional attentional and working memory demands of the random task resulted in greater negativity over frontal areas than the alternating task. Given the host of processes other than motor preparation proposed to contribute to the BP such as anticipation, attention, intention, motivation, effort, and timing it remains a mission for future research to design and employ experimental tasks that would disentangle the relative contributions of these processes to the latency, amplitude, and slope of the BP.

- *Models of BP generation*

Is the BP simply an electrophysiological correlate of motor and cognitive processes such as preparation, attention, timing, expectancy and anticipation or motivational states such as intention to act and effort, or does it have functional significance independent of these states? The circularity of explaining slow potentials such as the BP or CNV in terms of such psychological processes has been previously noted (Rockstroh et al 1982). More comprehensive models of why and how such slow potentials are generated are required. Despite its longevity, few models of how the BP is generated have been proposed.

Skinner & Yingling (1977) suggested that the thalamo-cortical system play a role in the regulation of attention and the EEG slow waves associated with it. According to this model, the mesencephalic reticular formation (MRF) and the mediothalamic frontocortcal system (MTFCS) play a prominent role in the regulation of attention and arousal. The MRF and the frontal cortex converge on the thalamic reticular nucleus, which exerts an inhibitory 'gating' function over the thalamic relay nuclei to the cortex. Skinner & Yingling (1977) showed that bilateral cryogenic blockade of the interconnections between the frontal cortex and medial thalamus abolished slow potentials of all origins in the frontal cortex, but did not affect the MRF-elicited ones. These results suggest that the activation of the MRF plays a role in the initiation of slow potentials over the frontal cortex through inhibition of the reticular nucleus. According to the dipole model proposed to explain the generation of negative slow potentials recorded over the scalp, the dendrites of pyramidal cells are activated by afferent fibres from the thalamus. This results in extracellular volume currents which in turn generate the negative potentials recorded in the surface EEG (Hari, 1980).

Rockstroh and colleagues (1982) proposed a model according to which slow potentials such as the CNV are directly related to the amount of "cerebral potentiality", defined as "a psychophysiological state necessary for any cerebral performance... which corresponds to a facilitation in neuronal firing or the increase in the likelihood of neuronal firing". It is proposed that neuronal excitability is increased during slow potentials and that this facilitation is produced by slow excitatory post synaptic potentials (EPSP) in apical dendrites and by glial contribution. As in the model of Skinner & Yingling (1977), the MRF and MTFCS are considered to be involved in the generation of slow potentials and in the regulation of cerebral potentiality. This 'cerebral potentiality' model was further elaborated and incorporated in the 'threshold regulation' model proposed by Elbert &

Rockstroh (1987), Birbaumer et al (1990), and Elbert (1993). According to the 'threshold regulation' model, "tuning of cortical excitability is achieved via control of the depolarisation in the apical dendrites which give rise to surface negative potentials in the EEG." (Elbert & Rockstroh, 1987). This tuning is proposed to be achieved by the thalamocortical fibres which synapse with the apical dendrites. Information about activity in the networks to be regulated is transmitted to the thalamus from the striatum which receives extensive information about cortical activity. Lowering the threshold for cortical excitability necessitated by anticipation and by self-paced movements is assumed to give rise to increased surface negativity, whereas conversely when thresholds are set high, surface positivity results. These are respectively considered to reflect a 'controlled' mode of processing when the EEG is under the control of the MTFCS and ERD occurs and an 'automatic' mode when threshold regulation is also automatic and the EEG is characterized by a predominance of oscillation in the alpha band.

While these models of slow potentials are not without their critics, they are nevertheless important in generating testable hypotheses and furthering our understanding of the BP. Future refinement of existing models and proposal of new models would be of value.

- ### *Brain plasticity and the BP*

The BP is not static and its latency, amplitude, and slope can be altered by disease or brain injury or conversely improve with clinical recovery, for example following recovery from depression or stroke or after thalamotomy or pallidotomy in patients with Parkinson's disease (see chapters by Westphal, Gerloff, Ikeda & Shibasaki, Praamstra, Jahanshahi & Rothwell). Such plasticity underlines the need for longitudinal follow-up studies of the BP in such disorders and also highlights that the mechanisms responsible for BP generation can show plastic changes. Such plasticity of negative movement-related potentials is also evident in the course of motor learning, whereby the amplitude of the BP or negative slow potentials recorded during movement execution are altered from the early phase of learning to the later phase of skilled performance (Lang et al, 1988). The latency of the LRP has also proved to be a sensitive index of implicit and explicit learning (Eimer et al, 1996; Russeler & Rosler, 2000) (see chapters by Lang, Eimer & Coles). Relatively little work has been done on learning or recovery of function using movement-related potentials such as the BP and the LRP. In light of the current interest in the functional anatomy of motor learning and the mechanisms of spontaneous or treatment-induced recovery of function following psychiatric or neurological illness, brain injury or stroke, this is likely to be a fruitful focus for future research endeavour.

- ### *Neurochemical and pharmacological influences on the BP*

While considerable work has been done on the neurochemical basis of slow cortical potentials such as the CNV in animals, pharmacological studies of these potentials in man are less extensive, particularly for the BP. Animal studies of neurochemical contributions to negative slow potentials have the advantage of precision and specificity of the sites and mechanisms of action, but this work has often been conducted on restrained animals under conditions which somewhat differ from those used for recording of the CNV and BP in man Conversely, studies on pharmacological influences on the CNV or BP in man suffer from systemic application and lack of specificity (Birbaumer et al, 1990). Pharmacological substances that stimulate excitatory neurotransmitters such as the cholinergic and serotonergic increase the amplitude of the CNV while those that operate on inhibitory neurotransmitter systems such as the GABAergic, reduce the amplitude of the CNV (Rockstroh et al, 1982). The individual's state of attention and arousal is also important in mediating drug effects on the CNV (Tecce et al,1978). Less information is available on pharmacological influences on the BP and the effect of dopamine agonists and antagonists and anticholinergics on the BP are discussed in the chapters by Praamstra, Jahanshahi & Rothwell, and Westphal. This is an area which clearly requires further investigation through animal experimentation and studies in man.

- ● *What is the function/value of the BP?*

Issues relevant to the functional significance and evolutionary value of the BP are discussed in the chapters by Deecke & Kornhuber, Lang, Libet, and MacKinnon. Since similar negative slow potentials as the BP are found in monkeys as well as man, the question of the evolutionary advantage or significance of the BP arises. If the BP is an index of anticipatory attention and motor preparation, then it is possible that these processes conferred some advantage or had some survival value for the animal, for example through mobilization of resources and priming of sensorimotor systems for impending action. Rockstroh et al (1982) speculate that slow brain potentials may constitute a third electrophysiological communication system in the brain operating in parallel with the fast-acting and 'fixed-wired' neuronal network and the more diffuse and slow-acting biochemical system. According to the threshold regulation model of Elbert & Rockstroh (1987), the increase and decrease in cortical activation reflected by negative and positive slow potentials would operate to dampen the danger of over-excitation in interconnected cell assemblies and neural networks.

What is the function of the BP? The work of Libet (1992) has shown that the conscious awareness (W) of an intention to act occurs some time, on average between 850 to 350 ms, after the onset of the BP, depending on whether the movements were perceived as pre-planned or spontaneous. From this finding two inferences are possible. First, that the decision or intention to act, the preparation for and initiation of movement as indexed by the onset of the BP occur unconsciously. Second, that the cortical negativity reflected by the BP needs to reach a certain amplitude or point in its time course before the prior 'intention to act' is consciously perceived. In this respect, it is of interest that movements that are performed automatically with limited conscious awareness are not preceded by a negativity, which has led some to propose that "slow negativities are a prerequisite for conscious processes" (Elbert, 1993). More recent evidence showing that in the Libet paradigm, the onset of the LRP was earlier on trials with early awareness than on trials with late awareness of the intention to act suggests that processes underlying the lateralised component of the BP, that is the LRP may give rise to awareness of movement initiation (Haggard & Eimer, 1999). Libet's results have implications for our conceptualisation of 'free will' and he proposes that this may primarily operate as a control mechanism selecting or vetoing actions through to completion. These results suggests that many of the brain's so-called 'voluntary' processes may be unconsciously initiated which may be a means of liberating conscious processes for more important control mechanisms (see chapters by Libet, Deecke & Kornhuber). These are pertinent issues that remain to be addressed.

REFERENCES

Ball, T., Schreiber, A., Feige, B., Wagner, M., Lücking, C. H. and Kristeva-Feige, R. (1999) The role of higher-order motor areas in voluntary movement as revealed by high-resolution EEG and fMRI. *Neuroimage* 10, 682-694.

Bauer, H., Korundka, Ch. and Leodolter, M. (1993) possible glial contribution in the electrogenesis of SP. In McCallum, W.C. and Curry, S.H. (Eds) Slow potential changes in the human brain., Plenum Press, New York, 23-34.

Benecke, R., Dick J.P.R., Rothwell, J.C., Day, B.C. and Marsden, C.D. (1985) Increase of the Bereitschaftpotential in simultaneous and sequential movements. *Neuroscience Letters* 62, 347-352.

Birbaumer, N., Elbert, T., Canavan, A.G.M. and Rockstroh, B. (1990) Slow potentials of the cerebral cortex and behavior. *Physiological Reviews.* 70, 1-41.

Botzel, K. and Schulze, S. (1996) Letter to *Brain.* 119: 1045-1048.

Boetzel, K., Plendl, H., Paulus, W. and Scherg, M. (1993) Bereitschaftspotential: is there a contribution of the supplementary motor area? *Electroencephalogr. Clin. Neurophysiol.,* **89,** 187-96.

Caspers, H. and Speckmann, E.J. (1974) Cortical DC shifts associated with changes of gas tension in blood and tissue. In *Handbook of Electroencephalography and Clinical Neurophysiology.* Edited by Reymond, A. Amsterdam: Elsevier, Vol 10A, p41-65.

Cheyne, D. and Weinberg, H. (1989) Neuromagnetic fields accompanying unilateral finger movements: pre-movement and movement-evoked fields. *Exp Brain Res* **78,** 604-612.

Chiarenza, G.A., Villa, M. and Vasile, G. (1995) Developmental aspects of Bereitschaftspotential in children during goal-directe behaviour. *International Journal of Psychophysiology,* **19,** 149-176.

Cunnington, R., Iansek, R., Bradshaw, J.L. and Phillips, J.G. (1995) Movement-related potentials in Parkinson's disease: Presence and predictability of temporal and spatial cues. *Brain* **118,** 935-950.

Cunnington, R., Iansek, R. and Bradshaw, J.L. (1999) Movement-related potentials in Parkinson's disease: External cues and attentional strategies. *Mov. Disord.* **14,** 63-68.

Cunnington, R., Windischberger, C., Deecke, L. and Moser, E. (2002) The preparation and execution of self-initiated and externally-triggered movement: a study of event-related fMRI. *NeuroImage* **15,** 373-385.

Coles, M. G. H. and Gratton, G. (1986) Cognitive psychophysiology and the study of states and processes. In: Hockey, G. R. J., Gaillard, A. W. K. and Coles, M. G. H. (Eds.) *Energetics and human information processing,* pp. 409-424. Dordrecht, The Netherlands: Martinus Nijhof.

Cui, R.Q., Huter, D., Lang, W., Lindinger, G., Beisteiner, R. and Deecke, L. (1996) Multichannel DC current source density mapping of the Bereitschaftspotential in the supplementary and primary motor area preceding differently loaded movements. *Brain Topography* **9**(2): 83-94.

Cui, R.Q., Huter, D., Lang, W. and Deecke, L. (1999) Neuroimage of voluntary movement: Topography of the Bereitschaftspotential, a 64-channel DC current source density study. *NeuroImage* **9,** 124-134.

Deecke, L. (1987) Bereitschaftspotential as an indicator of movement preparation in supplementary motor area and motor cortex. Porter, R. (Ed) *Motor Areas of the Cerebral Cortex. Ciba Found. Symp.,* **132,** Wiley, Chichester, 231-50.

Deecke, L., Kornhuber, H.H., Lang, W., Lang, M. and Schreiber, H. (1985) Timing function of the frontal cortex in sequential motor and learning tasks. *Human Neurobiol* **4,** 143-154

Deecke, L., Grozinger, B. and Kornhuber, H. H. (1976) Voluntary finger movement in man: cerebral potentials and theory. *Biol Cybern.* **23,** 99-119.

Deecke, L., Scheid, P. and Kornhuber, H.H. (1969) Distribution of readiness potential, pre-motion positivity and motorpotential of the human cerebral cortex preceding voluntary finger movements. *Exp Brain Res* **7,** 158-168.

Defebvre, L., Bourriez, J.L., Dujardin, K., Derambure, P., Destee, A. and Guieu, J.D. (1994) Spatiotemporal study of Bereitschaftspotential and event-related desynchronization during voluntary movement in Parkinson's disease. *Brain Topography* **6,** 237-244.

Deuschl, G., Toro, C., Matsumoto, J. and Hallett, M. (1995) Movement-related cortical potentials in writer's cramp *Annals of Neurology* **38**; 862-8.

Dick, J.P.R., Cantello, R., Buruma, O., Gioux, M., Benecke, R., Day B.L., Rothwell, J.C., Thompson, P.D. and Marsden, C.D. (1987) The Bereitschaftspotential, L-Dopa and Parkinson's disease. *EEG Clin Neurophysiol* **66,** 263-274

Dick, J.P.R., Rothwell, J.C., Day, B.L., Cantello, R., Buruma, O., Gioux, M., Benecke, R., Berardelli, A., Thompson, P.D. and Marsden, C.D. (1989) The Bereitschafts potential is abnormal in Parkinson's disease. *Brain* **112,** 233-244.

Dirnberger, G., Fickel, U., Lindinger, G., Lang, W. and Jahanshahi, M. (1998) The mode of movement selection: Movement-related cortical potentials prior to freely selected and repetitive movements. *Exp Brain Res,* **120,** 263-272.

Dirnberger, G., Reumann, M., Endl, W., Lindingere, G., Lang, W. and Rothwell, J.C. (2000) Dissociation of motor preparation from memory and attentional processes using movement-related cortical potentials. *Exp Brain Res* **135,** 231-240.

Eimer, M., Goschke, T., Schlaghecken, F. and Stürmer, B. (1996) Explicit and implicit learning of event sequences: Evidence from event-related brain potentials. *Journal of Experimental Psychology: Learning, Memory, and Cognition* **22,** 1-18.

Elbert, T. (1993) Slow cortical potentials reflect the regulation of cortical excitability. In WC McCallum and SH Curry (Eds) Slow potential changes in the human brain., Plenum Press, New York, p235-251.

Elbert, T. and Rockstroh, B. (1987) Threshold regulation- a key to understanding of the combined dynamics of EEG and event-related potentials. *J of Psychophysiology* **4,** 317-333.

Erdler, M., Beisteiner, R., Mayer, D., Kaindl, T., Edward, V., Windischberger, C., Lindinger, G. and Deecke, L. (2000) Supplementary motor area activation preceding voluntary movement is detectable with a whole-scalp magnetoencephalography system. *Neuroimage,* **11,** 697-707.

Fève, A.P., Bathien, N. and Rondot, P. (1992) Chronic administration of L-dopa affects the movement-related cortical potential in patients with Parkinson's disease. *Clin. Neuropharm.* **15,** 100-108.

Fève, A.P., Bathien, N. and Rondot, P. (1994) Abnormal movment-related potentials in patients with lesions of basal ganglia and anterior thalamus. *J. Neurol. Neurosurg. Psychiatry* **57,** 100-104.

Fuller, R., Nathaniel-James, D. and Jahanshahi, M. (1999). Movement-related potentials prior to self-initiated movements are impaired in patients with schizophrenia and negative signs. *Exp. Brain Res.* **126,** 545 – 555

Gaillard, A.W.K. (1977) The late CNV wave: preparation versus expectancy. *Psychophysiology* **14** 563-568.

Gerloff, C., Altenmüller, E. and Dichgans, J. (1996) Disintegration and reorganization of cortical motor processing in 2 patients with cerebellar stroke. *Electroencephalogr. Clin. Neurophysiol.,* **98,** 59-68.

Gerloff, C., Uenishi, N. and Hallett, M. (1998a) Cortical activation during fast repetitive finger movements in humans: Dipole sources of steady-state movement-related cortical potentials. *J. Clin. Neurophysiol.* **15,** 502-13.

Gerloff, C., Uenishi, N., Nagamine, T., Kunieda, T., Hallett, M. and Shibasaki, H. (1998b) Cortical activation during fast repetitive finger movements in humans: Steady-state movement-related magnetic fields and their cortical generators. *Electroenceph. Clin. Neurophysiol*, **109**, 444-53.

Green, J.B., Bialy, Y., Sora, E. and Ricamato, A. (1999) High-resolution EEG in poststroke hemiparesis can identify ipsilateral generators during motor tasks. *Stroke* **30**, 2659-65.

Haag, C., Kathmann, N., Hock, C., Günther, W., Vorderholzer, U. and Laakmann, G. (1994)Lateralization of the Bereitschaftspotential to the Left Hemisphere in Patients with Major Depression. *Biol Psychiatry* **36**, 453-457

Haggard, P. and Eimer, M. (1999). On the relation between brain potentials and the awareness of voluntary movements. *Exp. Brain Res.* **126**, 128-133.

Hari, R. (1980) *Sensory evoked sustained potentials in man.* Department of Physiology, University of Helsinki, Academic Dissertation.

Honda, M., Nagamine, T., Fukuyama, H., Yonekura, Y., Kimura, J. and Shibasaki, H. (1997) Movement-related cortical potentials and regional cerebral blood flow change in patients with stroke after motor recovery. *J. Neurol. Sci.* **146**, 117-26.

Ikeda, A., Shibasaki, H., Nagamine, T., Terada, K., Kaji, R., Fukuyama, H. and Kimura, J. (1994) Dissociation between contingent negative variation and Bereitschaftspotential in a patient with cerebellar efferent lesion. *Electroencephalogr. Clin. Neurophysio.*, **90**, 359-64.

Jahanshahi, M., Jenkins, I.H., Brown, R.G., Marsden, C.D., Passingham, R.E. and Brooks, D.J. (1995) Self initiated versus externally triggered movements. I. An investigation using measurement of regional cerebral blood flow with PET and movement-related potentials in normal and Parkinson's disease subjects. *Brain* **118**, 913-933.

Jahanshahi, M., Dirnberger, G., Filipovic, S. R., Jones, C., Fuller, R., Frith, C. D., and Barnes, G. (2000) Willed vs externally-triggered saccades. *Soc Neurosc Abstracts.*

Jahanshahi, M., Dirnberger, G., Liasis, A., Towell, A. and Boyd, S. (2001) Does the prefrontal cortex contribute to movement-related potentials? Recordings from subdural electrodes. *Neurocase, 7*, 495-501.

Johnson, K.A., Cunnington, R., Iansek, R., Bradshaw, J.L., Georgiou, N. and Chiu, E. (2001) Movement related potentials in Huntington's disease: movement preparation and execution. *Exp. Brain Res.* **138**, 492-499.

Keller, I. and Heckhausen, H. (1990). Readiness potentials preceding spontaneous acts: Voluntary vs. involuntary control. *Electroencephalography and Clinical Neurophysiology* **76**, 351-361.

Khanna, S., Mukundan, C.R., and Channabasavanna, S.M. (1989) Bereitschaftspotential in Melancholic Depression. *Biol Psychiatry* **26**, 526-529

Kitamura, J., Shibasaki, H. and Takeuchi, T. (1996) Cortical potentials preceding voluntary elbow movement in recovered hemiparesis. *Electroencephalogr. Clin. Neurophysiol*, **98**, 149-56.

Kitamura, J., Shibasaki, H., Takagi, A., Nabeshima, H. and Yamaguchi, A. (1993) Enhanced negative slope of cortical potentials before seqeuntial as compared with simultaneous extensions of two fingers *EEG Clin Neurophysiol*, **86**, 176-182.

Knight, R. T., Singh, J. and Woods, D. L. (1989) Pre-movement parietal lobe input to human sensorimotor cortex. *Brain Res.* **498**, 190-4.

Kornhuber, H. H. and Deecke, L. (1964) Hirnpotentialänderungen beim Menschen vor und nach Willkürbewegungen, dargestellt mit Magnetband-Speicherung und Rückwärtsanalyse. *Pflügers Arch.* **281**, 52.

Kornhuber, H. H. and Deecke, L. (1965) Hirnpotentialänderungen bei Willkürbewegungen und passiven Bewegungen des Menschen: Bereitschaftspotential und reafferente Potentiale. *Pflügers Arch.* **284**, 1-17.

Kristeva, R., Cheyne, D. and Deecke, L. (1991) Neuromagnetic fields accompanying unilateral and bilateral voluntary movements: topography and analysis of cortical sources. *EEG Clin Neurophysiol* **81**, 284-298.

Kutas, M. and Donchin, E. (1980) Preparation to respond as manifested by movement-related brain potentials. *Brain Research* **202**, 95-115.

Laming, P.R. (1989) Do glia contribute to behavior? A neuromodulatory review. *Comp Biochem. Physiol.* **94a**, 555.455.

Lang, L., Lang, M., Poddreka, M., Steiner, M., Uhi, F., Suess, E., Müller, Ch. and Deecke, L. (1988) DC-potentials shifts and regional cerebral blood flow reveal frontal cortex involvement in human visuomotor learning. *Exp Brain Res* **71**, 353-364.

Libet, B. (1992) Voluntary acts and readiness potentials, *Electroencephalography and Clinical Neurophysiology* **82**, 85-86.

Lemieux, L. Salek-Haddadi, A., Josephs, O., Allen, P., et al (2001) Event-related fMRI simultaneous and continuous EEG: Description of the method and initial case report. *NeuroImage* **14**, 780-787.

Lopes da Silva, F. (1996) Biophysical issues at the frontiers of the interpretation of EEG/MEG signals. *In RM Dashieff and DJ Vincent (Eds) Frontier Science in EEG: Continuous Waveform Analysis. EEG Clinical Neurophysiology, (EEG Suppl 45)* Elsevier, Amsterdam, pp1-7.

Mackinnon, C. D., Kapur, S., Hussey, D., Verrier, M. C., Houle, S. and Tatton, W. G. (1996) Contributions of the mesial frontal-cortex to the premovement potentials associated with intermittent hand movements in humans. *Human Brain Mapping*, **4**, 1-22.

McCallum, W.C. (1993) Human slow potential research. In McCallum, W.C. and Curry, S.H. (Eds) *Slow potential changes in the human brain.* Plenum Press, New York, p1-12.

Nagamine, T., Toro, C., Balish, M., Deuschl, G., Wang, B., Sato, S., Shibasaki, H. and Hallett, M. (1994) Cortical magnetic and electric fields associated with voluntary finger movements. *Brain Topogr.* **6**, 175-83.

Neafsey, E.J., Hull, C.D. and Buchwald, N.A. (1978) Preparation for movement in the cat. II Unit activity in the basal ganglia and thalamus. *EEG Clinical Neurophysiology* **44**, 714-723.

Neshige, R., Lüders, H. and Shibasaki, H. (1988) Recording of movement-related potentials from scalp and cortex in man. *Brain* **111**, 719-736.

Northoff, G., Pfennig, A., Krug, M., Danos, P., Leschinger, A., Schwarz, A. and Bogerts, B. (2000) Delayed onset of late movement-related cortical potentials and abnormal response to lorazepam in catatonia. Schizophrenia Research **44**, 193-211

Okano, K. and Tanji, J. (1987) Neuronal activities in the primate motor fields of the agranular frontal cortex preceding visually triggered and self-paced movement. *Exp. Brain Res.* **66**, 155-166.

Pedersen, J.R., Johannsen, P., Bak, C.K., Kofoed, B., Saermark, K. and Gjedde, A. (1998) Origin of human motor readiness field linked to left middle frontal gyrus by MEG and PET. *Neuroimage* **8**, 214-20.

Pfurtscheller, G., Aranibar, A. (1977) Event-related cortical desynchronization detected by power measurements of scalp EEG. *Electroenceph. Clin. Neurophysiol.* **42**, 817-826.

Praamstra, P., Cools, A.R., Stegeman, D.F. and Horstink, M.W.I.M. (1996a) Movement-related potential measures of different modes of movement selection in Parkinson's disease. *J. Neurol. Sci.* **140**, 67-74.

Praamstra, P., Stegeman, D. F., Horstink, M. W. and Cools, A. R. (1996b) Dipole source analysis suggests selective modulation of the supplementary motor area contribution to the readiness potential. *Electroencephalogr. Clin. Neurophysiol.,* **98**, 468-77.

Rektor, I., Féve, A., Buser, N., Bathien, N. and Lamarche, M. (1994) Intracranial recoroding of movement-related readiness potentials: an exploration in epileptic patients. *Electroenceph clin Neurophysiol* **90**, 273-283.

Rockstroh, B., Elbert, T., Birbaumer, N. and Lutzenberger, W. (1982) *Slow brain potentials and behavior.* Urban & Schwarzenberg, Munich.

Roitbak, A.L., Fanardjhayan, V.V., Melkonyan, D.S. and Melkonyan, A.A. (1987) Contribution of glia and neurons to the surface negative potentials of the cerebral cortex during its electrical stimulation. *Neuroscience* **20**, 1057-67.

Rüsseler, J. and Rösler, F. (2000) Implicit and explicit learning of event sequences: Evidence for distinct coding of perceptual and motor representations. *Acta Psychologica* **104**, 45-67.

Sasaki, K., Gemba, H., Hashimoto, S. and Mizuno, N. (1979) Influences of cerebellar hemispherectomy on slow potentials in the motor cortex preceding self-paced hand movements in the monkey. *Neurosci. Lett.,* **15**, 23-28.

Schultz, W. and Romo, R. (1992) Role of the primate basal ganglia and frontal cortex in the internal generation of movements. I Preparatory activity in the anterior striatum. *Exp Brain Res.* **91**, 363-384.

Shibasaki, H. (1975) Movement-associated cortical potentials in unilateral cerebral lesions. *J. Neurol.,* **209**, 189-98.

Shibasaki, H., Barrett, G., Halliday, E. and Halliday, A. M. (1980) Components of the movement-related cortical potential and their scalp topography. *Electroencephalogr. Clin. Neurophysiol.,* **49**, 213-26.

Shibasaki, H., Shima, F. and Kuroiwa, Y. (1978) Clinical studies of the movement-related cortical potential (MP) and the relationship between the dentatorubrothalamic pathway and readiness potential (RP). *J. Neurol.* **219**, 15-25.

Shima, K., Aya, K., Mushiake, H., Inase, M., Aizawa, H. and Tanji, J. (1991) Two movement-related foci in the primate cingulate cortex observed in signal-triggered and self-paced forelimb movements. *J of Neurophysiology* **65**, 188-202.

Simonetta, M., Clanet, M. and Rascol, O. (1991) Bereitschaftspotential in a simple movement or in a motor sequence starting with the same simple movement. *Electroenceph. Clin. Neurophysiol.* **81**, 129-134.

Singh, J. and Knight, R. T. (1990) Frontal lobe contribution to voluntary movements in humans. *Brain Res.* **531**, 45-54.

Singh, J. and Knight, R. T. (1993) Effects of posterior association cortex lesions on brain potentials preceding self-initiated movements. *J. Neurosci.* **13**, 1820-9.

Skinner, J.E. and Yingling, C.D. (1976) Regulation of slow potnetial shifts in nucleus reticularis thalami by the mesencepahic reticular formation and the frontal granular cortex. *EEG & Clinical Neurophysiology* **40**, 288-296.

Smid, H. G. O. M., Mulder, G. and Mulder, L. J. M. (1987) The continuous flow model revisited: Preceptual and central motor aspects. In: Johnson, R., Jr., Rohrbaugh, J.W. and Parasuraman, R. (Eds.) *Current trends in event-related potential research (EEG Suppl. 40),* pp. 270-278. Amsterdam: Elsevier.

Tarkka, I. M. and Hallett, M. (1991) The cortical potential related to sensory feedback from voluntary movements shows somatotopic organization of the supplementary motor area. *Brain Topogr.* **3**, 359-63.

Tarkka, I. M., Massaquoi, S. and Hallett, M. (1993) Movement-related cortical potentials in patients with cerebellar degeneration. *Acta Neurol. Scand.* **88**, 129-35.

Tecce J.J., Savignano-Bowman, J. and Meinbresse, D. (1978) Contingent negative variation and the distraction-arousal hypothesis. *EEG & Clinical Neurophysiology* **41**, 277-286.

Toro, C., Matsumoto, J., Deuschl, G., Roth, B. J. and Hallett, M. (1993) Source analysis of scalp-recorded movement-related electrical potentials. *Electroencephalogr. Clin. Neurophysiol,* **86**, 167-75.

Touge, T., Werhahn, K.J., Rothwell, J.C. and Marsden CD. (1995) Movement-related cortical potentials preceding repetitive and random-choice hand movements in Parkinson's disease. *Ann. Neurol.* **37**, 791-799.

Tsujimoto, T., Gemba, H. and Sasaki, K. (1993) Effect of cooling the dentate nucleus of the cerebellum on hand movements of the monkey. *Brain Res.* **629**, 1-9.

Uhl, F., Goldenberg, G., Lang, W., Lindinger, G., Steiner, M. and Deecke, L. (1990) Cerebral correlates of imanining colours, faces and a map - II. Negative cortical DC-potentials. *Neuropsychologia* **28**: 81-93

Van der Kamp, W., Rothwell, J.C., Thompson, P.D., Day, B.L. and Marsden, C.D. (1995) The movement-related cortical potential is abnormal in patients with idiopathic torsion dystonia. *Mov. Disord.* **10**, 630-633.

Verleger, R., Wascher, E., Wauschkuhn, B., Jaskowski, P., Allouni, B., Trillenberg, P. and Wessel, K. (1999) Consequences of altered cerebellar input for the cortical regulation of motor coordination, as reflected in EEG potentials. *Exp. Brain Res.* **127**, 409-422.

Walter, W.G., Cooper, R., Aldridge, V.J., McCallum, W.C. and Winter, A.L. (1964) Contingent Negative Variation: an electric sign of sensori-motor association and expectancy in the human brain. *Nature* **203**, 380-384.

Wessel, K., Verleger, R., Nazarenus, D., Vieregge, P. and Kömpf, D. (1994) Movement-related cortical potentials preceding sequential and goal-directed finger and arm movements in patients with cerebellar atrophy. *Electroenceph. Clin. Neurophysiol.* **92**, 331-341.

SURFACE RECORDINGS OF THE BEREITSCHAFTSPOTENTIAL IN NORMALS

Wilfried Lang
University Clinic of Neurology
Vienna
A-1090 Austria

INTRODUCTION

The BP was discovered by Kornhuber & Deecke (1964; 1965) during recordings made from electrodes placed over the scalp. In the intervening years, it is probably accurate to say that such surface recordings have been the most common means of measuring the BP and have been instrumental in providing key information about this movement-related potential. A first aim of this chapter is to examine the factors that influence the amplitude, slope or topography of the BP. Secondly, I will consider what information about BP-topography can contribute to our knowledge of the functional organization of cortical motor areas. Third, surface recordings of the brain potentials show a complex, spatio-temporal pattern shortly before and after movement onset. Their electrogenesis and functional significance will be discussed. In the final part of this chapter, I will review the evidence for and against separate components of the BP.

PROCESSES THAT CONTRIBUTE TO THE BP

In this section, the factors that have been most extensively investigated using surface recordings and shown to influence the amplitude, slope or togopgrahy of the BP will be reviewed. This information is important in unravelling some of the processes that may crucially contribute to the BP.

Awareness, intentionality, volition and readiness to move

In the original description, Kornhuber and Deecke (1965) reported that the amplitude of the BP depends on the level of intentional involvement in the task. When subjects get bored and perform the movements in a stereotyped and almost automatic manner, the amplitude of the BP decreases. This observation has been confirmed by later studies. Libet et al. (1982) classified self-initiated movements according to the subject's reports about their experiences of performing the movements. The shape of the BP depended on the type of experience associated with movements: Movements which were associated with "some general preplanning or preparation to act in the near future" had an earlier onset of the BP (about 1 s) as compared to movements when subjects were aware of the "urge to move" (about 500 ms). Subsequently, Keller and Heckhausen (1990) measured surface-negative potential shifts which preceded self-initiated movements in various conditions: Movements were either made (1) during a mental task (counting backwards), (2) while watching a hand on a computer monitor performing a clockwise motion (Libet's paradigm; Libet et al., 1983) or (3) while sitting relaxed with the instruction to look introspectively into their arms which performed the self-initiated movements. In experiments 1 and 3 movements were made which were not associated with introspective feelings of awareness of having moved. Slow surface-negative potentials shifts (SPS) were found to precede all types of movement. Maximal amplitude was found in recordings above the contralateral primary motor cortex (M1) with unconsciously performed movements, but above the SMA (FCz and Cz) for movements, which were associated with the feeling of intentionality and awareness of action (Fig. 1). Thus, we have to consider the fact that when subjects are invited to repetitively perform a simple motor task, such as a brisk isolated finger movement, the process of transducing the intention to move into action may or may not be present with a single movement (cf. experiment 3 of Keller and Heckhausen, 1990). The intentionality associated with a single movement in a sequence may even not be present in an all or none manner, but levels of intentionality in action may differ and have differential effects on the latency and the topography of the BP. When subjects participate in a study on repetitive single, isolated finger movements, the decisions of „what to do" or „how to do" are already settled in advance. During the experiments subjects are free (within certain limits) to make decisions about the time of performing the movement. The „Bereitschaftspotential", BP, has been associated with voluntary movements and is, therefore, restricted to those movements, which are not executed in an automatic, manner but which are executed with introspective feelings of the willful realization of the intention to move at a particular time (Kornhuber et al., 1984). Because of the fact that the information necessary for performance, that is the motor programs, were already available, the assumption of our group was that the BP, at least in the simple paradigm, does not reflect aspects of „programming" but the transitional process of the intention to act into action at a certain time. The involvement of this transitional process is assumed to reflect volition (Kornhuber et al., 1984). The decision about the time of performing the movement may be made contingent upon the existence of a „general preparatory state" of relevant motor areas (Gerloff et al. 1998). It has been shown that cortical excitability spontaneously changes. These changes are associated with surface-recorded potential shifts. It has been further shown that task presentation contingent upon the rise of negative surface-recorded potentials in relevant cortical areas facilitates task performance, whereas the presentation contingent on positive slow potential shifts disturbs task performance (Bauer and Nirnberger, 1981; Born et al., 1982; Rockstroh et al., 1989). Considering the consequences of overt motor behaviour for the individual it may be that the decision "when to act" is made contingent upon "spontaneous" rises of cortical excitability in relevant motor areas.

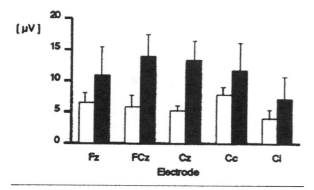

Figure 1. Comparison of the means of slow negative potential shifts (SPS, BP) with unconsciously (white columns) and consciously (black columns) performed motor acts. The bars give the standard error. Note, that there is a dissociative effect on SPS with larger amplitudes in Cc (recordings above the primary motor cortex contralateral to the movement) as compared to those in Cz (or FCz) with unconsciously performed motor acts but larger amplitudes in Cz (or FCz) as compared to those in Cc with consciously performed movements. Data from Keller and Heckhausen (1990), reprinted with permission.

Consequences of a movement

Voluntary movements have consequences for the individual. Consider, a waiter moving a glass from a table. The consequences of this movement are anticipated and postural adjustments are made in advance. Unfortunately, such movements performed in the course of daily living have not been tested so far. What has been assessed, so far, are paradigms in which subjects perform a movement which results in a subsequent task presentation. Here, surface-recorded negative potential shifts were associated with both, the organization and initiation of the movement and also the anticipatory preparation for the forthcoming task. In one study subjects were presented with behaviourally relevant visual stimuli in one hemi-field. The task was not only to initiate a simple finger movement (unilateral button press) but also to prepare for its consequence which was to respond to the visual cue with an adequate hand movement (Lang et al., 1984). In this situation, the BP was superimposed by a negative SPS which was specifically related to the forthcoming visually-guided task. There was an additional negative SPS in recordings above the occipito-temporal and inferior parietal cortex in the hemisphere contralateral to the hemi-field of vision where the visual cues appeared. This additional negative SPS was maximal in parietal recordings (Figure. 2).

Mountcastle and coworkers (Mountcastle et al., 1981; Lynch, 1981) have described neurons in the inferior parietal cortex which became active when monkeys expect the appearance of motivationally relevant objects within their extrapersonal space. In light of this observation and the clinical syndrome of hemineglect after lesions in the parieto-temporal cortex (Balint, 1909), we have considered this additional negative SPS to reflect mechanisms of directing attention towards the contralateral hemi-field of vision ("directed attention potential"). In other studies subjects were presented with demanding cognitive tasks (e.g. concept formation, selective auditory attention) following voluntary button pressing (Lang et al., 1987; Asenbaum et al., 1992). The execution of these tasks was associated with negative SPS above the prefrontal cortex. Because of the movement's consequences, the BP was superimposed by large negative SPSs in recordings over the prefrontal cortex (Fig. 3). Morgan et al. (1992) could demonstrate that the amplitude of these negative SPSs in prefrontal recordings correlated with the accuracy of performance in the cognitive task.

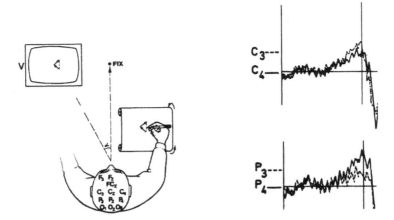

Figure 2: **Left part:** The subjects had to press down a pen held in the right hand. As a consequence of pressing down the pen, the subjects were presented with a moving visual target in the left hemi-field of vision which had to be tracked by the right hand.
Right part: In recordings above the primary motor cortices (C3, C4) BP was lateralized with larger amplitudes contralateral to the moving hand (C3 > C4). In parietal (as shown here) but also in occipital recordings the BP was lateralized with larger amplitudes above that hemisphere which primarily received the visual input (right hemisphere; P4 > P3). The additional negative potential shift in parieto-occipital recordings of the right hemisphere has been linked to mechanisms of directing sensory attention. Data from Lang et al. (1984), reprinted with permission. .

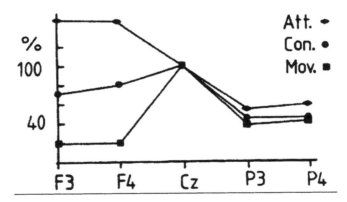

Figure 3. BP amplitudes (measured 1 s before movement onset) relative to Cz (100%) in three conditions: Single button press followed by selective attention task (Att.), single button press followed by concept formation task (Con.), and single button press without consequences (Mov.). Data from Lang et al. (1994), reprinted with permission.

Complexity of the movement

A concept proposed by Roland and coworkers (Roland et al., 1980; Roland, 1984) was that "the supplementary motor area (SMA) elaborates or retrieves from memory the necessary information to form a short sequence of motor commands in which the elementary movements to be executed are specified exactly. With an example from the motor sequence test one could say that the SMA specified: (1) the fingers to be moved in the near future, (2) which were the movements of the individual fingers and (3) the sequence of (1) and (2)". This hypothesis was tested in several studies. For example, Benecke et al. (1985) compared the BP preceding a single movement (A or B) to the BP preceding composite movements (sequential: A then B, simultaneous: A + B). Amplitudes of the BP were found to be significantly larger in composite simultaneous and sequential movements as compared to single movements. These observations were confirmed by Simonetta et al. (1991). In this study, subjects either performed a simple movement (flexion of the index finger) or a sequential motor task starting with the simple movement (flexion of the index finger, then, turning on and off a button located 20 cm away). They found not only larger BP amplitudes but also an earlier BP onset with complex movements (Fig. 4). Kitamura and coworkers (1993) found that the BP is larger prior to a sequential movement (A then B) as compared to a simultaneous movement (A + B).

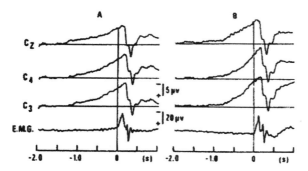

Figure 4. Grand averages of the BP recorded from Cz, C3 and C4 in the simple movement (A) & sequential motor tasks (B). Data from Simonetta et al. (1991), reprinted with permisison.

The findings of the studies examining the effect of movement complexity on the BP have two limitations. First, the spatial analysis of the BP was not sufficient to test the concept whether the effect of task complexity is regional (SMA?) or more widespread (SMA + M1). Second, manipulaton of the complexity of the motor task is associated with variation of the level of novelty, motivation and intentionality of the action all of which may also influence the BP. Furthermore, complex motor sequences may require learning which also affect the BP. We will consider the effect of skill on the BP next.

Level of skill and skill acquisition
        Taylor (1978) studied the BP when subjects learned the skill of pressing a sequence of buttons in a fixed order and found a transient increase of BP amplitude during learning, particularly in frontal recordings. After the acquisition of skill there was a decrease of amplitude. Lang and coworkers compared the BP preceding movement onset and slow negative potentials (SPS) accompanying the execution of motor sequences. Motor sequences consisted of the same motor elements, such as flexion and extension of the index finger, but the spatial and temporal order by which the fingers had to be moved were changed between conditions. In the most simple task, the two index fingers were moved without any constraints on temporal or spatial coordination (Fig. 5; UNKO). Other tasks required subjects to simultaneously move the fingers in the same or different directions (Fig. 5; SI-S and SI-D, resp.). The tasks became more complex, when subjects had to start the sequence with one hand and to follow with the other hand (sequential movements; Fig. 5; SE-S and SE-D). SPS which accompanied movement execution were far more sensitive to variations of the spatio-temporal pattern of the motor sequence than the BP preceding the task, in particular when motor sequences had to be performed for a longer period of time (Fig. 5; Lang et al., 1988, 1990).

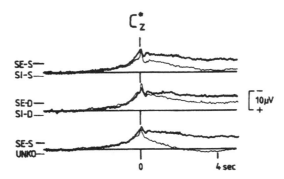

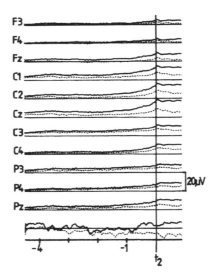

Figure 5: Grand averages of the BP preceding and accompanying the execution of different types of bimanual motor sequences. **SI-S:** SImultaneous movements of the two index fingers in Same directions. Starting from a middle finger position (P_0), (1) subjects flexed their index fingers simultaneously to reach P_{fl} (flexion position). (2) Ss made extensions to P_0, and subsequently (3) to P_{ex} (extension position). Then, (4) they flexed to reach P_0 and so on. **SE-S:** SEquential movements in the Same direction. As in SI-S but the right index finger started, the left finger followed. **SI-D:** Simultaneous movements in Different directions. The right index finger moved up, while the left index finger moved down, and so on. **SE-D:** SEquential finger movements in Different directions. **UNKO:** Uncoordinated movements of the two index fingers. Note, that the slow negative potential shifts accompanying the execution of the motor task vary to a large extent but the BP preceding the initiation of the task does not. Data from Lang et al. (1988), reproduced with permission.

In another study skilled musicians were trained to perform a bimanual tapping task. They started to tap with one side and had to bring in the other side either with the same rhythm or with a different rhythm. The amplitude of the BP preceding the start of the second motor sequence differed between the 'same rhythm' and the 'different rhythm' tasks. Differences started to be significant from 4 s before movement onset (Fig. 6).

Figure 6: Grand average of the slow negative potential shifts associated with a bimanual tapping task. In one task (thin, hatched line), trained musicians started to tap with the right index finger (about 2 movements/ 2 s) and had to bring in the left hand with the same rhythm after a period of time (about 4-6 s). In the other task (thick line), musicians started to tap with the right index finger (about 2 movements/ 2s) but had to bring in the other index finger with tapping of a different rhythm (about 3 movements/ 2s). EEG was averaged synchronized to the time (t_2) when the second finger was brought in. BP preceding the initiation of the movement of the second finger differed between the two tasks. Differences started to be significant from 4 s before t_2. Data from Lang et al. (1990) reprinted with permission.

The mode of movement selection

In 1991, a positron emission tomography (PET) study reported that the mode of movement selection, i.e. whether a movement is made in a pre-determined or self-determined manner, is associated with changes of SMA activity. During a sequence of randomly performed joystick movements, which had to be chosen out of a set of four possible movements - the so-called free mode of movement selection - the mean SMA activity was higher than during a sequence consisting of a single pre-determined movement that had to be performed repetitively ('fixed' mode of movement selection) (Deiber et al., 1991). Praamstra and Touge (Praamstra et al., 1995, Touge et al., 1995) employed this task to study the BP. Freely selected movements were found to be associated with a higher BP amplitude than pre-determined repetitive movements in a fixed mode. Praamstra made a spatio-temporal decomposition of the BP and succeeded to explain the surface-electric potentials by three current sources (Praamstra et al., 1996). One of them was located in the SMA and had a radial orientation, the others were located in the M1 of either hemisphere and had both, a tangential and a radial component. The most interesting result of this study from a functional point of view was that it was exclusively the current source in the SMA which was affected by the mode of movement selection (Fig. 7). Increased SMA acitivity in the free selection mode was considered to reflect the higher demands of planning before the execution of freely selected random movements (Praamstra et al., 1995, 1996).

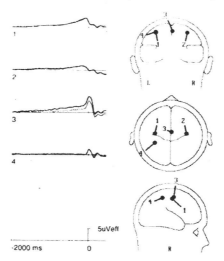

Figure 7: Topography of the BP preceding a finger movement is explained by three equivalent current dipoles (ECD): ECDs in the primary motor cortex (MI) of either hemisphere and an ECD in the SMA. Post-movement potentials are explained by a fourth ECD located in the primary somatosensory cortex. Activity of ECDs across time is shown in the left part of the figure. Note, that the time of activity of the SMA-ECD varies between the two conditions (free selection mode vs. fixed sequence). From Praamstra et al. (1996), reproduced with permission.

Using sequences of finger movements, Dirnberger et al. (1998) also showed that the BP is larger in a free selection than a fixed selection mode. But the mode effect on the spatial distribution of the BP was found to be more complex: There was a larger lateralization of the BP in the freely selected movements consistent with the assumption that the mode of movement selection had effects on M1-acitivity. Movements in the free mode are different from repetitive movements in more than one respect. In this study, free movements required a decision about which finger to move on each trial. Therefore, such freely-selected movements are accompanied by a higher rate of 'trial by trial novelty'. For repetitive movements or sequences, the BP and electric potentials present during movement execution were shown to habituate (Taylor, 1978; Lang et al., 1992; Dirnberger et al., 1998).

Another study, which has examined the effect of mode of movement selection on movement-related potentials is that of Jahanshahi et al. (1995). They recorded the EEG in two different paradigms which differed in terms of self- vs external selection of the *precise time of initiation* of a movement. In a first task ("self-initiated movements"), subjects made self-initiated movements on average once every 3 s. The movement involved a brisk lifting of the right index finger. A tone, which the subjects were told to ignore, was presented 100 ms after the self-initiated movement. Subjects were trained to produce movements at a target rate of about 3 s. In a second task ("externally triggered movements"), subjects made the same finger movements in response to a tone presented at an identical rate to that generated by the subjects in the self-initiated condition. Surface-negative slow potential shifts (SPS) were found in both conditions. Since the external stimulus was presented at a relatively regular rate between 3s and 4s, anticipation of its occurence and some kind of motor preparation or at least a state of motor responsiveness was possible. SPS in the "externally triggered" paradigm can be conceived to reflect the late component of the contingent negative variation (CNV; Walter et al., 1964). Self initiated movements were preceded by a BP. Amplitudes of the BP were found to be larger than the amplitudes of the SPS preceding externally triggered movements.

Sequence effects

Sequences of different finger movements offer the possibility to study specific phenomena, such as shifting movement initiation from one side of the body to the other. In a recent study in our laboratory (Dirnberger et al, 2001), subjects were free to either move the index or middle finger of either hand on a given trial. This design allowed us to compare the BP preceding a finger movement in two different situations: (a) when the subject had moved a finger of the same body side in the previous trial (absence of shift between the two sides of the body) or (b) when the subject had moved a finger of the contralateral side of the body in the previous trial (shift between the two sides). This factor (absence or presence of a shift) had a significant effect on the hemispheric assymmetry of the BP. There was additional negativity contralateral to the moving side when a shift of the side of movement initiation had taken place. This additional contralateral negative potential had a maximal amplitude in recordings above the premotor cortex, MI and the parietal cortex (Dirnberger et al., 2001). This finding led to the hypothesis that a change from one side of the body to the other for movement initiation involves mechanisms of (hemi-)spatial attention. Lesions of the parietal and premotor areas are known to cause the symptom of "hemineglect" which may affect the processing of sensory information, the initiation of a movement or even the awareness of one side of the body (Castaigne et al., 1970, 1972; von Giesen et al., 1994)

Different parts of the body performing the voluntary movement

The BP has been studied prior to movements of the fingers, toes (Boschert et al., 1983; Boschert and Deecke, 1986), and eyes (Becker et al., 1972) and prior to speech or vocalization (Grözinger et al., 1979, Deecke et al., 1986). Even more striking than differences among these conditions are the similarities. For all types of movement, the BP develops about 1.5 s before movement, has a widespread distribution and is bilaterally symmetrical in the initial period until about 500 ms before movement onset. Boschert et al. (1986) and Boschert and Deecke (1986) recorded the BP with unilateral movements of toes, foot, hip and index finger. Within the first second, the BP remained constant with movements of different parts of the body. Only within the last 500 ms did the spatial distribution of the BP vary depending on the part of the body executing the movement. Changes of BP distribution could be explained when assuming current sources within the centers of the MI representation areas (Fig. 8). But there are studies which show the confounding effects of task difficulty when varying the part of the body which performs the movement. BP amplitudes for movements of the middle finger are larger than those for index fingers. This difference was explained by middle-finger movements being, for anatomical reasons, more difficult and effortful than index-finger movements (Kitamura et al., 1993).

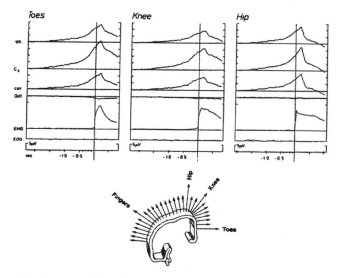

Figure 8: Upper part: Grand averages of the BP preceding plantarflexions of the left toes, flexions of the left knee and isometric extensions of the left hip (contractions of the gluteal muscles). Bipolar contralateral vs. ipsilateral recordings (Diff.) as calculated by computer.

Lower part: Schematic illustration of the orientation of electrical field vectors produced by different active sources in the MI representation areas. The tip of the arrows indicate the negative pole. Close to the mantle edge, the field vector stands upright in the saggital plane. Activity from this area does not impose any lateralization on BP topography. From Boschert and Deecke (1986), reproduced with permission.

Thus, BP topography varies with the part of the body performing the movement but this variation is small. The most remarkable finding is the similarity of the BP. Similarity was high in the early period of the BP (1.5 - 0.5 s); differences emerged in the later period of the BP (0,5 s to movement onset). These observations were taken as evidence for the concept that there is some general mechanism for the transitional process by which the intention to move becomes realized (Kornhuber et al., 1984).

Force and speed and precision of a movement

There are many experiments showing that factors such as force (Becker and Kristeva, 1980; Kutas and Donchin, 1974), inertial load (Kristeva et al., 1990) or speed (Becker et al., 1976; Hazemann et al., 1978) of a movement have an effect on the BP amplitude. Increasing force and inertial load imposed on the movement and increasing demands on the speed of the movement cause larger BP amplitudes. Increases of BP amplitude with increased voluntarily applied force start much earlier than increases of BP amplitude caused by higher levels of inertial load, where differences could only be found within the last 100 ms before movement onset (Kristeva et al., 1990). It has been hypothesized that in the case of subjectively applied force there is a greater degree of "effort" accompanying the preparation of the movement which is reflected in larger early BP amplitudes, whereas the addition of a physical load to the finger involves compensatory changes in the motor cortex more related to movement execution per se. Interestingly with bilateral finger movements, the addition of a physical load either to one side (right or left) or to both, did not change BP amplitudes. It has been suggested that bilateral movement organization involves "higher" aspects of motor control than those reflecting adjustment to conditions of inertial load as was the case with unilateral movements (Kristeva et al., 1990). Thus, there is evidence that "higher" aspects of motor control or the degree of "effort" required for a task have larger effects on the BP than adjustments of the motor program to different loads.

BP-TOPOGRAPHY AND ELECTROGENESIS

Three approaches have been used to get insights into the physiology of the BP: (a) functional dissociations of BP topography (b) topographical analyses of surface maps (amplitude maps of electric potentials or of current source densities, CSD), (c) biophysical modelling of current sources ("equivalent current dipoles").

Functional dissociations of BP topography: Assume that two spatially separated cortical areas (A1, A2) are activated in a motor task. Assume, furthermore, that variation of component C1 of a task selectively modulates activity of A1 but not A2, whereas variation of C2 activates A2 but not A1. In this situation variations of factors C1 and C2 should have differential effects on BP topography.

Topographical analyses of amplitude maps: Maps of BP amplitudes and of CSD on the scalp have been found useful to establish or test hyptheses about the cortical areas which may contribute to the BP (Nunez 1981; Lindinger et al. 1990; Lang et al. 1994; Cui et al. 1999). CSD maps have the advantage that they are reference independent and reduce spatial dispersion of electric potentials.

Biophysical modelling is based on the concept of decomposing the spatio-temporal patterns of the BP and "explaining" surface-recorded BP by a number of current sources ("equivalent current dipoles") which are defined by location, orientation, and time (Scherg 1990; Baumgartner 1993).

An example for a functional dissociation of BP topography is given by Praamstra et al. (1995). The mode of movement selection had an effect on the BP topography which is significantly different to the effect of the side of the performing hand. Mode of movement selection explained the variation of BP amplitude in recordings above the SMA while the side of the performing hand explained the variation of BP amplitudes in recordings above MI. Thus, there are two cortical areas, M1 and SMA, which are independently affected by two aspects of the motor task, side of the performing hand and task complexity, respectively. Analysis of variance (ANOVA) after normalization of amplitudes for each condition in order to exclude global amplitude effects may be the most approriate method for analysis (McCarthy and Wood 1985).

CSD maps of the BP showed that there are three areas with significant current sinks in the scalp, above the SMA, the contralateral and ipsilateral MI (Cui et al. 1999; see Fig. 8). Current sources appear above the frontopolar cortex. The pattern of current sinks and sources is consistent with the "dipole model" of Praamstra et al. (1996). They suggested that the surface-recorded BP can be explained by a radial equivalent current dipole (ECD) in the SMA and an ECD in MI of either hemisphere (even in unilateral movements) which has a radial and a tangential orientation. The radial SMA-ECD corresponds to the current sink above the SMA in CSD-maps. The M1-ECD is directed from posterior, lateral and superior to anterior, mesial and inferior which corresponds to a current sink slightly posterior and lateral to the MI hand area and a current source above the the fronto-polar cortex. This model of the electrogenesis of the BP is supported by analyses of the magnetic fields preceding movement onset (Cheyne et al. 1991; Kristeva et al. 1991; Lang et al. 1991; Beisteiner et al. 2000).

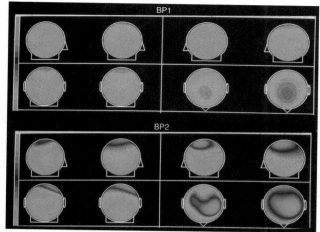

Figure 9: Upper part: Current source density maps (CSD) and amplitude maps of the BP at the time period from 1.6-1.5 s before movement onset in a single subject. Views of the head from the right side (left upper box), the left side (right upper box), from the front (left lower box) and from above (right lower box). CSD maps are at the left, amplitude maps of the BP at the right side in each box. Areas with radial current sinks or with negative BP amplitudes are coloured red, areas with current sources and positive BP amplitudes are coloured blue. Colour coded CSD values are linear and proporzional to radial current densities (colour gradings of CSD equivalents are indicated on the left side). Absolute values cannot be calculated since tissue resistance is not known. Note, there is a radial current sink in the area above the SMA.
Lower part: CSD maps and amplitude maps of the BP during the time period from 0.6-0.5 s before movement onset. Arrangements as in the upper part of the figure. Current sinks are widely distributed, from the area above the SMA to the area above motor cortices of the contralateral and the ipsilateral side. There is a current source in the prefrontal area (view from the front in the left lower box). From Cui et al. (1999), reproduced with permission.

With regards to the BP, there is controversy about whether the SMA and M1 are activated in parallel or whether SMA-activity precedes that of MI. Analysis of time activity of ECD in the model by Praamstra and coworkers (1996) suggested that SMA and MI are actived in parallel. Cui et al. (1999) studied the existence of a current sink in the areas above SMA and MI in an early (1.6 - 1.5 s) and a late (0.6 - 0.5 s) premovement period. On average across all subjects there was a significant current sink above the SMA but not above M1 in the early premovement period. Analysis of single subjects showed that there is a group of subjects (10 out of 17) who present an isolated current sink above the SMA in the early pre-movement period. Other subjects showed significant current sinks above the SMA and the M1 (n=3), above M1 (n=3) or neither above SMA nor above M1 (n=1). However, it is known that a certain area of the cortex needs to be activated in synchrony to produce an electric potential in surface-recordings. Thus, surface-recordings may be of limited value to solve the controversy. There is evidence from single cell recordings in subhuman primates that there is a higher prevalence of neurones in the SMA that start firing very early in the premovement period than in MI (Okano and Tanji 1987).

Analyses of BP topography and electrogenesis show that unilateral movements are not only preceded by contralateral but also by ipsilateral MI acitivity. Moreover, on a macroscopic level, the electrogenesis of contra- and ipsilateral MI-activity seems to be identical: ECDs in M1 of either side have an orientation which is consistent with the assumption of excitatory synaptic activity in apical dendrites and surface-to-depth current flow in large pyramidal tract cells. Time patterns of contra- and ipsilateral M1-activity do not differ and dipole strength of the ipsilateral M1 is about 50% - 75% of that of the contralateral M1 (Praamstra et al. 1996; Cheyne et al. 1991; Kristeva et al. 1991). This relation exceeds that of the relation between pyramidal tract fibres which do not cross (5%) as compared to those which cross to the other hemisphere. The spatial resolution of electric potentials or neuromagnetic fields is not sufficient to separate the contributions of the contra- and the ipsilateral SMA. There is indirect evidence in humans (Lang et al. 1991) but direct evidence from single cell recordings in sub-human primates (Brinkman et al. 1979) that the SMA of both hemispheres is active prior to unilateral movements.

ELECTRIC POTENTIALS AROUND MOVEMENT ONSET

There is a complex spatio-temporal pattern of electric potentials within the final 200 ms before and 200 ms after the onset of movement as detected by electromyographic activity (EMG onset). Electrogenesis is not only complicated with surface-recordings but also with epicortical recordings since the central sulcus which separates areas 4 and 3b has a complex structure. The disadvantage of magnetoencephalography (MEG) of being only sensitive to magnetic field changes produced by currents which are tangential to the scalp surface proves to be an advantage because it reduces complexity. MEG studies could clearly and consistently distinguish between a pre-movement ECD with the direction of current flow from surface-to-depth in the anterior bank of the central sulcus and a post-movement ECD with the opposite direction (Lang et al. 1990; Cheyne et al. 1991, 1992, Kristeva et al. 1991). This post-movement ECD had a maximum at about 110 ms after EMG onset and was found contralateral to the side of the performing hand. At about the same time, Lee et al. (1986) described a a bipolar pattern of electric brain potentials with negativity anterior to the central sulcus and positivity posterior to that line. Since a similar pattern was found with passive finger movements, it was argued that it should originate in the primary sensorimotor cortex (SI, area 3b) by surface-to-depth current flow. This bipolar pattern can also be found in surface recordings after voluntary finger movements (Fig. 9; see also Tarkka and Hallett 1991). Praamstra et al. (1996) explained the phenomenon with a ECD located posterior to the MI-Hand area and with current flow from anterior-mesial to posterior-lateral, a direction which is consistent with the geometry of the central sulcus. The fact, that the latency of post-movement brain potentials increases with toe movements as compared to finger movements supports the concept that these potentials are caused by stimulation of peripheral receptors (e.g. receptors of muscle spindles) at the onset of a movement. In summary, there is consistent evidence for the assumption that a bipolar pattern in surface recording with a maximum of negativity above the fronto-central midline and a positivity above the parietal cortex is caused by surface-to-depth current flow in area 3b and may reflect afferent input as caused by stimulation of peripheral receptors at movement onset.

However, additional electric events of the cortex may be hidden in surface-recorded brain potentials. Cheyne et al. (1992) studied the time activities of the pre-movement MI-ECD and the post-movement SI-ECD. After movement onset there is a decline of dipole strength of the MI-ECD and there may even be an "overshoot" i.e. a reversal of the direction of current flow. On the other hand, the SI-ECD was found to be already present prior to EMG onset and may have an opposite direction (depth-to-surface in area 3b) in the pre-movement period as compared to the post-movement period (surface-to-depth in area 3b). But, spatial resolution is not sufficient to disentangle these neural events. Single cell recordings in sub-human primates have shown that there is inhibition in area 3b prior to the onset of a voluntary movement (Nelson 1987) and it is known from psychophysiology that the threshold of perception of peripheral sensation is higher in a pre-movement period (Coquery 1978). Thus, it is not only the complex anatomy of the central sulcus but also complex functional interactions between MI and SI which make it difficult to distinguish between "motor" and "sensory" processes around EMG onset. Furthermore, EMG onset is defined by a threshold and it is possible to assume that there is sensory input preceding this event.

By analysis of waveform and topography of brain potentials, Tarkka and Hallett (1991) distinguished several electric events around EMG-defined movement onset: the PMP (Pre-Movement Positivity; Deecke et al. 1976) with a latency of about 50 ms prior movement onset (- 50 ms), isMP (initial slope of the Movement Potential), ppMP (parietal peak of the Motor Potential) and fpMP (frontal peak of the Movement Potential) with latencies of about -10 ms, 30 ms and 100 ms, respectively. The component fpMP, is to my understanding, the negative peak of the bipolar pattern which is casued by surface-to-depth current flow SI (SI-ECD). The electrogenesis and functional significance of PMP, isMP and ppMP are not established yet. Terms like "Motor Potential" (which has been introduced by Deecke et al. 1976) may be misleading for the reasons given above.

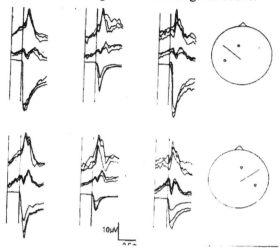

Figure 10: Brain potentials following the electromyographic onset of finger movements. Upper part: recordings from FC1 and CP5 together with rectified EMG for extension of the right index finger. Lower part: recordings from FC2 and CP6 together with the rectified EMG for extension of the left index finger. Three different subjects with three repetitive measurements in each subject. Note, there is a reversal of polarity consistent with the assumption of a surface-to-depth current flow in the posterior bank of the central sulcus after movement onset (unpublished data, W. Lang).

EVIDENCE FOR AND AGAINST SEPARATING BP COMPONENTS

Initially, research on event-related brain potentials was driven by the concept of finding associations between peaks (components) which emerge in surface-recordings and cognitive functions. With increasing numbers of paradigms and recording sites those components became more and more subdivided. The BP was found to have two phases, a slowly rising phase lasting from about 1.5 s to 500 ms before movement onset and a later, more rapidly rising phase lasting from about 500 ms to approximately the time of movement onset. Furthermore, it was observed that the BP is symmetrically distributed between the two hemispheres in the early phase whereas the BP gets asymmetrical in the later phase (500 ms to movement onset) (Deecke et al., 1976; Shibasaki et al., 1980). What is the physiological background for these changes of the BP pattern? Neither the physiology nor the functional significance of the change of steepness is currently known. The physiology behind hemispherical asymmetry of the BP in unilateral movements has been explained by asymmetries of activity in M1 (CSD-analysis; see Fig, 8; Cheyne et al., 1991; Kristeva et al., 1991; Praamstra et al., 1996). Given the established physiological background, BP-asymmetry (Lateralized Readiness Potential; LRP) became a valuable indicator for the time and the extent by which the motor system gets "biased" into one direction, i.e. to move with the right or the left hand (Coles et al., 1988). The start of BP-asymmetry may or may not coincide with the time when BP steepness changes. BP-asymmetry can easily be measured. Change of steepness of the BP is often quite clear in the grand average (average across subjects) but it is often not easy to determine in averages of single subjects. Furthermore, the time at which the BP steepness changes varies among recording sites.

Because of the difficulty of measuring the change of steepness of the BP, Dick et al (1989) suggested measuring the first component of the BP at 650 ms prior to movement onset and to calculate the second component of the BP from the period between 650 ms to movement onset. This procedure proved to be straightforward and helped to detect the reduction of the BP at 650 ms before movement onset in patients with Parkinson's disease. There is not only evidence from pathophysiology but also from physiology that the early part of the BP is different from the later part. Mappings of current source densities indicate that the initial and the later part of the BP may be generated by different neural sources with SMA contributing to the initial part but SMA and M1 to the second part (Cui et al. 1999). Thus, there are good arguments to hypothesize that their physiology and functional significance may be different for the first part of the BP (until about 500 ms before movement onset) and the second part of the BP (500 ms to movement onset). Since the BP pattern is abnormal in groups of patients (e.g. patients with Parkinson's disease; Dick et al. 1989, Touge et al. 1995; Jahanshahi et al. 1995) it is recommended that amplitudes in the initial and the second part of the BP should be measured.

SUGGESTIONS FOR FUTURE RESEARCH

We have learned a lot about the factors which affect the topography and the size of the BP but we still do not know why the BP starts so early even in simple movements. My hypothesis is that the movement is initiated at a time when the human motor cortices are in a general preparatory state. The existence of slow changes of cortical excitability have been demonstrated. Increase of surface-negativity in cortical areas was found to facilitate the execution of tasks which require the contribution of that cortical areas ("Brain Trigger Design"; Bauer and Nirnberger 1981; Born et al. 1982; Rockstroh et al. 1982). Contingencies between the background activity and event-related brain activity should further be studied.

When voluntary movements have consequences or are followed by other tasks, the BP-paradigm offers the possibility to study the timing, spatial pattern and amount of anticipatory activation of the brain in normals and in patients with disease. However, only a few studies in normals have been conducted. No other technique, such as PET or fMRI, has the temporal resolution to investigate the brain physiology related to the anticipation of a movement's consequences because of limited temporal resolution. Other areas of future research may be the acquisition of motor skills or to study the effects of transcranial magnetic stimulation (TMS) or stimulation with depth electrodes on the spatial pattern and amplitudes of the BP.

REFERENCES

Asenbaum, S., Lang, W., Eghker, A., Lindinger, G. and Deecke, L. (1992) Frontal DC potentials in auditory selective attention. *Electroenceph. Clin. Neurophysiol.* 82, 469-476.

Balint, R. (1909) Seelenlähmung des "Schauens", optische Ataxie, räumliche Störung der Aufmerksamkeit. *Monatsschr. Psychiat. Neurol.* 25, 51-81.

Bauer, H. and Nirnberger, G. (1981) Concept identification as a function of preceding negative or positive spontaneous shifts in slow brain potentials. *Psychophysiology* 18, 466-469.

Baumgartner, C. (1993) *Clinical electrophysiology of the somatosensory cortex.* Vienna: Springer.

Becker, W., Hoehne, O., Iwase, K. and Kornhuber, HH (1972) Bereitschaftspotential, prämotorische Positivierung und andere Hirnpotentiale bei sakkadischen Augenbewegungen. *Vision Research* 12, 421-436.

Becker, W., Iwase, K., Jürgens, R. and Kornhuber, HH (1976) Bereitschaftspotential preceding voluntary slow and rapid movements. In: W.C. McCallum and R.J. Knott, (Eds.) *The responsive brain,* pp. 99-102. Bristol: Wright.

Becker, W. and Kristeva, R. (1980) Cerebral potentials prior to various force deployments, in: Motivation, motor and sensory processes of the brain. In: H.H. Kornhuber and L. Deecke, (Eds.) Electrical potentials, behaviour and clinical use, pp. 189-194. Amsterdam: Elsevier.

Benecke, R., Dick J.P.R, Rothwell, J.C., Day, B.C. and Marsden, C.D. (1985) Increase of the Bereitschaftspotential in simultaneous and sequential movements. *Neuroscience Letters* 62, 347-352.

Born, J., Whipple, S.C. and Stamm, J. (1982) Spontaneous cortical slow-potential shifts and choice reaction time performance. *Electroenceph. Clin. Neurophysiol.* 54, 668-676.

Boschert, J. and Deecke, L. (1986) Cerebral potentials preceding voluntary toe, knee and hip movements and their vectors in human precentral gyrus. *Brain Res.* 376, 175-179.

Boschert, J., Hink, R.F., Deecke, L. (1983) Finger movement versus toe movement-related potentials: further evidence for supplementary motor area (SMA) participation prior to voluntary action. *Human Neurobiol* 52, 73-80.

Brinkman, C and Porter, R (1979) Supplementary motor area of the monkey: Activity of neurons during performance of a learned motor task. *Journal of Neurophysiology* 42, 681-709.

Castaigne, P., Laplane, D. and Degos, J.D. (1970) Trois cas de negligence motrice par lesion retrorolandique. *Rev. Neurol. Paris* 122(4), 233-242.

Castaigne, P., Laplane, D. and Degos, J.D. (1972) Trois cas de negligence motrice par lesion frontale prerolandique. *Rev. Neurol. Paris* 126, 5-15.

Cheyne, D., Kristeva, R. and Deecke, L. (1991) Homuncular organization of human motor cortex as indicated by neuromagnetic recordings. *Neuroscience Letters* 122, 17-20.

Cheyne, D., Kristeva, R., Deecke, L. and Weinberg, H. (1992) Spatio-temporal source modeling of sensorimotor cortex activation during voluntary movement in humans. In: M. Hoke, S.N. Erne, Y.C. Okada and G.L. Romani, (Eds) *Biomagnetism'91: Clinical aspects.* Amsterdam: Elsevier.

Coles, M.H., Gratton, G. and Donchin, E. (1988) Detecting early communi-cation: using measures of movement-related potentials to illuminate human information processing. *Biological Psychology* 26, 69-89.

Coquery, J.M. (1978) Role of active movement in control of afferent input from skin in cat and man. In: G. Gordon, (Ed.) *Active touch: the mechanisms of object manipulation: a multidisciplinary approach,* pp. 161-169. Oxford: Pergamon.

Cui, R.Q., Huter, D., Lang, W. and Deecke, L. (1999) Neuroimage of voluntary movement: topography of the Bereitschaftspotential, a 64-channel DC Current Source Density Study. *NeuroImage* 9, 124-134.

Deecke, L., Engel, M., Lang, W. and Kornhuber, H.H. (1986) Bereitschaftspotential preceding speech after holding the the breath. *Exp. Brain Res.* 65, 219-223.

Deecke, L., Grözinger, B. and Kornhuber, H.H. (1976) Voluntary finger movement in man: cerebral potentials and theory. *Biol. Cybern.* 23, 99-119.

Deiber, M.P., Passingham, R.E., Colebatch, J.G., Friston, K.J., Nixon, P.D. and Frackowiak, R.S.J. (1991) Cortical areas and the selection of movement: a study with positron emission tomography. *Exp. Brain Res.* 84, 393-402.

Dick, J.P.R., Rothwell, J.C., Day, B.L., Cantello, R., Buruma, O., Gioux, M., Benecke, R., Berardelli, A., Thompson, P.D. and Marsden, C.D. (1989) The Bereitschaftspotential is abnormal in Parkinson's disease. *Brain* 112, 233-244.

Dirnberger, G., Fickel, U., Lindinger, G., Lang, W. and Jahanshahi, M. (1988) The mode of movement selection: Movement-related cortical potentials prior to freely selected and repetitive movements. *Exp Brain Res.* 120, 263-72.

Dirnberger, G., Kunaver, C.E., Scholze, T., Lindinger, G. and Lang, W. (1992) The effects of alteration of effector and side of movement on movement-related cortical potentials, *Cogn Brain Res,* submitted

Erdler, M., Beisteiner, R., Mayer, D., Kaindl, T., Edward, V., Lindinger, G. and Deecke, L (2000) Supplementary motor area activation preceding voluntary movement is detectable with a whole scalp magnetoencephalography system. *NeuroImage* 11, 697-707.

Gerloff, C., Richard, J., Hadley, J., Schulman, A.E., Honda, M. and Hallett, M. (1998) Functional coupling and regional cortical motor areas during simple, internally paced and externally paced finger movements. *Brain* 121, 1513-1531.

Grözinger, B, Kornhuber, H.H. and Kriebel, J. (1979) Participation of mesial cortex in speech: Evidence from cerebral potentials preceding speech production in man. In: O. Creutzfeldt, H. Scheich and C. Schreiner (Eds.) *Hearing mechanisms and speech,* pp. 189-192. Berlin Heidelberg New York: Springer Verlag.

Hazemann, P., Metral, S. and Lille, F (1978) Influence of force, speed and duration of isometric contraction upon slow cortical potentials in man. In: D.A. Otto (Ed.) *Multidisciplinary Perspectives in Event-Related Brain Potential Research,* pp. 107-111. Washington: U.S. Environmental Protection Agency.

Jahanshahi, M., Jenkins, I.H., Brown, R.G., Marsden, C.D., Passingham, R.E., Brooks, D.J. (1995) Self-initiated versus externally triggered movements. I. An investigation using measurement of regional cerebral blood flow with PET and movement-related potentials in normal and Parkinson's disease subjects. *Brain* 118, 913-33.

Keller, I. and Heckhausen, H. (1990) Readiness potentials preceding spontaneous motor acts: voluntary vs. involuntary control. *Electroenceph. Clin. Neurophysiol.* 76, 351-361.

Kitamura, J., Shibasaki, H., Kondo T (1993) A cortical slow potential is larger before an isolated movement of a single finger than simultaneous movement of two fingers. *Electroenceph. Clin. Neurophysiol.* 86, 252-258.

Kitamura, J., Shibasaki, H., Takagi, A., Nabeshima, H., Yamaguchi, A. (1993) Enhanced negative slope of cortical potentials before sequential as compared with simultaneous extensions of two fingers. *Electroenc. Clin. Neurophysiol.* 86, 176-182.

Kornhuber, H.H. (1984) Mechanisms of voluntary movement. In: W. Prinz and A.F. Sanders (Eds.) *Cognition and motor processes,*.pp. 163-173. Berlin-Heidelberg-New York: Springer.

Kornhuber, H.H. and Deecke, L. (1965) Hirnpotentialänderung bei Willkürbewegungen und passiven Bewegungen des Menschen: Bereitschaftspotential und reafferente Potentiale. *Pflügers Arch. Ges. Physiologie* 284,1-17.

Kutas, M. and Donchin, E. (1974) Studies on squeezing: handedness, responding hand, response force and asymmetry of readiness potential. *Science* 186, 545-548.

Kristeva, R., Cheyne, D. and Deecke, L. (1991) Neuromagnetic fields accompanying unilateral and bilateral voluntary movements: topography and analysis of cortical sources. *Electroenceph. Clin. Neurophysiol.* 81, 284-298.

Kristeva, R, Cheyne, D., Lang, W., Lindinger, G. and Deecke, L. (1990) Movement-related potentials accompanying unilateral and bilateral finger movements with different inertial loads. *Electroenceph. Clin. Neurophysiol.* 75, 410-418.

Lang, M., Lang, W., Uhl, F., Kornhuber, A., Deecke, L. and Kornhuber, H.H. (1987) Slow negative potential shifts indicating verbal cognitive learning in a concept formation task. *Human Neurobiol.* 6, 183-190.

Lang, W., Lang, M., Heise, B., Deecke, L. and Kornhuber, H.H. (1984) Brain potentials related to voluntary hand tracking, motivation and attention. *Human Neurobiol.* 3, 235-240.

Lang, W., Lang, M., Uhl, F., Koska, Ch., Kornhuber, A. and Deecke, L. (1988) Negative cortical DC shifts preceding and accompanying simultaneous and sequential finger movements. *Exp. Brain Res.* 71, 579-587.

Lang, W., Obrig, H., Lindinger, G., Cheyne, D. and Deecke, L. (1990) Supplementary motor area activation while tapping bimanually different rhythms in musicians. *Exp. Brain Res.* 79, 504-514.

Lang, W., Torrioli, G., Pizzela, V., Romani, G.L. and Deecke, L. (1990) Movement-evoked reafferent fields. *Electroenceph. Clin. Neurophysiol.* 75, 62.

Lang, W., Cheyne, D., Kristeva, R., Beisteiner, R., Lindinger, G. and Deecke, L. (1991) Three-dimensional localization of SMA activity preceding voluntary movement - a study of electric and magnetic fields in a patient with infarction of the right supplementary motor area. *Exp. Brain Res.* 87, 688-695.

Lang, W., Beisteiner, R., Lindinger, G. and Deecke, L. (1992) Changes of cortical activity when executing learned motor sequences. *Exp. Brain Res.* 89, 435-440.

Lang, W., Höllinger, P., Egkher, A. and Lindinger, G. (1994) Functional localization of motor processes in the primary and supplementary motor areas. *J. Clin. Neurophysiol.* 11, 397-419.

Lee, B.I., Lüders, H., Lesser, R.P., Dinner, D.S. and Morris, H.H. (1986) Cortical potentials related to voluntary and passive finger movements recorded from subdural electrodes in humans. *Ann. Neurol* 20, 32-37.

Libet, B., Gleason, C.A., Wright, E.W. and Pearl, D.K. (1983) Time of conscious intention to act in relation to onset of cerebral activity (readiness potential). *Brain* 106, 623-642.

Libet, B., Wright, Jr., E.W. and Gleason, C.A. (1982) Readiness-potentials preceding unrestricted "spontaneous" vs. pre-planned voluntary acts. *Electroenceph. Clin. Neurophysiol.* 54, 322-335.

Lindinger, G., Baumgartner, C., Burgess, R., Lüders, H. and Deecke, L. (1994) Topographic analysis of epileptic spikes using spherical splines and spline laplacian (CSD). *J Neurol* 241, 276-277.

Lynch, J.C. (1981) The functional organization of posterior parietal association cortex. *Behav. Brain Sci.* 4, 485-534.

McCarthy, G. and Wood, C.C. (1985) Scalp distribution of event-related potentials: an ambiguity associated with analysis of variance models. *Electroencephal. Clin Neurophysiol* 69, 218-233.

Mountcastle, V.B., Lynch, J.C., Georgopoulos, A., Sakata, H. and Acuna, C. (1975) Posterior parietal association cortex of the monkey: command function for operations within the extrapersonal space. *J. Neurophysiol.* 38, 871-908.

Morgan, J.M., Wenzl, M., Lang, W., Lindinger, G. and Deecke, L. (1992) Fronto-central DC-potential shifts predicting behavior with or without a motor task. *Electroenceph. Clin. Neurophysiol.* 83, 378-388 (1992).

Nelson, R.J. (1987) Activity of monkey primary somatosensory cortical neurons changes prior to active movement. *Brain Res.* 406, 402-407.

Nunez P. (1981) *Electric fields of the brain, The neurophysics of EEG*, Oxford: Oxford University Press.

Okano, K. and Tanji, J. (1987) Neuronal activities in the primate motor fields of the agranular frontal cortex preceding visually triggered and self-paced movement. *Exp. Brain Res.* 66, 155-166.

Praamstra, P., Stegeman, D.F., Horstink, M.W.I.M., Brunia, C.H.M. and Cools, A.R. (1995) Movement-related potentials preceding voluntary movement are modulated by the mode of movement selection. *Exp. Brain Res.* 103, 429-439.

Praamstra, P., Stegeman, D.F., Horstink, M.W.I.M. and Cools, A.R. (1996) Dipole source analysis suggests selective modulation of the supplementary motor area contribution to the readiness potential. *Electroenceph. clin. Neurophysiol.* 98, 468-477.

Rockstroh, B., Elbert, T., Canavan, A., Lutzenberger, W. and Birbaumer, N. (1989) *Slow cortical potentials and behavior.* Baltimore: Urban & Schwarzenberg.

Roland, P.E, (1984) Organization of motor control by the human brain. *Human Neurobiol.* 2: 205-216.

Roland, P.E., Larsen, B., Lassen, N.A., Skinhoj, E. (1980) Supplementary motor area and other cortical areas in organization of voluntary movements in man. *J. Neurophysiol.* 43, 118-136.

Scherg, M. (1990) Fundamentals of dipole source potential analysis. In: F. Grandori, M. Hoke, Romani (Eds.) *Auditory evoked magnetic fields and electric potentials*, Advances in Audiology; vol 6, pp. 40-69. Basel: Karger.

Shibasaki, H., Barrett, G., Halliday, E. and Halliday, A.M. (1981) Cortical potentials associated with voluntary foot movement in man. *Electroencephalogr and Clin Neurophysiol* 52, 507-516.

Shibasaki, H., Barrett, G., Halliday, E., Halliday, A.M. (1980) Components of the movement-related cortical potential and their scalp topography. *Electroenceph. Clin. Neurophysiol.* 49, 213-26.

Simonetta, M., Clanet, M. and Rascol, O. (1991) Bereitschaftspotential in a simple movement or in a motor sequence starting with the same simple movement. *Electroenceph. Clin. Neurophysiol.* 81, 129-134.

Tarkka, I.M. and Hallett, M. (1991) Topography of scalp-recorded motor potentials in human finger movements. *J Clin Neurophysiol* 8, 331-341.

Taylor, M. (1978) Bereitschaftspotential during the acquisition of a skilled motor task. *Electroencephal. Clin. Neurophysiol.* 45, 568-576.

Touge, T., Werhahn, K.J., Rothwell, J.C. and Marsden, C.D. (1995) Movement-related cortical potentials preceding repetitive and random-choice hand movements in Parkinson's disease. *Ann. Neurol.* 37, 791-799.

von Giesen, H.J., Schlaug, G., Steinmetz, H., Benecke, R., Freund, H.J., Seitz, R.J. (1994) Cerebral network underlying unilateral motor neglect: evidence from positron emission tomography. *Journal of the Neurological Sciences* 125, 29-38 (1994).

Walter, W.G. (1964) The contingent negative variation. An electrical sign of significance of association in the human brain. *Science* 434.

THE BEREITSCHAFTS POTENTIAL (BP) AND THE CONSCIOUS WILL/INTENTION TO ACT

Benjamin Libet

Department of Physiology
University of California, San Francisco
San Francisco, CA 94143-0444
USA

INTRODUCTION

The discovery of the BP by Kornhuber and Deecke (1965) opened the way for further analyses of the volitional process, and it made possible my studies of the conscious will to act in relation to neural events in the brain.

Their finding that onset of BP preceded a voluntary act by about 1 sec (±) raised the question of whether the <u>conscious</u> initiation of the voluntary action also preceded the act by such long time intervals. This is a fundamental question for the nature of "free will". In the common view, free will was thought to initiate the neural processes in the brain that produce the voluntary act. That is, the conscious will would somehow order or command the brain to produce the voluntary act. Indeed, Sir John Eccles (1985) stated his belief that conscious will initiated the voluntary act in spite of the BP. But appearance of conscious will about a second before a freely voluntary act seemed to me intuitively unlikely. How could one test this issue?

Since the meaning assigned to the terms "voluntary action" and "will" can be quite complicated and are often related to one's philosophical biases, I shall attempt to clarify their usage here. In this experimental investigation and its analysis an act is regarded as voluntary and a function of the subject's will when (a) it arises endogenously, not in direct response to an external stimulus or cue; (b) there are no externally imposed restrictions or compulsions that directly or immediately control subjects' initiation and performance of the act; and (c) most important, subjects *feel* introspectively that they are performing the act on their own initiative and that they are *free* to start or not to start the act as they wish. The significance of point (c) is sharply illustrated in the case of stimulating the motor cortex (precentral gyrus) in awake human subjects. As described by Penfield (1958) and noted by others, under these conditions each subject regarded the motor action resulting from the cortical stimulation as something done to him by some external force; every subject felt that, in contrast to his normal voluntary activities, "he," as a self-conscious entity, had not initiated or controlled the cortically stimulated act.

Volitional processes may operate at various levels of organization and timing relative to the voluntary act. These may include consciously deliberating alternative choices as to what to do and when, whether or not to act, whether or not to comply with external orders or instructions to act, and so on. If any of these processes are to result in the motor performance of a voluntary act, they must somehow work their way into a "final common motor activation pathway" in the brain. Without an overt motor performance, volitional deliberation, choosing, or planning may be interesting for its mental or psychological content, but it does not constitute *voluntary action*. It is specifically this overt performance of the act that was experimentally studied by us.

EXPERIMENTAL STUDY OF 'WILL TO ACT'

We conducted a series of studies (Libet et al., 1982; 1983) in which the subjects were free to choose to perform a simple action at any time of desire, urge, decision, and will should arise in them. (They were also free not to act out any given urge or initial decision to act; and each subject indeed reported frequent instances of such aborted intentions.) The freedom of the subject to act at the time of his choosing actually provides the crucial element in this study. The objective was in fact to compare the time of the conscious will with the time of onset of associated cerebral processes. The specific choice of what act to perform was not material to the question being asked.

The subject is also instructed to allow each such act to arise "spontaneously," without deliberately planning or paying attention to the "prospect" of acting in advance. The subjects did indeed report that the inclination for each act appeared spontaneously ("out of nowhere"), that they were consciously aware of their urge or decision to act before each act, that they felt in conscious control of whether or not to act, and that they felt no external or psychological pressures that affected the time when they decided to act (Libet et al. 1982; Libet, Gleason, Wright & Pearl 1983).

Initially, I thought it would be impossible to measure the time of appearance of the conscious will (W) to act. W is a subjective phenomenon and we can only know about it from an introspective report by the subject. But such a report involves motor actions, in speaking or otherwise indicating the time of W. That kind of report would be distorted by the latency for making it and, more importantly, by a BP that precedes the voluntary report itself. Such a BP would contaminate the BP related to the voluntary act about which the report is being made.

A workable way to obtain the introspective report of W, without the difficulties cited above, finally occurred to me. It simply involved having the subject observe a fast moving "clock", without moving the eyes, and to note the position of the clock when he/she became first aware of the wish or urge to act. (Fig. 1). That clock position was to be reported after the subject performed the voluntary act under study (Libet et al. 1983). There are potential difficulties with this method, as raised by a number of commentators (Libet, 1985). I shall not repeat my reply to those arguments. I shall only point out two relevant experimental findings that supported the validity of the W reports (the time of appearance of the conscious will to act now).

For recording suitable RP, ("readiness potential" for BP) the average for 40 acts was obtained, with a report of W given by the subject after each event. There was, of course a distribution of the 40 Ws. But the S.E. for the mean value of W (close to -200 msec), in each set of 40 trials, was close to ±20 msec, for all subjects. That is, the reported values of W were moderately consistent, not wildly random.

Secondly, a·control series of 40 trials was carried out in each session of study. In this, the subject was asked not to move (confirmed by the EMG recording). Instead, a near-threshold effective electrical stimulus was delivered to the back of the subject's hand. This skin stimulus was delivered at a random time not known to the subject, but objectively known to the observer. The subject was instructed to note the "clock time" at which he/she felt the stimulus, and to report that clock time after the trial was over. Subjects reported stimulus times that averaged -50 msec, before the actual stimulus. This procedure mimicked that for reporting W times with voluntary acts. The apparent negative bias for the skin stimulus reports could be subtracted from the mean W of -200 msec, giving a corrected W of -150 msec.

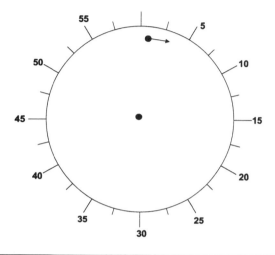

Figure 1: Oscilloscope 'clock'. Spot of light revolves around periphery of screen, once in 2.56 sec. (instead of 60 sec. for a sweep-second hand of a regular clock). Each marked off 'second' (in the total of 60 markings) represents 43 msec. of actual time here. The subject holds his gaze to the center of the screen. For each performed quick flexion of the wrist, at any freely chosen time, the subject was asked to note the position of the clock spot when he/she first became aware of the wish or intention to act. This associated clock time is reported by the subject later, after the trial is completed.

Preplanned versus fully spontaneous acts

We found that the recorded RPs tended to fall into one of two groups. (Libet et al. 1982). RPs with earlier onsets (of about -1 sec) and greater peak amplitudes (RP I) were distinguishable from RPs with later onsets (about -550 msec) and smaller peak amplitudes (RP II) (Fig. 2). Upon interrogating the subjects we found that RP I was associated with experiences of preplanning to act at some time in the next second or two. Some series with such reports occurred even though we asked each subject to try to let each voluntary act to appear "on its own", i.e. spontaneously. In the majority of 40 trials series subjects reported having no experiences of preplanning; these produced RP IIs.

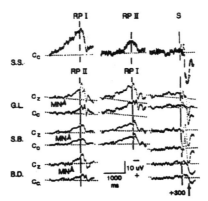

Figure 2: Readiness potentials (RP) preceding self-initiated voluntary acts. Each horizontal row is the computer-averaged potential for 40 trials, recorded by a DC system with an active electrode on the scalp, either at the midline-vertex (C_z) or on the left side (contralateral to the performing right hand) approximately over the motor/premotor cortical area that controls the hand (C_c).

When every self-initiated quick flexion of the right hand (fingers or wrist) in the series of 40 trials was (reported as having been) subjectively experienced to originate spontaneously and with no pre-planning by the subject, RPs labeled type II were found in association. (Arrowheads labeled MN indicate onset of the 'main negative' phase of the vertex recorded type II RPs in this figure; see Libet et al. 1982). When an awareness of a general intention or preplanning to act some time within the next second or so was reported to have occurred before some of the 40 acts in the series, type I RPs were recorded (Libet et al., 1982). In the last column, labeled S, a near-threshold skin stimulus was applied in each of the 40 trials at a randomized time unknown to the subject, with no motor act performed; the subject was asked to recall and report the time when he became aware of each stimulus in the same way he reported the time of awareness of wanting to move in the case of self-initiated motor acts.

Evoked potentials following the stimulus are seen regularly to exhibit a large positive component with a peak close to $=300$ ms. (arrow indicates this time); this P300 event-related potential had been shown by others to be associated with decisions about uncertain events (in this case, the time of the randomly delivered stimulus), and it also indicates that the subject is attending well to the experimental conditions.

The solid vertical line through each column represents 0 time, at which the electromyogram (EMG) of the activated muscle begins in the case of RP series, or at which the stimulus was actually delivered in the case of S series. The dashed horizontal line represents the DC baseline drift.

It seems likely that most RP studies by others are of the RP I type, with onsets close to 1 sec roughly. Most studies do not give instructions for subjects to try to act spontaneously. In addition, some restrictions of when to act are often imposed on the subject. Since hundreds of voluntary acts are often averaged to produce a RP, the investigator attempts to keep the total experimental session to a reasonable time by asking subjects to perform their "self-paced" movements within a time period not to exceed, say, 6 sec. Such an instructions may lead the subject to preplan when to perform his "self-paced" movements.

The distinction between our RP I and RP II is an important one. The component of preplanning, in RP I, introduces a process that is distinctly different from that involved in producing the final "act now" phase of a voluntary act. RP II would seem to be more representative of this "act now" phase. One point in support of this was our finding that the mean W values were closely similar (about -200 msec) for both the RP I and RP II series. That is, the final awareness of wanting to act came at similar times, independent of the preplanning process.

As a general point, processes in deliberating about choices of *"whether to act"* and of *"when to act"* (as in preplanning), are different from the final decision to *"act now"*. One can of course deliberate about an action all day long without ever acting at all. It is the final process in deciding to "act now" that constitutes the volitional process for action.

Unconscious initiation of a voluntary act

We found a clear and consistent gap of time between onsets of RP II and the reported times for W (first awareness of wanting to act now). Onsets of RP II preceded Ws by about 350 msec (see Fig. 3). Or, if Ws are corrected for the -50 msec bias seen in reporting awareness of a skin stimulus, (making W= - 150 msec) the gap becomes 400 msec. One could only conclude that brain activity involved in initiating the volitional process occurs unconsciously, before one is consciously aware of the wish or intention to act.

This sequence of events and the conclusion were confirmed by Keller and Heckhausen (1990) and Haggard and Eimer (1999). Keller and Heckhausen added an experiment in which slight spontaneous movements were made unconsciously while the subject concentrated on counting backwards in steps of 3. They recorded some sort of RP's with onsets about –460 msec before such acts. They concluded that RP's could occur even with unconsciously performed acts, but I suggested that the mental task could have disrupted the memory of those acts (Libet, 1992). Haggard and Eimer made additional studies relating reports of early and late RP respectively to the W reports. They concluded that awareness of conscious will (W) was produced in the pre-motor area of the cortex rather than RP processes. Their results were probably affected by large components of pre-planning by their subject, because of their instructions. That calls into question the validity of their conclusion about the locus of W (a discussion of these issues (Haggard and Libet, 2001)

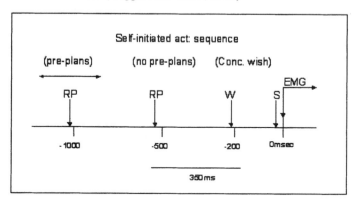

Figure 3: Diagram of sequence of events, cerebral and subjective, that precede a fully self-initiated voluntary act. Relative to 0 time, detected in the electromyogram (EMG) of the suddenly activated muscle, the readiness potential (RP)(an indicator of related cerebral neuronal activities) begins first, at about -1050 ms., when some pre-planning is reported (RP I), or about -550 ms. with spontaneous acts lacking immediate pre- planning (RP II). Subjective awareness of the wish to move (W) appears at about -200 ms., some 350 ms. after onset even of RP II; however, W does appear well before the act (EMG). Subjective timings reported for awareness of the randomly delivered S (skin) stimulus average about -50 ms. relative to actual delivery time. (From Libet, 1989.)

We should note that the neural activities represented by the recorded RP II may or may not be the first initiator of the voluntary process. It seems likely that the supplementary motor area (SMA), thought to be the source of RP II, may itself be activated from elsewhere in the brain. If so, the gap between cerebral initiating processes and W would be even larger than 400 msec. How unconscious initiation of the volitional process affects our views of "free will" will be discussed below.

The gap of about 400 msec, before W appears, is surprisingly close to the values (0.5s) for durations of brain activity found to be necessary to elicit awareness of a sensory stimulus (see Libet 1973; 1993). This suggests that to elicit an endogenous awareness also required activities lasting about 0.5 sec.

Some critiques of our findings

Following my review article in *Behavioral and Brain Sciences* (Libet, 1985) twenty-five different commentaries appeared, together with my reply. It may be of interest to summarize some of this exchange.

Some commentators felt the conscious will (W) must be developing gradually until a threshold is reached for reportability. The basis for this could be thought to lie in signal detection studies, in which the degree of detection rises gradually from the zero level of a signal. But detection should not be confused with conscious subjective experience; detection can occur *unconsciously* (Libet et al. 1991). Our subjects insisted they were not aware of anything before their reported W time.

Some commentators argued that there would be variable delays in becoming visually aware of the "clock-time," and so reported W's may not accurately reflect the actual time of first conscious will. But one should not confuse *what* the subject reports with *when* he/she actually becomes aware of that. Additionally, the control series with skin stimulus, instead of movements (see above), showed the subjects reporting the clock time to within 50 msec of the actual stimulus time. In this series, all the worries of the commentators are potentially present, with no significant affect.

There were various arguments about the meaning of the recorded RP. For example, does the RP really represent process of preparation to perform the voluntary act. The fact is that the RP only appears in association with a *preparation* to move. A motor response to an unwarned stimulus is not preceded by a RP. Acts preformed "automatically," without any conscious intention, appear with little or no RP preceding the act.

There were arguments that the conscious veto of an act must itself be developed by preceding unconscious processes, I have made a full rebuttal to that (in Libet, 1999.) It is chiefly based on viewing the conscious veto as a "control" function. It is not in the same category as conscious awareness, and it need not require the time for developing an awareness.

There is also a later critique by Gomes (1998), which is too lengthy and complicated to summarize here. I analyzed it fully in a reply paper (Libet, 2000).

IMPACT ON NATURE OF FREE WILL

An accepted view of free will would require that a "free" decision to act be made consciously. If the voluntary preparation to "act now" is made unconsciously it would be difficult to see how an individual could be held responsible for his wish too, when he has no conscious control over the initiating event.

Our finding that a freely voluntary act is initiated unconsciously, well before an individual is consciously aware of the wish or decision to act, means that free will does not initiate the freely voluntary process. (I am of course extrapolating from our findings with a simple voluntary act, flexion of the hand and fingers, to voluntary acts generally). Is there then some potential role for free will in voluntary actions?

For this, a crucial feature of our findings must be noted. Although W does not occur until about 400 msec after the brain initiates the process unconsciously, W does appear about 150 msec before the voluntary motor act. That leaves time for the conscious function to have an effect on the outcome of the volitional process. The obvious possible effect can be for the conscious will to block or "veto" the process, so that no motor act actually occurs (Libet 1985).

Our subjects at times reported having had an urge or wish to act but they vetoed it. The consequent absence of an EMG meant there was no trigger to the computer to back track and record a RP. To obtain a RP with a veto we asked the subject to pre-plan to move at a pre-set time but to veto that intention when the clock reached 100 to 200 msec before that preset time. We could then apply a proper signal to the computer at the known pre-set time, and record any RP even with no motor act. The RP that started before the pre-set time resembled a RP I, but tended to flatten at the normal rise about 200 msec before the pre-set time in accord with the veto.

The existence of a potential conscious veto is actually well known and not in doubt. We have all had the experience of becoming aware of an urge or intention to perform an act and of blocking/vetoing the performance of the act. Such vetoes often occur when the particular urge is viewed as socially or personally unacceptable. For example, you may have a conscious urge to kick the professor (in relation to a nasty quiz by him), but you normally veto that urge. One may propose, then, that free will could function as a control agency; it could either allow (or foster) a given volitional process to go to completion, or it could veto it and abort the act.

In the overall picture, one may visualize the brain unconsciously "burbling up" initiatives for various urges or intentions to act, with the conscious function selecting and controlling which such initiatives will be consummated in an actual motor act. Such a control role for free will is actually in accord with religious instructions to "control yourself" in your behaviors. Most of the Ten Commandments are orders not to do certain things.

DOES THE CONSCIOUS VETO ITSELF ARISE UNCONSCIOUSLY?

If the conscious veto is developed by preceding unconscious processes, as is the case for the appearance of W, then free will acting as a control agent would itself not be a bona fide independent conscious function. This issue is discussed at some length by Libet (1999), and will only be briefly outlined here.

The conscious veto is a control function, different from simply becoming aware of the wish to act. There is no logical imperative in any mind-brain theory, even identity theory, that requires specific neural activity to precede and determine the existence of a conscious control function. And, there is no experimental evidence against the possibility that the control process may appear without development by prior unconscious processes.

The possibility is not excluded that factors, on which the decision to veto (control) is based, do develop by unconscious processes that precede the veto. However, one could consciously accept or reject the programme offered up by the whole array of preceding brain processes. The conscious decision to veto could still be made without the direct specification of that decision by the preceding unconscious processes.

DETERMINISM AND FREE WILL

There remains a deeper question about free will that the foregoing considerations have not addressed. What we have achieved experimentally is some knowledge of how free will may operate. But we have not answered the question of whether our consciously willed acts are fully determined by natural laws that govern the activities of nerve cells in the brain, or whether acts and the conscious decision to perform them can proceed to some degree independently of natural determinism. The first of these options would make free will illusory. The conscious feeling of exerting one's will would then be regarded as an epiphenomenon, simply a by-product of the brain's activities but with no causal powers of its own.

The philosophy of materialist determinism holds that everything in the world, including thought, will and feeling, can be explained only in terms of the natural laws that appear to govern the behavior of matter.

First, it may be pointed out that free choices or acts are not predictable, even if they should be completely determined. The 'uncertainty principle' of Heisenberg precludes our having a complete knowledge of the underlying molecular activities. Quantum mechanics forces us to deal with probabilities rather than with certainties of events. And, in chaos theory, a random event may shift the behavior of a whole system, in a way that was not predictable. However, even if events are not predictable in practice, they might nevertheless be in accord with natural laws and therefore determined.

Let us re-phrase our basic question as follows: *Must* we accept determinism? Is non-determinism a viable option? We should recognize that both of these alternative views (natural law determinism vs. non-determinism) are unproven theories, i.e. unproven in relation to the existence of free will. Determinism has on the whole, worked well for the physical observable world. That has led many scientists and philosophers to regard any deviation from determinism as absurd and witless, and unworthy of consideration. But there has been no evidence, or even a proposed experimental test design, that definitively or convincingly demonstrates the validity of natural law determinism as the mediator or instrument of free will.

There is an unexplained gap between the category of physical phenomena and the category of subjective phenomena. As far back as Leibniz, it was pointed out that if one looked into the brain with a full knowledge of its physical makeup and nerve cell activities, one would see nothing that describes subjective experience. The whole foundation of our own experimental studies of the physiology of conscious experience (beginning in the late 1950s) was that externally observable and manipulable brain processes and the related reportable subjective introspective experiences must be studied simultaneously, as independent categories, to understand their relationship. The assumption that a deterministic nature of the physically observable world (to the extent that may be true) can account for subjective conscious functions and events is a speculative *belief*, not a scientifically proven proposition.

Non-determinism, the view that conscious will may, at times, exert effects not in accord with known physical laws, is of course also a non-proven speculative belief. The view that conscious will can affect brain function in violation of known physical laws, takes two forms. In one it is held that the violations are not detectable, because the actions of the mind may be at a level below that of the uncertainty allowed by quantum mechanics. (Whether this last proviso can in fact be tenable is a matter yet to be resolved). This view would thus allow for a non-deterministic free will without a perceptible violation of physical laws. In a second view it may be held that violations of known physical laws are large enough to be detectable, at least in principle. But, it can be argued, detectability in actual practice may be impossible. That difficulty for detection would be especially true if the conscious will is able to exert its influence by minimal actions at relatively few nerve elements; these actions could serve as triggers for amplified nerve cell patterns of activity in the brain. In any case, we do not have a scientific answer to the question of which theory (determinism or non-determinism) may describe the nature of free will.

However, we must recognize that the almost universal experience that we can act with a free, independent choice provides a kind of *prima facie* evidence that conscious mental processes can causatively control some brain processes (Libet, 1994). As an experimental scientist, this creates more difficulty for a determinist than for a non-determinist option. The phenomenal fact is that most of us feel that we do have free will, at least for some of our actions and within certain limits that may be imposed by our brain's status and by our environment. The intuitive feelings about the phenomenon of free will form a fundamental basis for views of our human nature, and great care should be taken not to believe allegedly scientific conclusions about them which actually depend upon hidden ad hoc assumptions. A theory that simply interprets the phenomenon of free will as illusory and denies the validity of this phenomenal fact is less attractive than a theory that accepts or accommodates the phenomenal fact.

In an issue so fundamentally important to our view of who we are, a claim for illusory nature should be based on fairly direct evidence. Such evidence is not available; nor do determinists propose even a potential experimental design to test the theory. Actually, I myself proposed an experimental design that could test whether conscious will could influence nerve cell activities in the brain, doing so via a putative 'conscious mental field' that could act without any neuronal connections as the mediators (Libet, 1994). This difficult though feasible experiment has, unfortunately, still to be carried out. If it should turn out to confirm the prediction of that field theory, there would be a radical transformation in our views of mind-brain interaction.

My conclusion about free will, one genuinely free in the non-determined sense, is then that its existence is at least as good, if not a better, scientific option than is its denial by determinist theory. Given the speculative nature of both determinist and non-determinist theories, why not adopt the view that we do have free will (until some real contradictory evidence may appear, if it ever does.) Such a view would at least allow us to proceed in a way that accepts and accommodates our own deep feeling that we do have free will. We would not need to view ourselves as machines that act in a manner completely controlled by the known physical laws. Such a permissive option has also been advocated by the neurobiologist Roger Sperry (see Doty, 1998).

I close, then, with a quotation from the great novelist Isaac Bashevis Singer that relates to the foregoing views. Singer stated his strong belief in our having free will. In an interview (Singer, 1968) he volunteered that "The greatest gift which humanity has received is free choice. It is true that we are limited in our use of free choice. But the little free choice we have is such a great gift and is potentially worth so much that for this itself life is worthwhile living".

REFERENCES

Doty, R.W. (1998) Five mysteries of the mind, and their consequences. *Neuropsychologia* **36,**(10).1069-1076.

Eccles, J.C. (1985) Mental summation: The Timing of Voluntary intentions by cortical Activity. *Behavioral and Brain Sciences* **8,** 542-543.

Gomes, G. (1998) The timing of conscious experience. A critical review and reinterpretation of Libet's research.. *Consciousness and Cognitive* **7,** 559-595.

Haggard, P. & Eimer, M (1999). On the relation between brain potentials and the awareness of voluntary movements. *Exp. Brain Res.* **126,** 128-133.

Haggard, P & Libet, B. (2001) Conscious intention and brain activity. *J Consciousness Studies* **8,** 47-63

Keller, I & Heckhausen, H (1990). Readiness potentials preceding spontaneous acts: Voluntary vs. involuntary control. *Electroencephalography and Clinical Neurophysiology* **76,** 351-361.

Kornhuber, H.H. and Deecke, L. (1965), Hirnpotentialanderungen bei
    Willkurbewegungen und passiven Bewegungen des Menschen: Bereitschaftspotential und reafferente Potentiale, *Pfluegers Arch Gesamte Physiol Menschen Tiere* **284**, 1-17.

Libet, B. (1973), Electrical stimulation of cortex in human subjects, and conscious
    sensory aspects. In : *Handbook of Sensory Physiology*: vol II, *Somto-sensory system*, ed. A. Iggo (Berlin: Springer-Verlag).

Libet, B. (1989), Conscious subjective experience vs. unconscious mental functions: A
    theory of the cerebral processes involved. In: *Models of brain function* ed. R.M.J. Cotterill (Cambridge University Press).

Libet, B. (1985), Unconscious cerebral initiative and the role of conscious will in
    voluntary action. *Behavioral and Brain Sciences* **8**, 529-66.

Libet, B. (1992) Voluntary acts and readiness potentials, *Electroencephalography and Clinical Neurophysiology* **82**, 85-86.

Libet, B. (1993), The neural time factor in conscious and unconscious mental events, in
    Ciba Foundation Symposium #174, *Experimental and Theoretical Studies of Consciousness* (Chichester: Wiley).

Libet, B (1994), A testable field theory of mind-brain interaction. *J. Consciousness Studies* **1**, 119-26.

Libet, B. (1999) Do we have free will? *J. Consciousness Studies* **6**, 47-57.

Libet, B. (2000) Time factors in conscious processes. A reply to Gilberto Gomes.
    Consciousness and Cognition **9**, 1-12.

Libet, B., Gleason, C.A., Wright, E.W. and Pearl, D.K. (1983), Time of conscious
    intention to act in relation to onset of cerebral activity (readiness potential): The unconscious initiation of a freely voluntary act. *Brain* **106**, 623-42.

Libet, B., Pearl, D.K., Morledge, D.G., Gleason, C.A., Hosobuchi, Y., and
    Barbaro, N.M. (1991), Control of the Transition from Sensory Detection to Sensory Awareness in man. The cerebral "time on" factor. *Brain* **114**, 1731-57.

Libet, B. Wright, E.W. and Gleason, C. A. (1982), Readiness potentials preceding
    unrestricted spontaneous pre-planned voluntary acts. *Electroenceph. & Clin. Neurophysiology* **54**, 322-5.

Singer, I.B. (1968), Interview by H. Flender, in *Writers at Work* (1981), ed. G. Plimpton
    (New York: Penguin Books).

GENERATOR MECHANISMS OF BEREITSCHAFTSPOTENTIALS AS STUDIED BY EPICORTICAL RECORDING IN PATIENTS WITH INTRACTABLE PARTIAL EPILEPSY

Akio Ikeda[1] and Hiroshi Shibasaki, M.D.[1,2]

Department of Neurology[1], and Human Brain Research Center[2],
Kyoto University Graduate School of Medicine,
Shogoin, Sakyo-ku, Kyoto, 606, JAPAN

INTRODUCTION

Since the initial description of Bereitschaftspotental (BP) by Kornhuber and Deecke in 1965, the cortical generators of BP in humans have been investigated mainly by the use of electroencephalographic (EEG) source analysis and magnetoencephalography (MEG), and also invasively in patients with intractable partial epilepsy. The BP is considered to provide an important clue to the understanding of preparatory cortical functions in association with voluntary movements in normal subjects, pathogenesis of movement disorders like dystonia and parkinsonism, and psychogenic movement disorders. Subdural recordings in epilepsy surgery have been shown to be useful not only for identifying the precise focus of epileptic discharges but also for assessing function of the cortical areas around the epileptogenic zone, mainly with the aid of cortical stimulation (Hahn and Lüders,1987; Lüders et al., 1987). Our group performed direct recording of pre-movement slow potentials from the cerebral cortex by using chronically implanted subdural electrodes in those patients during presurgical evaluation. One of the most important current topics with regard to cortical generators of BP is the role of supplementary motor area (SMA). Symmetric distribution of the initial slow potentials at the vertex area with unilateral hand movement could reflect either the activities of bilateral SMAs or a simple summation of the activities arising from bilateral primary motor cortices (MIs). Since clear distinction of pre-SMA and SMA proper has been proposed based on monkey experiments (Luppino et al., 1991; Matsuzaka et al., 1992), the presence of rostral and caudal SMAs in humans has drawn an increasing attention, and thus the previous results of cortical potentials recorded from the mesial frontal area (Ikeda et al., 1992a,b; Shibasaki and Ikeda,1996) have been reappraised. Another important concern is whether, in addition to MI, the lateral rostral non-primary motor area generates the BP or not. Recent invasive studies played a pivotal role in answering those essential questions about the generator mechanism of the BP.

The Bereitschaftspotential
Edited by Jahanshahi and Hallett, Kluwer Academic/Plenum Publishers, New York, 2003

LATERAL PREMOTOR VS. PRIMARY MOTOR CORTICES

In early studies, Neshige et al (1988) recorded the BP associated with self-paced finger movements from subdural electrodes placed over the lateral frontal convexity in epilepsy patients. The BP starting at –1 to –1.5 sec (1 to 1.5 sec before the onset of finger movement), the NS' starting at –250 to -400 msec, and the MP (Motor Potential) were all highly localized at the part of MI and the part of the primary somatosensory area (SI) each somatotopically corresponding to the moving finger, particularly at the part of MI where high frequency cortical stimulation elicited positive motor responses such as finger twitching. With unilateral finger movements, the BP was recorded from bilateral MI-SI hand areas, and the NS' was predominantly from the contralateral MI-SI hand area, while the MP was recorded only from the contralateral MI-SI hand area. In that report, those potentials were surface negative in polarity at all recorded sites, and no phase reversal was observed across the central sulcus **(Figure 1)**.

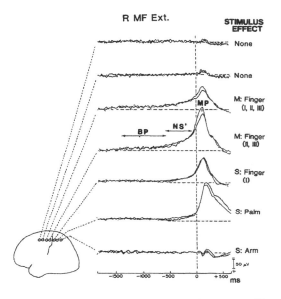

Figure 1 Movement-related cortical potentials in association with self-paced extension of the right middle finger recorded from the subdural electrodes placed across the left central sulcus in an epilepsy patient. BP, NS' and MP are well localized to the finger motor area, and to a lesser degree, to the sensory area. M=motor, S=sensory. (from Neshige et al., 1988 with permission)

The data suggested that the potentials arising only from a radial dipole, most likely generated by the crown of the precentral gyrus, were recorded by those subdural electrodes, and that the tangential dipoles generated from the anterior or posterior bank of the central sulcus were either absent or not recorded.

In the self-paced tongue protrusion task, clear pre-movement slow potentials were recorded from the lower perirolandic area where high frequency cortical stimulation produced surface-positive motor responses in the tongue (Ikeda et al, 1995a). However, the polarity of the pre-movement potentials along the central sulcus varied among the subjects, and a total of four different patterns were observed **(Figure 2)**. Of interest, the polarity reversed across the central sulcus in most subjects. In view of the intracortical field potential studies in monkeys suggesting that the pre-movement slow potentials represent excitatory postsynaptic potentials generated by the thalamo-cortical projections in the superficial layers of apical dendrites of cortical pyramidal neurons (Sasaki and Gemba, 1991a,b), the possible cortical generators along the central sulcus were postulated to explain the potential distribution over the scalp in association with self-paced tongue protrusion (**Figure 2**). The anterior bank of the central sulcus and the crown of the precentral gyrus may not always generate pre-movement slow potentials simultaneously. The post-movement potentials were always observed as surface-positive potentials on the postcentral gyrus. In order to explain these findings, common activation of both the posterior bank of the central sulcus and the crown of the postcentral gyrus is plausible. Histologically in human brain, the tongue area at the crown of the precentral gyrus corresponds to area 6a rather than area 4 (**Figure 3**). Therefore, at least as far as self-paced tongue movements are concerned, not only area 4 but also area 6 might generate pre-movement potentials. When stimulated with a train of high frequency (50 Hz) electric pulses for functional cortical mapping in epilepsy surgery, the crown of the precentral gyrus elicits positive motor responses, and thus is regarded as an eloquent, indispensable cortical area that projects to the corticospinal tracts like area 4. Therefore, at least in the perirolandic area, it is most likely that both the area 4 and a part of area 6a , from both of which corticospinal tracts originate, generate pre-movement slow potentials.

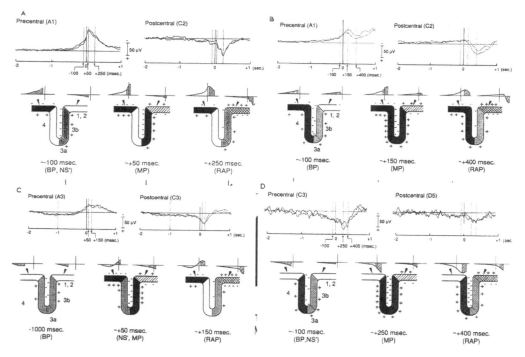

Figure 2 Assumed generator location of movement-related cortical potentials across the central sulcus in tongue protrusion (A,C,D) and in vocalization (B) in epilepsy patients.
A: Surface-negative BP at the precentral and positive BP at the postcentral area are seen.
B: Only negative potentials are seen across the central sulcus with vocalization.
C: The potential distribution is similar to A, except that the initial part of BP at the precentral area is positive in polarity.
D: Positive BPs are seen at the pre- and postcentral areas.
(from Ikeda et al., 1995a with permission)

It was expected that the MP might represent the activity of pyramidal neurons that project to the corticospinal tracts as final motor output to the spinal cord whereas BP and NS' could represent preparatory activity of the neurons in MI and the adjacent areas. As shown in **Figure 4**, epicortical recording of pre-movement potentials showed that BP, NS' and MP were not necessarily generated from the common electrodes, but from different electrodes within MI. In the example shown in **Figure 4**, BP is maximal at A3, NS' and MP at A1 and B1, and MP at A2 and B2.

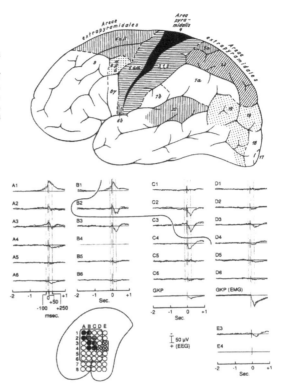

Figure 3 Cytoarchitectonic map of human brain adapted by Foerster from the Vogts.
The crown of the precentral gyrus in the lower rolandic area is little occupied by area 4 (filled area), but mainly occupied by area 6a α (shaded area). (cited from Fig.3a in O.Bumke and O. Foerster, Handbuch der Neurologie, Bd.6, 1936 with permission)

Figure 4 Movement-related cortical potentials in association with self-paced tongue protrusion recorded from the subdural electrodes placed across the left lower central sulcus in an epilepsy patient.
Large negative BP, NS' and MP are seen in the tongue motor area (A1 and B1). A3 shows only BP whereas A2 and B2 show only MP.
(from Ikeda et al., 1995a with permission)

When recording pre-movement slow potentials in association with self-paced movements of various parts of the body, the distribution of the potentials along the central sulcus was consistent with the somatotopy in the precentral gyrus. Therefore, BP analysis by subdural electrodes is clinically useful for functional mapping of the so-called "primary motor area" (MI) (which is defined by positive motor responses to cortical stimulation) in epilepsy surgery. Recording of BP is especially useful when the functional area is involved by epileptogenicity, because electric stimulation of that area easily produces seizures even with the conventional stimulus intensity used in the clinical mapping protocol (Yazawa et al.,1997; Ikeda et al., 2001). Furthermore, we recently observed that, whichever part of the boby was voluntarily moved, the lateral frontal area at or adjacent to the lateral negative motor area (LNMA) (Lüders et al., 1995), located just rostral to the MI face area, generated surface-positive pre-movement slow potentials (Kunieda et al., 1999). It still remains to be solved whether this potential is generated from the prefrontal area (areas 44 and 45) anterior to the precentral sulcus or from the ventral premotor area (area 6aα posterior to that sulcus. So far, no other cortical generators of pre-movement slow potentials have been identified in the lateral frontal area rostral to the positive motor area.

electrodes placed along the central sulcus. The RAP was surface positive in the postcentral and negative in the precentral area with phase reversal across the central sulcus. It was similar in morphology to the second tangential component (P30 or P2) of the somatosensory evoked potentials (SEPs) to median nerve stimulation recorded in the same session. With self-paced tongue movements, robust RAPs were likewise recorded in the postcentral area (activities shown at the postcentral parts of **Figure 2A-D**), and the distribution was also similar to the SEPs to lip stimulation recorded in the same patients (Ikeda et al., 1995a).

SMA PROPER AND PRE-SMA

a) Activity of the SMA proper

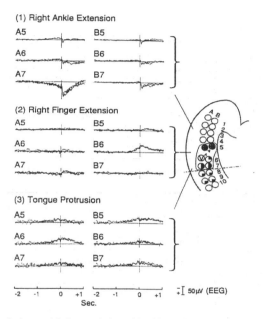

Figure 5 Movement-related cortical potentials in association with self-paced movements of the right finger, right foot and tongue recorded from the left SMA in an epilepsy patient. The figure illustrates the somatotopic distribution of the BP within the SMA proper, consistent with the results of cortical stimulation.
The broken line shows the boundary between the primary motor foot area posteriorly (lower part in the right figure) and the SMA proper anteriorly (middle part in the right figure).
Symbols in the figure show the results of electric cortical stimulation (A5 & B5 = positive motor response in the face and negative motor response of the bilateral hands; A6 = rhythmic vocalization; B6 = tonic motor response of the right hand; A7 and B7 = positive motor response of the right hand and foot; A8, B8, A9 & B9 = clonic motor response of the right foot)
(cited from Ikeda et al., 1992a with permission)

The SMA has been regarded as a supramotor area for programming and execution of voluntary movements based on clinical observations (Penfield and Welch, 1951) and on the pioneering regional cerebral blood flow studies (Orgogozo and Larsen, 1979). Since the initial component of slow cortical potentials before voluntary movements was always symmetrically distributed with the vertex maximum, it was strongly hypothesized that the SMA might participate in its generation. Neshige et al (1988) reported one epilepsy patient in whom very small pre-movement slow potentials were recorded from subdural electrodes chronically implanted on the mesial frontal cortex. The recorded potentials were quite small probably because those were recorded through the falx cerebri from electrodes inserted in the interhemispheric fissure of the other side. In this study, therefore, a rather widely distributed and small BP was picked up as the result of the epidural recording condition. They interpreted that the SMA generated a much smaller BP than MI did.

We further investigated the contribution of the SMA to the generation of BP (Ikeda et al., 1992a). By using subdural strip or grid electrodes directly placed on the mesial frontal cortex, the following findings were obtained. (1) In unilateral finger or foot movements, bilateral SMA proper generated clear BP and NS' with some contralateral predominance in amplitude, (2) MP was observed mainly on the contralateral SMA proper, and (3) pre-movement slow potentials for movements of different parts (tongue, finger, foot, etc) followed a somatotopic distribution within the SMA proper, which was consistent with somatotopic organization of the SMA proper defined by high frequency electric stimulation **(Figure 5)**.

b) Activity of the pre-SMA

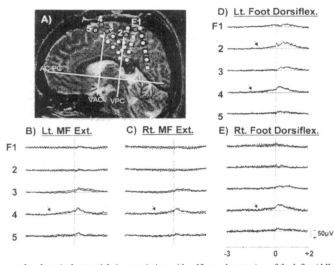

Figure 6 Movement-related cortical potentials in association with self-paced extension of the left middle finger (B) and the right middle finger (C), and dorsiflexion of the left foot (D) and the right foot (E), recorded from the right mesial frontal cortex in an epileptic patient. In all tasks, negative BP was observed at F4, located anterior to the VAC line, being most likely pre-SMA. F2 just on the paracentral lobule (most likely MI foot area) generated BP only in association with the left foot movements.
VAC = a line on the anterior commissure perpendicular to AC-PC line, VPC = a line on the posterior commissure perpendicular to AC-PC line.
(from Yazawa et al., 2000 with permission)

The pre-SMA is believed to be delimited as the rostral part of the conventional SMA, typically anterior to the vertical anterior commissural (VAC) line in humans. According to the recent neuroimaging studies of humans and invasive animal studies (Wiesendanger,1986; Luppino et al., 1991; Matsuzaka et al., 1992; Tanji, 1994; Picard and Strick, 1996), pre-SMA was significantly activated in association with higher order functions of motor control such as intrinsic movement selection, motor learning, complex movements, Go/NoGo trials and so on, independent of body part involved. The pre-SMA could correspond, at least in part, to the so-called "supplementary negative motor area" (SNMA) (Lüders et al., 1995), where high frequency cortical stimulation interrupts voluntary tonic muscle contraction or rapid alternating movements without loss of consciousness.At some electrodes located at or just adjacent to SNMA, usually rostral to the VAC line, slow negative potentials were consistently recorded in association with movements of various parts of the body (Yazawa et al., 2000). By contrast, SMA proper, i.e., caudal SMA, generated pre-movement slow potentials in accordance with its somatotopy as described above. The amplitude of BP at the pre-SMA was usually smaller than that at the SMA proper, but the onset times of BP in the two areas were not different **(Figure 6)**. Since those electrodes generating the BP are located at or just adjacent to SNMA, it is most likely that a part of the pre-SMA is a cortical generator of BP, but the precise relation between pre-SMA and SNMA still remains to be clarified.

Among four components of movement-related cortical potentials (BP, NS', MP and RAP), the pre-SMA seems to generate only BP, followed by only an ill defined NS', and no MP or RAP. This strongly suggests a functional, hierarchical difference among pre-SMA, SMA proper and MI in preparation for voluntary movements **(Table)**. Allison et al (1996) also investigated the BP from subdural electrodes placed in the mesial frontal area in patients with intractable partial epilepsy, and they concluded that BP on the mesial frontal surface was not so well localized as to explore functional motor maps, although they did not distinguish the pre-SMA and SMA proper.

Table 1: Characteristics of subcomponents of movement-related cortical potentials in different cortical areas.

	Pre-SMA	SMA proper	MI/SI (hand)	MI/SI (foot)
BP	small, bilat.	large, bilat.	large, bilat.	large, contralat.
NS'	absent	large, bilat.	large, contralat.	large, contralat.
MP	absent	large, contralat.	large, contralat.	large, contralat.
RAP	absent	absent	large, contralat.	large, contralat.

(MI=primary motor area, SI=primary sensory area, bilat.=bilateral, contralat.=contralateral)

c) Comparison of BP between the primary sensorimotor area and SMA proper

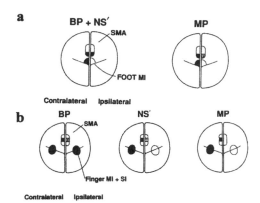

Fig.7: Schematic representation of the cortical generators of BP, NS' and MP for foot movements (**a**: upper half) and hand movements (**b**: lower half) viewed from the top.
The degree of darkness of the shading in the SMA and MI is approximately proportional to the amplitude of the corresponding potentials.
MI=primary motor area, SI= primary sensory area
(cited from Ikeda et al., 1992a with permission)

We have compared the temporal evolution pattern of pre-movement slow potentials in association with finger and foot movements in the MI-SI and SMA proper (Ikeda et al.,1992a,1995b; Ohara et al., 2000).

For foot movements **(Figure 7a)** (Ikeda et al.,1992a,1995b), the paracentral lobule and the mesial part of the superior frontal gyrus were simultaneously investigated, so that the slow potentials arising from the MI foot area and the foot area of the SMA proper could be compared. The MI foot area clearly generated BP, NS' and MP for the contralateral foot movements, whereas for the ipsilateral foot movements essentially no preceding potentials were observed. The foot area in the SMA proper generated clear BP and NS', and also MP to a lesser degree, equally for the contralateral and ipsilateral foot movements. The time difference of BPs between the MI foot area and the foot area of the SMA proper was not significant, and amplitudes of the potentials were not significantly different between the two areas **(Figure7a)**. RAPs were also clearly different between the SMA proper and the MI foot area, which are located very closely to each other. Only the MI foot area clearly generated transient large RAPs with the earliest latency of 145 msec after the movement onset. Furthermore, on more than half of occasions, the recorded BP and NS' were surface positive in polarity, but no counterparts of negative potentials were recorded adjacent to those areas. Thus the presence of a tangential dipole buried in the sulcus is unlikely. Positive polarity with foot movements might represent a surface positive dipole, in contrast to a surface negative dipole with hand or tongue movements (Sasaki, personal communication), and further studies on this generator mechanism are needed.

The opportunity to study both the MI-SI hand area on the lateral convexity and the mesial frontal area of the same hemisphere simultaneously is extremely rare in clinical situations. Therefore, very limited data have been available for a direct comparison so far (Ikeda et al., 1993, Ohara et al, 2000). The MI-SI hand area generated BP, NS' and MP for the contralateral movements, and less BP and no NS' or MP for ipsilateral movements. The hand area in the SMA proper bilaterally generated as large BP and NS' as the contralateral MI hand area **(Figure 7b)**. The onset time of BP in the two areas was not different. Similar temporal evolution of BPs in the SMA and the lateral frontal area was described also for voluntary saccades (Sakamoto et al., 1991). Similar findings for hand movements were reported from another laboratory (Rektor et al., 1994, 1998). By contrast, analysis of event-related desynchronization of the background rhythmic activity clearly showed that SMA proper started showing desynchronization of the 18 to 22 Hz cortical activity much earlier (-3.4 $\pm$ 0.5 sec) than MI (-1.7 $\pm$ 0.7 sec) with respect to the movement onset, whereas BP was not significantly different between the two areas (Ohara et al., 2000). The physiological significance of this discrepancy between the two phenomena remains to be clarified.

d) How much does SMA contribute to the scalp-recorded BP ?

Of scalp-recorded pre-movement slow potentials, the initial component (BP) is symmetrically distributed with the maximum amplitude at the vertex starting at 1.5 sec before the voluntary muscle contraction. It is followed by a steeper negative slope (NS' by Shibasaki et al.,1980) and MP with the predominance at the contralateral central area arising from about 300 msec before the movement onset. Based on the distribution of each component on the scalp, the BP was thought to be represented by the bilateral SMA activity, and the NS' and MP were mainly by the contralateral MI-SI. As described above based on the subdural recordings, the SMA clearly generates BP and NS'. However, it still remains unsolved how much the scalp-recorded pre-movement potentials reflect the potentials arising from the SMA. As shown in **Figure 8**, the SMA is located in the relatively deep, mesial frontal cortex whose dipoles with

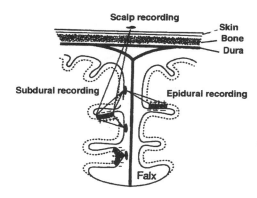

Figure 8: Schematic representation of the dipoles demonstrating the generators of BP in the mesial frontal surface. Dipoles with a surface negative pole could be oriented in different directions depending on whether they are generated in the crown of the gyrus, or in the superior or inferior bank of the sulcus. The scalp electrodes would only record summated potentials arising from several generators of both hemispheres, and are also significantly attenuated by different volume conductors of relatively high resistance like bone, dura and skin.
(cited from Ikeda et al., 1992a with permission)

negative polarity on the cortical surface can be directed tangential or to any other direction with respect to the scalp surface at the vertex area, and furthermore, two tangential dipoles arising from the left and right SMAs can cancel out each other.

When BP was recorded from patients with a unilateral mesial frontal lesion (Deecke,et al.,1987), a large negative BP was recorded at the vertex, and a relatively symmetric distribution of BP was observed in association with unilateral hand movements contralateral to the lesion. Therefore, it was assumed that initial slow potentials with maximum at the vertex and symmetric distribution might result mainly from a summation of the potentials arising from bilateral MIs. Furthermore, based on the dipole analysis study, the contribution of the BP arising from the bilateral SMAs located in the mesial frontal area to the scalp-recorded BP would be small, if any (Toro et al., 1993).This notion was also supported by the fact that MEG failed to identify any pre-movement slow field at the vertex, partly because the tangential current dipoles arising from the bilateral SMAs might be cancelled out (Lang et al., 1991). However, recent dipole analysis suggested active contribution of the SMA as one of the dipole sources to explain the scalp distribution of the BP (Praamstra et al.,1996: Urbano et al.,1996).

The SMA proper is usually located on the mesial aspects of the superior frontal gyrus, but not infrequently it is also found on the lateral aspect of the superior frontal gyrus as demonstrated by high frequency cortical stimulation in epilepsy surgery (Morris et al., 1988). It can produce a part of the radial dipoles for the scalp recording at the vertex area. We also observed in one patient (Patient 1 in Ikeda et al.,1999) that an electrode located rostral to the VAC line, on the boundary between the mesial and lateral part of the superior frontal gyrus, was one of the areas generating the largest pre-movement slow potentials, and thus it was regarded as pre-SMA. Luppino et al (1991) also showed that the cytoarchitectonically and electrophysiologically defined SMA-proper and pre-SMA (F3 and F6 by those authors, respectively) in monkey partly extended from the mesial surface over to the dorsal surface of the frontal lobe.

SUBCOMPONENTS OF PRE-MOVEMENT SLOW POTENTIALS

Several subcomponents of the scalp-recorded pre-movement slow potentials were described previously (Shibasaki et al., 1980), and their cortical generators were postulated as follows. Symmetric initial slow component at the vertex area, BP, would be essentially generated by SMA, while NS' and MP, predominantly contralateral to the hand movement (and maximum at the vertex for foot movements), would represent the activity of the MI-SI. Current understanding of the subcomponents based on direct cortical recording in epilepsy patients can be summarized as follows; (1) BP and NS' are clearly observed not only in MI but also in the SMA proper, and (2) clear MP is observed at MI. Furthermore, (3) premotion positivity (PMP) was described in the scalp recording as a transient positivity occurring around 50 msec before the movement onset on the ipsilateral hemisphere, but it has never been observed in the cortical recording, and thus it is regarded as a composite waveform observed in the scalp recording condition (Neshige et al., 1988). (4) Intermediate slope (IS) was described as a slow potential occurring between BP and NS' (starting about 900 msec before the movement onset and lasting for 500 to 600 msec), and based on the left-sided dominance regardless of the side of movement, it was thought to represent the activity of the premotor area (Barrett et al., 1986). However, subdural recording has not confirmed this slope as a consistent finding.

FACTORS AFFECTING THE LATENCY, AMPLITUDE AND SLOPE OF THE BP IN SUBDUAL RECORDINGS

There have been many studies investigating the factors affecting the latency, amplitude and slope of BP based on the scalp recording in normal subjects. As regards the complexity of the movements, the sequential movements of two fingers of one hand produced a larger and earlier BP at the vertex and at the bilateral precentral areas than the simultaneous movements of those two fingers (Kitamura et al., 1993). In order to clarify those affecting factors, several studies were conducted by subdural recording as follows.

(1) Single finger movements and sets of rapid repetitive movements of the same finger were compared in relation to BP generation from the SMA proper. There was no significant increase of BP amplitude arising from the SMA proper before the rapid repetitive movement task as compared with the single movement task (Ikeda et al., 1993).

(2) Voluntary movements can be produced not only by muscle contraction but also by its relaxation, and both conditions were associated with clear BP (Terada et al.,1995). Scalp-recorded BP before the onset of muscle relaxation was as large and early as that for muscle contraction, and small surface-positive slow potentials were observed in the frontal to frontopolar area in some subjects only with muscle relaxation task, suggesting a possibility about the additional activity of SNMA or pre-SMA for generation of BP in the muscle relaxation. However, as the result of subdural recordings done so far, the SMA proper and pre-SMA generated similar BPs for both muscle contraction and relaxation tasks (Yazawa et al., 1998).

(3) Rapid-rate repetition of voluntary movements for recording the BP with scalp electrodes was shown to be clinically useful for exploring motor cortex function in patients with movement disorders (Gerloff et al., 1997). Pre-SMA and SMA proper generated clear BPs in the slow-rate repetition of voluntary movements, but little BP in the rapid-rate repetition of movements, whereas MI generated clear BP equally for both slow- and rapid-rate repetition of movements (Kunieda et al.,2000). In the rapid-rate repetition of movements, voluntary

movements may be conducted as a kind of "automatic" movements, while in the slow-rate repetition of movements each movement might be executed discretely, each of which may involve distinct motor control mechanisms.

(4) Language-related slow potentials in association with vocalization were investigated by scalp recording. Asymmetry of scalp-recorded slow potentials could not exclude the possibility of slow shift artifacts caused by glossokinetic potentials, galvanic skin responses or any slow artifacts related to contraction of facial, laryngeal and pharyngeal muscles (Grözinger et al., 1980). Since the subdural recording condition is essentially free from these kinds of artifacts as far as one of the subdural electrodes placed distant from the region of interest is chosen as a reference electrode, we investigated the BP directly from the peri-rolandic area and language area in association with vocalization, and compared it with simple tongue protrusion (Ikeda et al., 1995a). There were no slow pre-movement potentials before vocalization at either Broca's or Wernicke's area. BPs for tongue protrusion and vocalization were commonly seen at a part of area 4: either anterior bank of the central sulcus or the crown of the precentral gyrus. For vocalization, a whole area 4, i.e., both anterior bank of the central sulcus and the crown of the precentral gyrus, were active in BP generation.

SUBCORTICAL GENERATOR OF BP

In monkey studies by intracortical field potential recording, the BP disappeared on the hemisphere after contralateral cerebellar hemispherectomy (Sasaki et al., 1979). In patients with cerebellar lesions, especially involving the dentate nucleus and its efferent pathways, scalp-recorded BP and NS' were often absent (Shibasaki et al., 1986). Both suggest that BP is generated by the feedforward activation of the motor cortices through the cerebrocerebellar communication (Allen and Tsukahara, 1974). There has been no direct invasive study to explore these neuronal connections further in humans.

COMPARISON OF THE BP AND CNV

Contingent negative variation (CNV) is a slow negative brain potential occurring between two successive stimuli only when the two stimuli are associated with or contingent on each other (Walter et al, 1964). In general, the first stimulus (S1) serves as a warning signal for the second imperative stimulus (S2), to which a motor response is required. Thus, CNV, especially the late CNV component, may share common generator mechanisms with BP at least partially (Rohrbaugh et al., 1976; Tecce et al., 1993), although BP is obtained before self-paced, spontaneous voluntary movements whereas CNV represents sensorimotor integration or association and preparation of voluntary movements closely related to the cue stimuli.

The scalp-recorded late CNV component shows little laterality with the maximum at the vertex to the frontal area regardless of the type of motor task, and larger amplitude as compared with BP (Tecce et al., 1993; Ikeda et al., 1996a; Lai et al., 1997). In patients with cerebellar efferent lesions, the scalp-recorded BP was absent whereas the CNV preserved its normal amplitude (Ikeda et al., 1994). Invasive recording of CNV demonstrated that late CNV components in a choice-reaction time paradigm were generated from the lateral and mesial prefrontal areas and non-primary motor areas (pre-SMA, SMA proper, lateral premotor areas) (Ikeda et al., 1996a, 1996b, 1999), and also from the parietal and temporal association areas (Hamano et al.,1997). This result was also consistent with a report from another laboratory

(Lamarche et al., 1995).

TECHNICAL CONSIDERATIONS FOR INVASIVE RECORDING

The BP is usually recorded with a long time constant setting of the amplifier of longer than 3 sec, sometimes longer than 10 sec, or even with direct current (DC) setting. For scalp recording, electrodes made of silver/silver chloride (Ag/AgCl) are recommended to minimize unnecessary artifacts and maximize slow potential recording, as compared with stainless steel or silver (Ag) electrodes. For invasive recording, however, Ag/AgCl electrodes are toxic for the tissue and thus can not be used, and therefore, stainless steel or platinum electrodes are currently used. For recording the low frequency activities sufficiently, use of platinum electrodes, electrodes with large recording surface, and large input impedance of the amplifier of more than 200MΩ should be carefully adopted, and the recording conditions in which the waveforms are obtained should always be taken into account when interpreting the results (Ikeda et al., 1998).

PROBLEMS TO BE ADDRESSED IN FUTURE STUDIES

Epilepsy surgery has provided us with the opportunity to investigate the generator mechanism of BP, and the results thus obtained provide clinically useful information for functional motor mapping for each individual patient. Future problems to be solved in relation to the invasive recording of BP are listed as follows.
(1) It is apparent that multiple cortical areas play different roles in generating BP for voluntary motor control, and thus the temporal evolution pattern among those generators is to be clarified for specifying the functional importance of various motor areas.
(2) Whenever we have the opportunity to record epicortical potentials in patients with intractable partial epilepsy by subdural electrodes implanted chronically at or close to the possible epileptogenic area, the functional area that generates the BP might be located at or very close to the epileptogenic area. Therefore, as done in our previous studies, we have to pay special attention to avoid any potentially misleading results caused by those pathological conditions in terms of the distribution, polarity, onset time of potentials and amplitude.
(3) From the clinical point of view, in presurgical evaluation of epilepsy patients by using subdural grid electrodes, we currently make motor mapping not only by high frequency cortical stimulation but also by recording BP/NS' for motor tasks involving different parts of the body, or for the specific motor task, whose cortical generators along the central sulcus are clinically needed. All the obtained data are then taken into account for the final functional mapping. To promote the clinical usefulness of BP in presurgical evaluation, more practically available tasks like rapid-rate movement are to be considered, and the sensitivity and specificity of the results of BP should be further clarified as compared with the conventional high frequency electric cortical stimulation.
(4) Scalp-recorded BP showed that the activity at the parietal area was selectively diminished in patients with focal stroke at the posterior parietal area (Knight et al., 1989). Since the disturbance in the posterior parietal area would result in ideomotor or ideational apraxia, the possibility that this area also might generate or modify the BP remains to be elucidated.

ACKNOWLEDGMENT

This study was supported by Grants-in-Aid for Scientific Research on Priority area (C)-Advanced Brain Science Project 12210012, (B2)-13470134, (C2)-13670460 from Japan Ministry of Education, Science, Sports and Culture, and Research for the Future Program from the Japan Society for the Promotion of Science JSPS-RFTF97L00201.

REFERENCES

Allen, G.I., Tsukahara, N. (1974) Cerebrocerebellar system, *Physiol Rev* 54, 857-1006.

Allison, T., McCarthy, G., Luby, M., Puce, A., Spencer, D.D. (1996) Localization of functional regions of human mesial cortex by somatosensory evoked potential recording and by cortical stimulation. *Electroenceph clin Neurophysiol* 100, 126-140.

Barrett, G., Shibasaki, H., Neshige, R. (1986) Cortical potentials preceding voluntary movement: evidence for three periods of preparation in man. *Electroencephalogr Clin Neurophysiol* 63, 327-339.

Deecke, L., Lang, W., Heller, H.J., Hufnagi, M., Kornhuber, H.H. (1987) Bereitschaftspotential in patients with unilateral lesions of the supplementary motor area. *J Neurol Neurosurg Psychiatry* 159, 1430-1434.

Gerloff, C., Toro, C., Uenishi, N., Cohen, L.G., Leocani, L., Hallett, M. (1997) Steady-state movement-related cortical potentials: a new approach to assessing cortical activity associated with fast repetitive finger movements. *Electroenceph clin Neurophysiol* 102, 106-113.

Grözinger, B., Kornhuber, H.H., Kriebel ,J., Spirtes, J., Westphal, K.T.P. (1980) The Bereitschaftspotentials p receding the act of speaking. Also an analysis of artifacts. *Prog Brain Res* 54, 798-804.

Hahn, J.,,Lüders, H. (1987) Placement of subdural grid electrodes at the Cleveland Clinic. In: Engel, J.Jr., (Ed.) Surgical treatment of the epilepsies. New York: *Raven Press*, 621-627.

Hamano, T., Lüders, H.O., Ikeda ,A., Collura, T., Comair, Y.G., Shibasaki, H. (1997) The cortical generators of the contingent negative variation in humans: a study with subdural electrodes. *Electroenceph clin Neurophysiol* 104, 257-268.

Ikeda, A., Lüders, H.O., Burgess, R.C., Shibasaki, H. (1992a) Movement-related potentials recorded from supplementary motor area and primary motor area: role of supplementary motor area in voluntary movements. *Brain* 115, 1017-1043.

Ikeda, A., Shibasaki, H.(1992b) Invasive recording of movement-related cortical potentials in humans. *J Clin Neurophysiol* 9, 509-520.

Ikeda, A., Lüders, H.O., Burgess, R.C., Shibasaki, H. (1993) Movement-related potentials associated with single and repetitive movements recorded from human supplementary motor area. *Electroenceph clin Neurophysiol* 89, 269-277.

Ikeda, A., Shibasaki, H., Nagamine, T., Terada, K., Kaji, R., Fukuyama, H., Kimura, J. (1994) Dissociation between contingent negative variation and Bereitschaftspotential in a patient with cerebellar efferent lesion. *Electroenceph clin Neurophysiol* 90, 359-364.

Ikeda, A., Lüders, H.O., Burgess, R.C., Sakamoto, A., Klem, G.H., Morris, H.H., Shibasaki, H. (1995a) Generator locations of movement-related potentials with tongue protrusions and vocalization: subdural recording in human. *Electroenceph clin Neurophysiol* 96, 310-328,.

Ikeda, A., Lüders, H.O., Shibasaki, H., Collura, T.F., Burgess, R.C., Morris, H.H., Hamano, T. (1995b) Movement-related potentials associated with bilateral simultaneous and unilateral movement recorded from human supplementary motor area. *Electroenceph clin Neurophysiol* 95, 323-334.

Ikeda, A., Lüders, H.O., Collura, T.F., Burgess, R.C., Morris, H.H., Hamano, T., Shibasaki, H. (1996a) Subdural potentials at orbitofrontal and mesial prefrontal areas accompanying anticipation and decision making in humans: a comparison with Bereitschaftspotential. *Electroenceph clin Neurophysiol* 98, 206-212.

Ikeda, A., Lüders, H.O., Shibasaki,H. (1996b) Generation of contingent negative variation (CNV) in the supplementary sensorimotor area. In: Lüders, H.O., (Ed.) Supplementary sensorimotor area. Advances in Neurology, vol.70, New York: *Lippincott-Raven*, 153-159.

Ikeda, A., Nagamine, T., Yarita, M., Terada, K., Kimura, J., Shibasaki,H.(1998) Reappraisal of the effects of electrode property on recording slow potentials. *Electroenceph clin Neurophysiol* 107, 59-63.

Ikeda, A., Yazawa, S., Kunieda, T., Ohara, S., Terada, K., Mikuni, N., Nagamine, T., Taki, W., Kimura, J., Shibasaki, H. (1999) Cognitive motor control in human pre-supplementary motor area studied by subdural recording of discrimination/ selection-related cortical potentials. *Brain* 122, 915-931.

Ikeda, A., Miyamoto, S., Shibasaki, H. (2001) Cortical motor mapping in epilepsy patients: information from subdural electrodes in presurgical evaluation. *Epilepsia* (in press).

Kitamura, J., Shibasaki, H., Takagi, A., Nabeshima, H., Yamaguch, A.(1993). Enhanced negative slope of cortical potentials before sequential as compared with simultaneous extension of two fingers, *Electroenceph clin Neurophysiol* 86, 176-182.

Knight, R.T., Singh, J., Woods, D.L. (1989) Pre-movement parietal lobe input to human sensorimotor cortex.

    Brain Res 498, 190-194.

Kornhuber, H,H,, Deecke, L. (1965) Hirnpotentialänderungen bei Willkürbewegungen und passiven
    Bewegungen des Menschen: Bereitschaftspotential und reafferente Potentiale. *Pflügers Archiv* 284: 1-
    17.

Kunieda, T., Ohara, S., Ikeda, A., Taki, W., Baba, K., Yagai, K., Hashimoto, N., Shibasaki, H. (1999)
    Bereitschaftspotentials arising from the human primary negative motor area in humans. *Jap J
    Electroenceph Electromyograph* 27, 168 (abstract) (written in Japanese)

Kunieda, T., Ikeda, A., Ohara, S., Yazawa, S., Nagamine, T., Taki, W., Hashimoto, N., Shibasaki, H. (2000)
    Different activation of pre-supplementary motor area (pre-SMA), SMA-proper and primary sensorimotor area
    depending on the movement repetition rate in humans. *Exp Brain Res* 135, 163-172.

Lamarche, M., Louvel, J., Buser, P., Rector, I. (1995) Intracerebral recording of slow potentials in a contingent
    negative variation paradigm: an exploration in epileptic patients. *Electroenceph clin Neurophysiol* 95,
    268-276.

Lai, C., Ikeda, A., Terada, K., Nagamine, T., Honda, M., Xu, X., Yoshmura, N., Howong, S., Barrett, G.,
    Shibasaki, H. (1997) Event-related potentials associated with judgment: comparison of S1- and S2-
    choice conditions in a contingent negative variation (CNV) paradigm. *J Clin Neurophysiol* 14, 394-405.

Lang, W., Cheyne, D., Kristeva, R., Beisteiner, R., Lindinger, G., Deecke, L. (1991) Three dimensional
    localization of SMA activity preceding voluntary movement. *Exp Brain Res* 87, 688-695.

Lee, B.I., Lüders, H., Lesser, R., Dinner, D.S., Morris, H.H. (1986) Cortical potentials related to voluntary and
    passive finger moevements recorded from subdural electrodes in humans. *Ann Neurol* 20, 32-37.

Lüders, H., Lesser, R.P., Dinner, D.S., Morris, H.H., Hahn, J.F., Friedman, L., Skipper, G., Wyllie, E.,
    Friedman, D. (1987) Commentary: Chronic intracranial recording and stimulation with subdural
    electrodes. In: Engel, J.Jr. (Ed.) Surgical treatment of the epilepsies. New York: *Raven Press*, 297-321.

Lüders, H.O., Dinner, D.S., Morris, H.H., Wyllie, E., Comair, Y.G. (1995) Cortical electric stimulation in
    humans: the negative motor areas. In: Fahn, S., Hallett, M., Lüders, H.O., (Eds.) Negative motor
    phenomena. Advances in Neurology, vol.67, New York: *Lippincott-Raven*, 115-129.

Luppino, G., Matelli, M., Camarda, R., Gallese, V., Rizzolatti, G. (1991) Multiple representations of body
    movements in mesial area 6 and the adjacent cingulate cortex: an intracranial microstimulation study in
    the macaque monkey. *J Comp Neurol* 311, 463-82.

Luppino, G., Matelli, M., Camarda, R., Rizzolatti, G. (1993) Corticocortical connections of area F3 (SMA-
    proper) and area F6 (pre-SMA) in the macaque monkey. *J Comp Neurol* 338, 114-40.

Morris, H.H. III., Dinner, D., Lueders, H., Wyllie, E., Kramer, R. (1988) Supplementary motor seizures: clinical
    and electroencephalograhic findings. *Neurology* 38, 1075-1082.

Matsuzaka, Y., Aizawa, H., Tanji, J. (1992) A motor area rostral to supplementary motor area (presupplementary
    motor area) in the monkeys: neuronal activity during a learned motor task. *J Neurophysiol* 68, 653-62.

Neshige, R., Lüders, H., Shibasaki, H. (1988) Recording of movement-related potentials from scalp and cortex
    in man. *Brain*. 111, 719-36.

Ohara, S., Ikeda, A., Kunieda, T., Yazawa, S., Baba, K., Nagamine, T., Taki, W., Hashimoto, N., Mihara, T.,
    Shibasaki, H. (2000) Movement-related change of electrocorticographic activity in human SMA proper.
    Brain 123, 1203-1215.

Orgogozo, J.M., Larsen, B. (1979) Activation of the supplementary motor area during voluntary movement in
    man suggests it works as a supramotor area. *Science* 206, 847-850.

Penfield, W., Welch, K. (1951) The supplementary motor area of the ecerebral cortex: a clinical and
    experimental study. *Arch Neurol Psychiat* 66, 289-317.

Picard, N., Strick, P. (1996) Motor area of the mesial wall: a review of their location and functional activation.
    Cerebr Cortex 6, 342-353.

Praamstra, P., Stegeman, D.F., Horstink, M.W., Cools, A.R. (1996) Dipole source analysis suggests selective
    modulation of the supplementary motor area contribution to the readiness potential. *Electroenceph clin
    Neurophysiol* 98, 468-477.

Rektor, I., Féve ,A., Buser, N., Bathien, N., Lamarche, M. (1994) Intracranial recoording of movement-related
    readiness potentials: an exploration in epileptic patients. *Electroenceph clin Neurophysiol* 90, 273-283.

Rektor, I., Louvel, J., Lamarche, M. (1998) Intracranial recording of potentials accompanying simple limb
    movements: a SEEG in epileptic patients. *Electroenceph clin Neurophysiol* 107, 277-286.

Rohrbaugh, J.W., Syndulko, K., Lindsley, D.B. (1976) Brain wave components of the contingent negative
    variation in human. *Science* 191, 1055-1057.

Sakamoto, A., Lüders, H., Burgess, R. (1991) Intracranial recoording of movement-related potentials to
    voluntary saccades. *J Clin Neurophysiol* 8, 223-233.

Sasaki, K. (1979) Cerebro-cerebellar interconnection in cats and monkeys. In: Massioon, J., Sasaki, K. (Ed.)
    Cerebro-cerebellar interaction. Amsterdam: *Elsevier*, 105-124.

Sasaki, K., Gemba, H. (1991a) How do the different cortical motor areas contribute to motor learning and
    compensation following brain dysfunction ? In: Humphrey, D.R., Freund, H.J., (Eds.) Motor control:
    concepts and issues. New York; *John Wiley & Sons*, 445-461.

Sasaki, K., Gemba, H. (1991b) Cortical potentials associated with voluntary movements in monkeys.
    Electroenceph clin Neurophysiol Suppl 42, 80-96.

Shibasaki, H., Barrett, G., Halliday, E., Halliday, A.M. (1980) Components of the movement-related cortical potentials and their scalp topography. *Electroenceph clin Neurophysiol* 49, 213-26.

Shibasaki, H., Barrett, G., Neshige, R., Tomoda, H. (1986) Volitional movement is not preceded by cortical slow negativity in cerebellar dentate lesion in man. *Brain Res* 368, 361-165.

Shibasaki, H., Ikeda, A. (1996) Generation of movement-related potentials in the supplementary sensorimotor area. In: Lüders, H.O. (Ed.) Supplementary sensorimotor area. Advances in Neurology, vol.70, New York: *Lippincott-Raven*, 117-125.

Tanji, J. (1994) The supplementary motor area in the cerebral cortex *Neurosci Res* 19. 251-268.

Terada, K., Ikeda, A., Nagamine, T., Shibasaki, H. (1995) Movement-related potentials associated with voluntary muscle relaxation. *Electroenceph clin Neurophysiol* 95, 335-345.

Tecce, J.J., Cattanach, L. (1993) Contingent negative variation. In: Niedermeyer, E., Lopes da Silva, F., (Eds.) Electroencephalography. Basic principles, clinical applications and related fields. Baltimore: *Wiliams & Wilkins*, 887-910.

Toro, C., Matsumoto, J., Deuschl, G., Roth, B.J., Hallett ,M. (1993) Source analysis of scalp-recorded movement-related electric potentials. *Electroenceph clin Neurophysiol* 86, 167-175.

Urbano, A., Babiloni, C., Onorati, P., Babiloni, F. (1996) Human cortical activity related to unilateral movements. A high resolution EEG study. *NeuroReport* 8, 203-206.

Walter, W.G., Cooper, R., Aldridge ,V.J., McCallum, W.C., Winter, A.L. (1964) Contingent negative variation: an electrical sign of sensorimotor association and expectancy in the human. *Nature* 203, 380-384.

Wiesendanger, M. (1986) Recent development in studies of the supplementary motor area of primates. *Rev Physiol Biochem Pharmacol* 103, 1-59.

Yazawa, S., Ikeda, A., Terada, K., Mima, T., Mikuni, N., Kunieda, T., Taki, W., Kimura, J., Shibasaki, H. (1997) Subdural recording of Bereitschaftspotential is useful for functional mapping of the epileptogenic motor area: a case report. *Epilepsia* 38, 245-248.

Yazawa, S., Ikeda, A., Kunieda, T., Mima, T., Nagamine, T., Ohara, S., Terada, K., Taki, W., Kimura, J., Shibasaki, H. (1998) Human supplementary motor area is active in preparation for both voluntary muscle relaxation and contraction: subdural recording of Bereitschaftspotential. *Neuroscience* Lett, 244: 145-148.

Yazawa, S., Ikeda, A., Kunieda, T., Ohara, S., Mima, T., Nagamine, .T, Taki, W., Hori, T., Shibasaki, H. (2000) Human pre-supplementary motor area is active before voluntary movement: subdural recording of Bereitschaftspotentials from mesial frontal cortex. *Exp Brain Res* 131, 165-177.

INTRACEREBRAL RECORDINGS OF THE BEREITSCHAFTSPOTENTIAL AND RELATED POTENTIALS IN CORTICAL AND SUBCORTICAL STRUCTURES IN HUMAN SUBJECTS

Ivan Rektor

1st Department of Neurology, Masaryk University, St. Anne's Hospital
Brno, Czech Republic

More than thirty-five years after its discovery, the Bereitschaftspotential (BP) remains a very productive research tool in both clinical and cognitive neurophysiology. Despite fruitful research results, there are several basic questions about BP that remain unanswered:

1. Where in the brain is BP generated?

2. What is the relation of BP to some other cerebral phenomena that share common features with BP, such as movement preparation and execution (contingent negative variation, CNV), and even changes in power during an identical testing protocol (event related desynchronization and synchronization, ERD/ERS)?

3. What physiological activity is represented by BP?

Intracerebral recordings may help to answer these questions. There are some medical reasons for human subjects to have electrode contacts that record subcortically and from the cortex. When the electrodes are placed, it is possible for the depth electrodes to provide direct information from cortical and subcortical structures. Intracerebral recordings can provide information not readily available through other means.

METHODS

Twenty-six patients examined in Paris (Hôpital Ste Anne, Service de Neurochirurgie; INSERM, U 97) were implanted with orthogonal cortical electrodes. BP was tested in seventeen patients, CNV was tested in fourteen patients, and both protocols were tested in seven patients. Thirty-one epilepsy surgery candidates were examined in Brno (Centre for

Epilepsy, St. Anne's Hospital). They had orthogonally implanted depth electrodes; subdural electrodes were used infrequently. In addition to the orthogonal or subdural electrodes, twelve of these patients had also one or two diagonal electrodes, which reached the amygdala and hippocampus via the basal ganglia. Depth electrodes were placed in sites corresponding to the electroclinical characteristics of their seizures. The patients were candidates for epilepsy surgery who had remained unresponsive to conventional forms of therapy and who were recommended by a special commission for stereotactic exploration. Three patients who had been implanted because of chronic pain had electrodes in the posterior thalamus.

Microdeep (DIXI Besançon) intracerebral 5-15 contact stainless steel and later platinum electrodes were used. The electrode diameter was 0.8 mm, the contact length was 2 mm, and the distance between contacts was 1.3 mm. The electrodes were implanted stereotactically. For subdural recordings, Radionics platinum electrode strips and grids were used. The position of the electrodes was verified by MRI, and their functionality was verified by electrical stimulation. The video-SEEG (stereoelectroencephalographic) recordings were performed over a period of 5-10 days. Regions with clearly pathological SEEG activity or with MRI lesions were excluded from the BP testing.

The data acquisition and averaging were performed using Nihon Kohden Neuropack 4 and 8 set, the band was 0.01-500 Hz for the intracerebral recordings. Final recordings were performed with a 96-channel M&I neurophysiological device. All recordings were taken against a biauricular reference. In every session, the number of trials was between 40 and 60. Every session was repeated at least once.

Slightly modified versions of classical BP and CNV testing protocols were used (Kornhuber and Deecke 1965; Walter et al. 1964). For the CNV tests, the French recordings used auditory signals for both the warning and the imperative stimuli; the Czech recordings used S1 auditory and S2 visual stimuli (Bareš et al. 1999), with interstimulus intervals of 1.5 and 3.0 seconds, respectively. We investigated slow potentials appearing before (BP and CNV) and during (movement accompanying potential, MAP) a simple repetitive distal limb movement, namely a hand or foot flexion. In one series, a more complex movement (turning pages in a book) was used (see below).

When the BP exhibited two distinct slopes (BP and NS´), the intersection point of these slopes was considered as the starting point of the second (NS´) component. Absolute amplitudes were measured from the baseline, which was defined as the electrical activity that occurred from 3000-3500 ms before the movement. The distance from the electrode to the generator heavily influences the amplitude of intracerebrally recorded potentials, and thus the differences of amplitude can only be compared intraindividually. The intracerebral potentials occurred with both positive and negative polarities. This was due to variances in the positions of the electrode contact and of the dipole generator. Intracerebrally, both positive and negative locations are possible. The Contingent *Negative* Variation can be positive with depth electrode recording.

Intracerebrally, we have primarily studied the brain potentials elicited in the well-known BP and CNV test protocol. We have also studied the potential recorded in the BP and CNV protocol during the movement performance. We call this potential MAP, for "Movement Accompanying Potential" (referred to in an earlier paper as MASP -Movement Accompanying Slow Potential, Rektor et al. 1998). In the cortex, MAP may follow BP or it may be recorded separately in areas where BP was not displayed. MAP was found to be broadly distributed in the prefrontal and motor regions in the cortex, and in some parietal and temporal regions. More details concerning the patients, methods and results have already been published (Rektor et al.1994, 1998, 2001a, 2001b; Rektor 2000a, 2000b; Lamarche et al. 1995). Consent was obtained from each patient for the electrophysiological testing, about which they were amply informed. The research was approved by the local ethics committee.

INTRACRANIAL GENERATORS OF BP

BP is generated in several cortical and subcortical structures that are known to be directly or indirectly linked with motor control. Cortical sources of BP (Fig. 1), localised with intracranial recordings, were displayed contralaterally to the movement in the primary motor cortex and somatosensory cortex, and bilaterally in the supplementary motor area (SMA), in the preSMA and in the cingulate cortex (Lee et al. 1986; Ikeda et al. 1992; Rektor et al. 1994, 1998; Lamarche et al. 1995; Yazawa et al. 2000). More generators may remain to be revealed in structures that have not yet been sufficiently explored. Experimental data obtained from animals in unit and field studies reveal potentials prior the movement by several hundreds of milliseconds in the somatosensory, primary motor, and premotor cortices, and in the SMA, always bilaterally (Gemba and Sasaki 1984; Arezzo and Vaughan 1975). However, the very restricted localisation of cortical BP in human subjects differs from the data obtained in monkeys.

Subcortical generators of BP were found in the basal ganglia (putamen, pallidum and caudate head: Rektor et al. 2001c) and in the thalamus (posterior thalamus and the VIM: Rektor et al. 2001b; Fève 1993). In light of previous observations of BP in the brainstem and the motor thalamic nuclei (Haider et al.1981; McCallum 1993), we presume that there are generators of BP in various subcortical structures. There are many nuclei that have never been tested for the occurrence of slow potentials.

Cortical BP

Primary motor cortex

We recorded BP in twelve patients with electrodes in the primary motor cortex (MC), and in two patients with subdural electrodes over the MC. In two of the patients with electrodes implanted quasi-symmetrically in both hemispheres, recordings could be simultaneously performed ipsilaterally and contralaterally to the movement. In all but two cases, BP appeared only in the MC contralateral to the movement (Fig.2). A patient with a bilateral BP appearance had a large porencephalic cyst and an EMG revealed a synkinetic movement of the non-tested hand. Our results with BP recorded with electrodes in the MC did not substantially differ from the results with BP recorded with subdural electrodes over the MC. Neshige et al. (1988) recorded BP1 but not BP2 (NS') in the contralateral as well as in the ipsilateral MC. NS' was seen exclusively on the contralateral side. Ikeda et al. (1992) recorded BP with foot movements (as opposed to hand movements) only over the MC contralateral to the movement side. This unilateral occurrence of BP indicates that the physiological events expressed by BP are absent in the ipsilateral cortex. However, it does not exclude the possibility of the participation of the ipsilateral MC in motor preparation by other mechanisms (e.g. Chen et al. 1997).

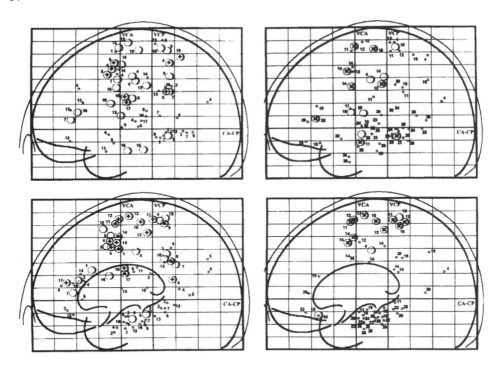

Figure 1. The cortical recording sites in the Bereitschaftspotential protocol (left) and in the Contingent Negative Variation protocol (right) in twenty-six patients. Upper part: Lateral view of the brain. Lower part: Sagittal aspect. In the supracallosal part, most of the internal contacts of multilead electrodes reached the mesial wall. In the lower part of the diagram, some of the recording points correspond to sites located more laterally. This schema presents the results in one or in both hemispheres.

The numbers correspond to the individual patients. The large circles indicate the MAP occurrence. They are marked "+" when the BP preceded the MAP. The "X" indicates the occurrence of CNV. The small circles indicate electrodes that recorded no slow potential. Reprinted from: Rektor I. et al. Electroencephalogr clin Neurophysiol, 1998; 107: 277–286, with permission from Elsevier Science.

SMA

We recorded BP in twelve of thirteen patients with electrodes in or over SMA. In five of the patients with electrodes implanted quasi-symmetrically in both hemispheres, recording could be simultaneously performed ipsilaterally and contralaterally to the movement. We found bilateral BP in all but one recording. The amplitude was higher on the contralateral side (Fig. 4). In several patients, simultaneous recordings in the SMA and MC revealed the almost identical duration of BP in both structures. The average duration of BP in the SMA and BP in the MC did not differ significantly. Ikeda et al. (1992) observed a bilateral appearance of BP in the SMA, with the BP more prominent on the contralateral side. Recently, independent BP generators in the SMA and in the preSMA have been noted by Yazawa et al. (2000) in three patients. Unlike the somatotopic occurrence of BP in the SMA, the preSMA displayed BP in conjunction with the movement of any limb. We also observed that BP in the SMA is elicited in some cases only by the movement of upper limbs or lower limbs, while in other cases BP preceded any limb movement (Rektor et al. 1994). However, the distribution of these responses did not adhere to the VPA line that is considered to represent the border between SMA and preSMA. It is true that we did not perform extensive electrical stimulation of this region, and that we have studied the movements of all limbs in only a few patients. The fact that we observed responses typical for preSMA in posterior parts of the SMA, and responses typical for SMA in the preSMA, leads us to believe that more recordings are needed for a functional delimitation of SMA and preSMA.

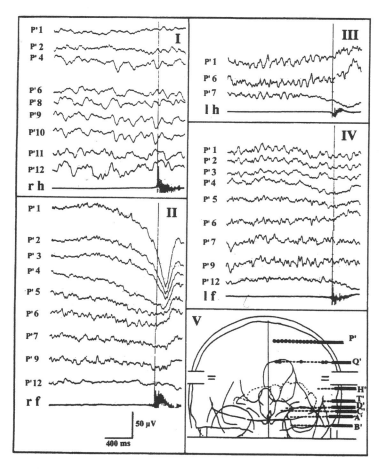

Fig. 2. Patient 10. Left side electrode P′ passing from the parietal convexity cortex (P′ 12) into the MC foot area on the mesial wall (P′1); recorded using several successive leads. BP on electrode P′ were tested in connection with a movement of the right hand (I, r h), the right foot (II, r f), the left hand (III, l h), and the left foot (IV, l f). Bottom right indicates the P′ and Q′ electrode positions; frontal view with the recording leads marked by dots. Recordings were made from other electrodes (H′, T′, C′, B′, A′), but they were virtually silent (not shown). Notice the presence of intracranial BP in the MC foot area, in response to contralateral foot movement only.

Adapted from Rektor et al. Electroenceph clin Neurophysiol 1994: 90, 273-283, with permission from Elsevier Science.

Cingulate gyrus

The consistent appearance of BP was observed in the anterior caudal cingulate (ACC) in four patients (Rektor et al 1998). BP is also generated in other parts of the anterior and posterior cingulate gyrus (CG), although less consistently than in the motor cortices (Štreitová et al. 1999).We studied the appearance of BP in the CG, as well as in the "motor" ACC, in eleven patients (some of them with two electrodes). Two recording were obtained from the posterior CG, area 31, with one BP displayed. All other recordings were obtained from the supracallosal part of the anterior CG, specifically from the paralimbic (areas 24c and 32) and limbic (areas 24a and 24b) portions. In the limbic area, BP appeared in all four recordings; in the paralimbic portion, it appeared in five of the eight sites. The results indicate that BP is generated in the CG, even outside its "motor" ACC. However, an important variable is that BP appearance is less consistent in the CG than in the MC or SMA. This could be the result of individually different cognitive strategies linked with movement preparation and performance. In a PET study, Paus et al. (1998) found significant regional differences in the ACC linked to task difficulty, motor output and recent memory requirement. In the ACC, the effects of these variables may combine.

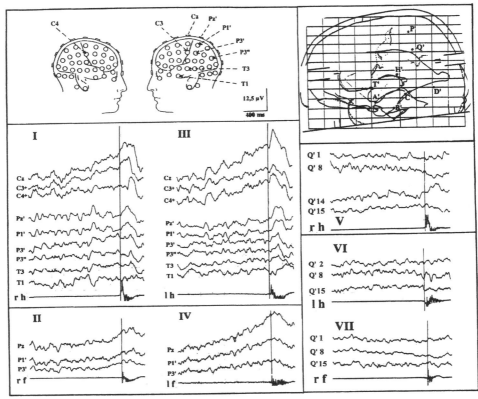

Fig. 3. Patient 10 (as in Fig. 2). On the left scalp recorded BP preceding movement of all 4 limbs. The location of the electrodes is indicated above ("primed" 10-20 system). I, II, movements of the right hand (r h) and the right foot (r f), respectively; III, IV, the left hand (l h) and the left foot (l f), respectively. Scalp electrodes P_1, P''_3, T_3 and T_1 were positioned at the sites selected for future implantation of depth electrodes (depth electrodes: sagittal view top right, frontal view in Fig. 2). On the right side, recording from intracerebral electrode Q' passing from the parietal convexity (Q' 15), into the posterior cingulate (Q'1). Movements of the right hand (r h – V), left hand (l h – VI) the right foot (r f – VII), respectively. Compare the scalp recordings (I-III) with the depth recordings (V-VII and with Fig. 2). Note that while intracranial electrode Q was silent, the scalp recording electrode just above it (P'₃) displays some BP. The same holds true when comparing the intracerebral electrode P'12 (in Fig. 2) with scalp electrode P'₁ (for foot flexion).

Adapted from Rektor et al. Electroenceph clin Neurophysiol 1994: 90, 273-283, with permission from Elsevier Science.

Further possible cortical generators

The possibility of other cortical generators cannot be fully discounted. There may be a small BP generator in the dorsolateral prefrontal cortex (area 9, 46), but the recording results are not yet sufficiently consistent. We have observed a small bilateral BP in these regions from four electrodes in three patients; however, in four other recordings, no BP was recorded. It is necessary to make further recordings. There are other structures that are usually not explored in epilepsy surgery candidates, e.g. the posterior parietal cortex, the occipital lobe and the insula. Moreover, BP is sometimes generated in regions from which it is absent in the majority of cases. While in many recordings, no BP generator could be identified in the lateral premotor cortex (Rektor et al. 1994, 1998, Ikeda et al. 1992), in one case, we found BP in the bottom of a premotor sulcus that is usually inaccessible to such exploration (Fig. 4). As it is the only recording from this area, it is not possible to know whether this is an individual variation or a new source of BP. Such individual BPs in areas where BP is usually not displayed are occasionally observed. Therefore, some new generators might be described in the future, but this should be based on consistent recordings from several patients.

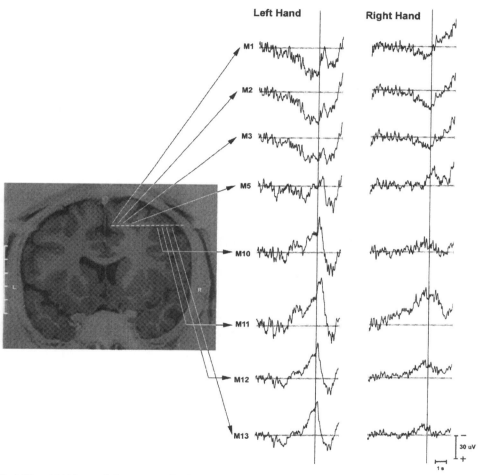

Fig.4 BP and MAP recorded on consecutive leads of an electrode passing from the SMA (contacts M1 and 2, and possibly M3), through the white matter (M4 to M9; only M5 is shown), through the premotor cortex in the bottom of an vertically oriented sulcus (M11) into the premotor cortex of the convexity (M 13).

Left: MR imaging of the brain with a superimposed reconstruction of the electrode position based on MR artefacts caused by the electrode (not shown). Middle: recording triggered by the movement of the contralateral (left) hand. Right: recording triggered by the movement of the ipsilateral (right) hand.

Note the BP with maximal amplitude (a generator) in the SMA (M2) and in a part of the premotor cortex (M11), although no generator was recorded in the premotor cortex of the convexity. No BP or only small far field potentials were recorded in the white matter (M5; other contacts not shown). The amplitude of all BP is higher with contralateral than with ipsilateral hand movement.

Subcortical BP

In an attempt to explain the discrepancy between the wide distribution of BP on the scalp and its restricted distribution in the cortex, we suggested the possibility of some additional generators in subcortical structures (Rektor et al. 1994). We wondered if BP, which is a potential time-locked with the preparation of a movement, might be generated in some subcortical structures other than the cortex. Theoretically, structures other than the cortex that also participate in the motor execution loop *(Alexander 1994)*, namely the basal ganglia and the thalamus, could be involved in the motor act preparation. BP is influenced by cerebellar (Sasaki et al. 1979) and basal ganglia lesions. In some patients undergoing intracranial exploration, an electrode may have contacts in subcortical structures. Below are details from our recordings of BP in the basal ganglia and in the posterior thalamus (Rektor et al. 2001c, Rektor et al. 2001b).

The basal ganglia

The Bereitschaftspotential test protocol was studied in twelve patients. In all the patients, one or two diagonal electrodes passed through or touched the basal ganglia to reach the amygdala and the hippocampus. In seven patients, several depth electrodes were implanted orthogonally, in the temporal, fronto-orbital and prefrontal cortex. In four patients, subdural strip electrodes were used for the exploration of the fronto-temporal convexity. No BP was recorded in these areas. No contacts were placed in the central region known to generate cortical BP.The putamen was explored in eleven patients (in three of them bilaterally); the caudate head was explored in two patients; and the pallidum in two patients. BP was observed in the caudate in all recordings, and in the pallidum in one patient. It was recorded in the putamen in eight out of the eleven explored patients. BP was displayed contralaterally to the movement nine times in thirteen explorations, ipsilaterally four times in nine explorations. In the basal ganglia, BP preceded the onset of movement by 500-1500 ms. The shape of BP resembled the BP shape in the cortex and on the scalp. Given the amplitude gradients on neighbouring contacts, it is evident that the electrodes in the basal ganglia were actually located in the immediate vicinity of the generators. Movement Accompanying Potentials (MAP) were also present in all three explored structures. The electrophysiological characteristics of MAP differed from that of BP, indicating separate generators.

During most of the premovement period, as well as during the execution of movement, BP and MAP were displayed synchronously in the cortex and in the basal ganglia. When comparing the average duration of BP in the basal ganglia (1086 ± 330 ms) with that measured in other patients in the primary motor and supplementary motor cortices (1264 ± 393 ms and 1401 ± 451 ms, respectively: Rektor et al. 1994), it seems that the BP in the basal ganglia starts slightly later. This could be demonstrated in patients with simultaneous recordings from cortical and subcortical structures.

There are more indications of the influence of the basal ganglia on the occurrence of BP. Scalp BP was found to be disturbed in patients with basal ganglia lesions (Deecke 1985; Dick et al. 1989; Fève et al. 1991). The BP preceding a first step is larger than that preceding a simple dorsiflexion of the ankle in controls; however, this is not true for patients with an impairment of the basal ganglia function that is manifested by gait ignition failure (Vidailhet et al. 1995). The recordings of BP in the basal ganglia confirm a cognitive role of the basal ganglia in connection with motor regulation. Jahanshahi and Frith (1998) proposed that the mechanisms for motor deficit (akinesia) and for cognitive deficit (executive function) in willed action in Parkinson's disease are similar. In a PET study (Krams et al. 1998), there was significant activation in the putamen either during the preparation for or the execution of a movement. The activation was greater when there was a period of preparation before the execution than when the execution was immediate. Experimental data demonstrated that 20% of monkey striatal neurones were activated at least three seconds before a self-initiated movement (Schultz and Romo 1992).

BP is generated in subcortical structures in addition to the cortex. The cortico-basal ganglia-thalamo-cortical circuitry seems to be involved not only in movement execution but also in its preparation. From this point of view, the disturbance of scalp BP in some extrapyramidal diseases (eg Deecke 1985, Fève et al. 1991), and perhaps also in schizophrenia (Singh et al 1992), is not surprising.

The posterior thalamus

Recordings were taken from three patients with intractable chronic pain who underwent analgesic electrical stimulation of the contralateral thalamus. An intracerebral electrode was targeted into the ventroposterolateral (VPL) nucleus of the thalamus. The multilead electrode made it possible to record from several thalamic nuclei. The recordings were done in three patients from the VPL, and in one case from the nucleus posterolateralis, from the pulvinar, and from the internal capsule in the vicinity of the ventral anterior nucleus. The protocols used enabled us to study simple information processing (SEP and middle latency evoked potentials in the CNV protocol), as well as some functions linked with motor regulation and cognitive functions (BP, CNV, MAP, auditory oddball P3).

The results obtained from the VPL recordings were in concurrence with its established role as a relay nucleus, processing somatosensory information to the primary somatosensory cortex. The VPL generated the "thalamic" SEP, which was the only potential regularly recorded in this nucleus. Neither BP nor CNV appeared in the VPL. The absence of BP from the VPL was also observed in one patient by Fève (1993). From the pulvinar, only a visually evoked potential was recorded. Interesting results were obtained from the nucleus posterolateralis (PL) region, located just dorsally to the VPL. The PL is connected with parietal association cortex. The PL generated several electrophysiological potentials in tests involving cognitive operations. The oddball P3, the CNV, and the BP were generated here. It seems that there is a generator of BP in the PL that is more active with contralateral movement than with ipsilateral movement (the amplitude of the ipsilateral BP was lower and consisted only of the second slope). The BP generator is not identical to the MAP generator. The MAP recorded here was probably a far field potential. The recording of middle and long latency potentials evoked by auditory and visual stimuli in the CNV protocol showed that the PL also processes various modalities of afferent information. Unfortunately, we were able to perform the electrophysiological studies in the PL in only one patient. Therefore, the PL should be a candidate for further studies.

THE SOURCES OF BP RECORDED ON THE SCALP

The Bereitschaftspotential on the scalp is broadly distributed above the vertex, central, prefrontal and parietal areas (Kornhuber and Deecke 1965, Brunia and Damen, 1988, Shibasaki, et al. 1981). BP has generally been regarded as a cortical phenomenon. A term often used synonymously with BP is Movement Related Cortical Potential (MRCP). In fact, several generators of BP were localised in the cortex (see above). However, there are also BP generators in the putamen, the head of the caudate nucleus, in the pallidum and in several thalamic nuclei (posterior thalamus, motor thalamic nuclei, and midline thalamic nuclei) (Rektor et al. 2001b, 2001c, Fève 1993). McCallum et al. (1975) observed a BP rostral to the thalamic region and in the brainstem, i.e. in the pes peripedunculi, nucleus peripeduncularis, pulvinar, and medial geniculate (for older reviews, see McCallum 1993,

Haider et al.1981). BP may also be generated in some other subcortical structures that have not yet been explored. We do not know the extent to which the BP recorded on the scalp is influenced by the cortical or subcortical generators. According to theoretical models (Birbaumer et al. 1990), the slow waves must have cortical sources with occasional input from subcortical generators. Our observations do not explain the generation of scalp BP by the contribution of either cortical or subcortical sources alone. Intracranial cortical recordings are in contradiction to the wide distribution of scalp-recorded BP. Topographical studies of scalp BP distribution revealed BP in a large region of the head, encompassing both sides, with only slight asymmetry in the later BP components (Kornhuber and Deecke 1964, Shibasaki et al. 1980). The cortical dipoles that are produced in very restricted locations, exhibiting amplitudes of about 50-100 μV, are unlikely to be the only source of scalp BP. These highly restricted locations of cortical generators probably cannot explain the wide bilateral distribution of BP on the scalp. In our studies, scalp BP was recorded directly over scalp regions whose underlying brain areas displayed no BP. A typical example is presented in Fig. 3. In this example (with a foot movement), nearly all known BP dipoles were generated in the mesial cortex, while the appearance of BP on the scalp was bilateral above the convexities. BP appeared on the scalp exactly at a site above the cortex where an intracerebral electrode revealed the absence of BP. The reasons for this discrepancy remain obscure. However, it is also improbable that a dipole produced in subcortical structures (with the highest amplitude about 30-50 μV) could reach the electrodes on the scalp that have an approximate amplitude of 25-50 μV. A reproducible correlation between the amplitudes of deep and surface recordings of interictal epileptic activity generated in the mesiotemporal structures suggested that the ratio of deep to surface activity is around 1:2000 (Alarcon et al. 1994). On the other hand, a generator located in the depth of the hemisphere could be an additional source for potentials that are displayed broadly on the scalp. Cerebral electromagnetic activity can be recorded on the scalp when it is widely synchronised. We presume that in the case of BP, the deep dipoles might reach the scalp, as they are produced by a relatively huge mass of subcortical neuronal tissue. For example, BP can be recorded (with changing amplitude) in all contacts passing through the putamen. This is in contrast to epileptic spikes, which are produced in the epileptogenic zone that is composed of a network of more or less asynchronously spiking generators (Rektor and Švejdová, 1995). Based on intracranial recordings, we strongly suspect that the BP recorded on the scalp represents a summation of potentials that are generated simultaneously in several cortical as well as in several subcortical structures. McCallum et al. (1975) have concluded that "electrophysiological phenomena held to be manifestations of higher cortical functions are also to be found as deep as the midbrain level..." The surface BP is a spatial composite of several dipoles. We believe that it is not an exclusively cortical phenomenon.

THE RELATION OF BP TO ERD/ERS, CNV AND MAP

Several electrophysiological phenomena are linked with motor preparation and execution. We studied the appearance of slow potentials (BP and MAP) and of rhythmic activities (ERD/ERS) in a self-paced protocol, and we studied the appearance of slow potentials (CNV and MAP) in a protocol with cued movements. Although the movement used was identical, we observed that the location of generators varies according to the type of recording (BP/MAP compared with ERD/ERS) and according to the experimental method (BP/MAP compared with CNV/MAP). We conclude that before and during a movement, various processes occur that are presented by various electrophysiological phenomena.

Event-Related Synchronisation (ERS) and Desynchronisation (ERD)

Movement-related potentials – Bereitschaftspotential (BP) and alpha (8-12 Hz) and beta (16-24 Hz) ERD/ERS rhythms in self-paced hand and foot movements - were analysed in eight epilepsy surgery candidates (Sochůrková et al. 2000). We explored the primary motor and sensory cortex, the supplementary motor area, the premotor and cingulate cortices, the mesial temporal (amygdala, hippocampus, parahippocampal gyrus, fusiform gyrus) and lateral temporal structures, and the insula and orbital gyrus. Of the structures, the middle frontal gyrus, the cingulate gyrus and the amygdala were explored in both hemispheres and during both left and right upper limb movements. In the BP and ERD/ERS off-line evaluations, we used the same EEG recordings. In evaluating ERD/ERS, we used software designed by Stančák et al. in 1997 (smooth window of one second; overlap interval of 125 ms; FFT). Our preliminary results indicate that the ERD/ERS of spontaneous rhythms in alpha and beta frequency bands is more widespread than BP in the intracranial recordings. In contrast to BP, ERD/ERS frequently occurs in the mesiotemporal structures, suggesting a different functional significance for both electrophysiological phenomena.

Contingent Negative Variation

The appearance of intracerebral CNV was broader than that of BP. CNV was present in two large areas: the central region (premotor and motor cortex, supplementary motor area, cingulate and postcentral cortex) and the temporal lobe. In the motor structures, the somatotopy was less strict with CNV than with BP. For example, in the SMA of one patient, BP preceded foot (but not hand) movements, while the CNV was present in the same leads, regardless of the limb in motion.

Based on recordings made from the scalp, CNV has been separated into two distinct phases. The later phase has been associated with BP by several authors (e.g. Grünewald et al. 1979). The anatomical distribution of CNV revealed by intracerebral recordings tends to support an alternative opinion (Ruchkin et al. 1986; Frost et al. 1988), which holds that the later CNV phase cannot only be identified as BP. The second phase of CNV may share critical features with motor preparation (Brunia and Damen 1988) but it is not identical to BP (Lamarche et al. 1995).

Movement Accompanying Potential

MAP is defined as a brain potential appearing during a simple repetitive movement. MAP is monophasic or polyphasic, and starts before or during the movement. MAP is self-paced in the BP protocol, and is triggered by an external signal in the CNV protocol (Rektor et al. 1998). In BP protocol, MAP is recorded in the cortex and in the basal ganglia in two situations: following BP, or independently. Our recording showed a widespread distribution, bilaterally involving SMA, and the premotor, prefrontal, parietal (area 5, 7, 40) and midtemporal cortices. In contrast with the CNV protocol, the cingulate cortex was heavily involved, while in the primary motor cortex, MAP appeared only contralateral to the movement. In some areas (primary motor cortex, cingulate gyrus), the occurrence of MAP depended on the task context; it was present only in one of the two tested protocols. In monkeys (Romo et al. 1992), segregated populations of striatal neurones are engaged in the generation of self-paced and stimulus-triggered movement.

Electrophysiological characteristics, such as polarity and amplitude gradient, are not identical in BP and MAP. Despite a close relationship, they are produced by distinct generators. These observations concur with the theory that different neuronal populations are active during the preparation for and the execution of a motor act. Kropotov and Etlinger (1999) reported segregated neuronal populations in human basal ganglia responsible for the preparation of a motor response and for motor acts. Following the cortical study of MAP with self-paced and cued motion, we suggested that MAP is not, or is not only, the expression of movement generation in the corticospinal system. MAP might also at least partially represent some kind of cognitive activity (Rektor et al. 1998). The composite nature of MAP indicates that it is also a complex phenomenon.

THE PHYSIOLGICAL ACTIVITY REPRESENTED BY THE BP

It is presumed that movement-related brain potentials are generated by the coherent synaptic activity of cortical neurones (Fuster, 1984; Lang et al. 1994), perhaps with participation of glia in producing the slow shifts (Bauer et al. 1993). However, there is a discrepancy between the recording of the neuronal activity in premotor and frontal cortices (Mauritz and Wise, 1986), and the absence of slow waves.

BP preceding a complex motor task. What does "self-paced" mean?

BP starts about one, two, or sometimes three seconds before the onset of a self-paced movement. The slow waves are composed of several elements, reflecting the notion that the slow activation is a composite phenomenon. Their cerebral distribution is determined more by the task condition preceding the movement than by the movement itself. Several events occur during this period, including some acts that could be considered as cognitive: the timing; the decision to perform the movement; the preparation for, and the initiation of, the movement. Most of these events are unconscious (Libet, 1985).

The exact functional significance expressed by the BP is not known. Available evidence suggests that the underlying process is related to the subsequent movement (Libet, 1985). This process is self-paced in the sense that no external cue is necessary to trigger it. But what exactly does "self-paced" mean? The processes associated with attention, cognition and expectancy could not solely account for BP. BP is generated by processes specifically involved in the preparation for performing or the intention to perform a movement (Libet et al. 1982). It is difficult to investigate what happens during the BP, because the underlying process is mostly unconscious.

Our intention was to study the BP by comparing a standard BP with the BP elicited by a familiar activity (looking at pictures, deciding to act, performing a motor action). While exploring eleven epileptic patients with intracerebral electrodes, we recorded the BP preceding a complex motor task (Rektor et al. 2001a). The movement triggering the recording was the turning of a page in an architectural book. Each patient performed the task under two conditions. The first condition required that the patient turn the pages of the book in a self-paced manner, without looking at the pictures. He was instructed only to maintain an interval of at least five seconds between movements, without estimating the elapsed time by counting. The second condition required that the patient carefully inspects a particular picture, e.g. to determine whether the building was a villa or a factory, what type of roof it had, etc., and to turn the page at the moment he decided that he had seen it enough, and was ready to look at another picture. The patient then turned and successively inspected the subsequent pages (showing different pictures). The patients were instructed to make their own differentiation in performing both tasks, i.e., during the first task, they were asked to ignore the book and only to look straight ahead, turning the page at a certain moment, and during the second task, they were asked to look at the picture as long as necessary in order to make judgements about it.

BPs were present in the contralateral primary sensorimotor cortex and the bilateral SMA, and in the anterior caudal cingulate cortex. No difference between the cortical topography of the BP preceding a simple motor task (Rektor et al. 1994, 1998, Ikeda et al. 1992) and the topography of the BP occurring in connection with a complex movement was observed. In the two conditions (a simple motor task and a complex movement), the BP did not significantly differ, either in its location in the brain, or in its shape, duration, and amplitude. Evidently, the cognitive activity linked with looking at a picture had no influence on the generation of the BP. In the experiments by Libet et al. (1982), the cognitive activity related to exogenous stimuli could not solely account for the BP either. There is one simple explanation for this lack of difference. It is possible that, despite the protocol differences, the behavioural events related to the BP were identical in both conditions. This means that in the classic self-paced protocol, the BP was related to the succession of actions performed to accomplish an early implicit task (the measurement of the inter-movement period by an internal clock; the decision to move; the preparation and initiation of the movement itself). Our suggestion, that the long unconscious period is associated with preparation to act in the near future, while the process with an onset 0.5 sec before an act is associated with voluntary choice and the endogenous urge to act, is supported by the results of Libet et al. (1982). If this is the case, then "self-paced" does not mean "spontaneous". Instead, this notion covers an internal program related to the given task.

Slow parallel information processing in the motor system

We have observed that variations in movement-related potentials (BP, CNV, and MAP) occur simultaneously in multiple cortical and subcortical structures. The topography of simultaneously active cortical regions is task-related. During identical movements, the areas with slow activation differed, depending on the kind of pacing. The slow potentials (MAP) in the self-paced protocol were present in the primary motor cortex, only contralateral to the movement, but they were bilateral in the CNV protocol. The cingulate cortex was frequently involved in the self-paced movements, but rarely in the cued movements. In the self-paced protocol before the movement (BP), only the regions known to be involved in motor activity were involved. In the CNV protocol with auditory stimulation, the slow waves (CNVs) also appeared beside the central regions in the acoustic temporal zone (Lamarche et al. 1995).

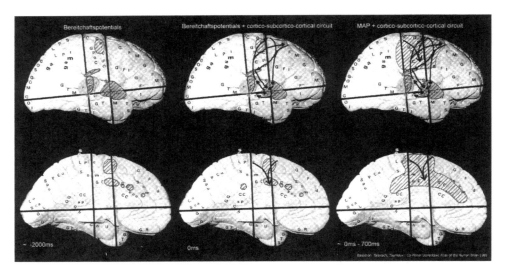

Fig 5. Schematic presentation of a long-lasting simultaneous activation of cortical and sub-cortical structures in the motor system before (BP) and during (MAP) self-paced movement of the contralateral hand. Left: period starting approximately 2000 ms before the movement. Middle: the moment of the movement onset. In conjunction with the slow parallel activity, there is rapid sequential processing of information in the cortico-basal ganglia-thalamo-cortical circuitry. Right: the movement is executed. The rapid sequential processing continues while the slow activation spreads over more cortical areas

Striped: slow parallel activity. Arrows: rapid sequential activity.

According to the model of Alexander (1994), a motion is based on a sequential activity of structures in a cortico–basal ganglia–thalamo-cortical circuitry that are modulated by several sub-cortical loops. Our results indicate that in addition to sequential processing of information, there is also a simultaneous, slow activation of cortical and sub-cortical structures, lasting from hundreds of milliseconds to seconds (Fig. 5). These results concur with the concept that sensorimotor processing is an operation distributed in a parallel manner, conducted simultaneously at several brain sites.

We conclude that in addition to the rapid sequential processing of information, there is a long-lasting simultaneous activation of cortical and sub-cortical structures in the motor system when a movement is prepared and executed. Which regions are simultaneously active is dependent on the context of the task.

The BP similar to other slow potentials represents a slow event linked to certain tasks. The location of the BP in the brain indicates that the areas generating the BP participate in the task, but their exact role remains unclear. The role played by individual areas may be varied, i.e. the output of some areas might be excitatory, while that of other areas might be inhibitory. Some structures might be control structures, while the function of other structures could be closer to the execution of a given task. An example could be a PET study for language translation and switching: The basal ganglia and anterior cingulate are suggested as parts of control circuits controlling the cortical circuits (Price et al. 1999). However, in the case of the BP, the meaning of activity in individual structures remains speculative.

It is not clear which kind of brain function the slow potentials represent. The appearance of BP, CNV and MAP is not always directly linked with the preparation for and the performance of a movement. This has been demonstrated both in our studies and in studies of cases with basal ganglia lesions, where slow potentials did not appear while the movements were being performed (Deecke, 1985; Fève et al. 1991a,b). Motor areas are active both when executing, as well as when imagining, movement (Lang et al. 1994;

Beisteiner et al. 1995). Prolonged voluntary muscle contraction alone does not cause any cerebral negativity, neither before nor after movement, if the action is not goal-directed, or has no motivational significance (Tecce et al. 1984; quoted in Birbaumer et al. 1990). The presence or absence of MAP is, at least in some brain regions, dependent on the protocol tested, and not on the execution of movement. BP and MAP do not (or do not only) reflect the neuronal discharges of structures involved in generation of descending corticospinal volleys and the arrival of sensory feedback information related to the movement itself.

The slow potentials in question could be an expression of the spread of information through the relevant structures. The significance of the slow potentials studied here may be a "readiness to subsequently act" and "attention to action" (Posner and Dehaene, 1994). The role of slow potentials seems to be linked to states of preparedness, receptiveness, and possibly resource mobilisation (McCallum, 1993).

ACKNOWLEDGEMENTS

This research was supported by the Yamanouchi European Foundation and by the MŠ ČR 112801 research program. The author wishes to thank the co-authors of previous studies: M. Bareš, N. Bathien, P. Buser, A. Fève, P. Kaňovský, D. Kubová, M. Lamarche, J. Louvel, M. Kukleta, I. Rektorová, D. Sochůrková, and A. Stancák, Jr. Additional thanks are due to Mme L. Olive, Mr. P. Daniel, and Ms. Y. Břenková for their help with the preparation of figures and the manuscript.

REFERENCES

Alarcon, G., Guy, C.N., Binnie, C.D., Walker, S.R., Elwes, R.D., Polkey, C.E. (1994) Intracerebral propagation of interiactal activity in partial epilepsy: implications for source localisation. J. Neurol. Neurosurg. Psychiatry 57 (4), 435-449.

Alexander, G.E. (1994) Basal ganglia-thalamocortical circuits: their role in control of movements. J. Clinical. Neurophysiol. 11, 420-431.

Arezzo, J., Vaughan, Jr., H.G. (1975) Cortical potentials associated with voluntary movements in the monkey. Brain Research 88, 99-104.

Bareš, M., Rektor, I. (1999) Basal ganglia are involved in the generation of auditory and visual evoked potential components – a SEEG study of a contingent negative variation paradigm. Clin. Neurophysiol. 110 (1), 220.

Bauer, H., Korunka, Ch., Leodolter, M. (1993) Possible glial contribution in the electrogenesis of SPs. In: McCallum, W.C. and Curry, S.H., (Eds.) Slow Potential Changes in the Human Brain, pp. 23-34. New York: Plenum Press.

Beisteiner, R., Hoellinger, P., Lindinger, G., Lang, W., Berthoz, A. (1995) Mental representations of movements. Brain potentials associated with imagination of hand movements. Electroencephalogr. clin. Neurophysiol. 96, 183-93.

Birbaumer, N., Elbert, T., Canavan, A.G.M., Rockstroh, B. (1990) Slow Potentials of the Cerebral Cortex and Behavior. Phys. Rev. 70, 1-40.

Brunia, C.H.M., Damen, E.J.P. (1988) Distribution of slow brain potentials related to motor preparation and stimulus anticipation in a time estimation task. Electroenceph. clin. Neurophysiol. 69, 234-243.

Chen, R., Gerloff, C., Hallett, M., Cohen, L.G. (1997) Involvement of the ipsilateral motor cortex in finger movements of different complexities. Ann. Neurol. 41 (2), 247-254.

Deecke, L. (1985) Cerebral potentials related to voluntary actions: parkinsonian and normal subjects. In: Delwaide PJ, Agnoli A., (Eds.) Clinical Neurophysiology in Parkinsonism. Amsterdam, Elsevier 91-105.

Dick, J.P.R., Rothwell, J.C., Day, B.L., Cantello, R., Buruma, O., Gioux, M., Benecke, R., Berardelli, A., Thompson, P.D., Marsden, C.D. (1989) The Bereitschaftspotential is abnormal in Parkinson's disease. Brain 112, 233-344.

Fève, A.P., Bathien, N., Rondot, P. (1991a) Evolution des potentiels corticaux lies au mouvements chez patients parkinsoniens, avant et apres traitement par la levodopa. Neurophysiol. Clin. 21, 105-119.

Fève, A.P., Bathien, N., Rondot, P. (1991b) Les potentiels corticaux lies au mouvement de l'homme age. Neurophysiol. Clin. 21, 281-291.

Fève, A., Bathien, N., Rondot, P. (1991) Evolution des potentiels corticaux liés au mouvement chez les patients parkinsoniens, avant et après traitement par la lévodopa. Neurophysiol. Clin. 21, 105-119.

Fève, A.P. (1993) Origine sous corticale des potentiels pre-moteurs (movement-related-potentials) chez l'homme. Thèse de Doctorat. Université Paris 6. Paris.

Frost, B.G., Neill, R.A., Fenelon, B. (1988) The determinants of the non-motoric CNV in a complex, variable foreperiod, information processing paradigm. Biol. Psychol. 27, 1-21.

Fuster, J.M. (1984) Behavioral electrophysiology of the prefrontal cortex. TINS 408-414.

Gemba, H., Sasaki, K. (1984) Distribution of Potentials Preceding Visually Initiated and Self-Paced Hand Movements in Various Cortical Areas of the Monkey. Brain Res. 306, 207-214.

Grünewald, G., Grünewald-Zuberbier, E., Netz, J., Hömberg, V., Sander, G. (1979) Relationships between the late component of the contingent negative variation and the Bereitschaftspotential. Electroenceph. clin. Neurophysiol. 46, 538-545.

Haider, M., Knapp, E.G., Ganglberger, J.A. (1981) Event related slow (DC) Potentials in the Human Brain. Res. Physiol. Biochem. Pharmacol. 88, 125-197.

Ikeda, A., Luders, H.O., Burgess, R.C., Shibasaki, H. (1992) Movement-related potentials recorded from supplementary motor area and primary motor cortex. Brain 115, 1017-1043.

Jahanshahi, M. and Frith, C.D. (1998) Willed action and its impairments. Cognitive Neuropsychology 15, 483-534.

Kornhuber, H.H., Deecke, L. (1964) Hirnpotentialaenderung beim Menschen vor und nach Willkuerbewegungen, dargestellt mit Magnetbandspeicherung und Rueckwaertsanalyse. Pfluegers Archiv 281, 52.

Kornhuber, H.H., Deecke, L. (1965) Hirnpotentialanderungen bei Willkurbewegungen und passiven Bewegungen den Menschen: Bereitschaftspotential und reafferente Potentiale. Pflugers Archiv 284, 1-17.

Krams, M., Rushworth, M.F.S., Deiber, M-P., Frackowiak, R.S.J., Passingham, R.E. (1998) The preparation, execution and suppression of copied movements in the human brain. Exp. Brain Res. 120, 386-398.

Kropotov, J.D. and Etlinger, S.C. (1999) Selection of actions in the basal ganglia-thalamocortical circuits: review and model. Intern. J. Psychophysiol. 31, 197-217.

Lamarche, M., Louvel, J., Buser, P., Rektor, I. (1995) Intracerebral recordings of slow potentials in a contingent negative variation paradigm: an exploration in epileptic patients. Electroenceph. clin. Neurophysiol. 95, 268-276.

Lang, W., Hollinger, P., Eghker, A., Lindinger, G. (1994) Functional Localization of Motor Processes in the Primary and Supplementary Motor Areas. J. Clin. Neurophysiol. 11, 397-419.

Lee, B.I., Lüders, H., Lesser, R.P., Dinner, D.S., Morris, H.H. (1986) Cortical potentials related to voluntary and passive finger movements recorded from subdural electrodes in humans. Ann. Neurol. 20, 32-37.

Libet, B., Wright, E.W. Jr, Gleason, C.A. (1982) Readiness-potentials preceding unrestricted "spontaneous" vs. pre-planed voluntary acts. Electroenceph. clin. Neurophysiol. 54, 322-335.

Libet, B. (1985) Unconscious cerebral initiative and the role of conscious will in voluntary action. Behav. Brain Sci. 8, 529-566.

Mauritz, K.H., Wise, S.P. (1986) Premotor cortex of the rhesus monkey: neuronal activity in anticipation of predictable environmental events. Exp. Brain Res. 61, 229-244.

McCallum, W.C. (1975) Behavioural and clinical correlates of brain slow potential changes. Proc R Soc Med 68, 3-6.

McCallum, W.C. (1993) Human Slow Potential Research: A review. In: McCallum, W.C. and Curry, S.H., ((Eds.) Slow Potential Changes in the Human Brain, pp. 1-11. New York: Plenum Press.

Neshige, R., Lüders, H., Shibasaki, H. (1988) Recording of movement related potentials from scalp and cortex in man. Brain 111, 719-736.

Paus, T., Koski, L., Caramos, Z., Westbury, Ch. (1998) Regional differences in the effects of task difficulty and motor output on blood flow response in the human anterior cingulate cortex: a review of 107 PET activation studies. Neuroreport 9 (9) 37-47.

Posner, M.I., Dehaene, S. (1994) Attentional networks. TINS 17, 75-79.

Price, C.J., Green, D.W., von Studnitz, R. (1999) A functional imaging study of translation and language switching. Brain 122, 2221-2235.

Rektor, I., Fève, A., Buser, P., Bathien, N., Lamarche, M. (1994) Intracerebral recording of movement related readiness potentials: an exploration in epileptic patients. Electroenceph. clin. Neurophysiol. 90, 273-283.

Rektor, I., Švejdová, M. (1995) Spatiotemoporal analysis of interictal spikes. A stereoelectroencephalographic study. Neurophysiol. Clin. 25, 12-18.

Rektor, I., Louvel, J., Lamarche, M. (1998) Intracerebral recording of potentials accompanying simple limb movements. A SEEG study in epileptic patients. Electroenceph. clin. Neurophysiol. 107, 277–286.

Rektor, I. (2000a) Long-lasting simultaneous activation of cortical and subcortical structures in movement preparation and execution. Clinical neurophysiology at the beginning of the 21ˢᵗ century. Suppl. Clin Neurophysiol. 53, 192-195.

Rektor, I. (2000b) Cortical activation in self-paced versus externally cued movements: a hypothesis. Parkinsonism and Related Disorders 6, 181-184.

Rektor, I., Bareš, M., Kaňovský, P., Kukleta, M. (2001a) Intracerebral recording of readiness potential induced by a complex motor task. Movement Disorders 16, 698-704.

Rektor, I., Kaňovský, P., Bareš, M., Louvel, J., Lamarche, M. (2001b) Evoked potentials, ERP, CNV, readiness potential, and movement accompanying potential recorded from the posterior thalamus in human subjects. A SEEG study. Neurophysiologie clinique/Clinical Neurophysiology 31, 1- 9.

Rektor , I., Bareš, M., Kubová, D. (2001c) Movement related potentials in the basal ganglia: a SEEG readiness potential study. Clin Neurophysiol, in press

Romo, R., Scarnati, E., Schultz, W. (1992) Role of primate basal ganglia and frontal cortex in the internal generation of movements. II. Movement-related activity in the anterior striatum. Exp. Brain Res. 91, 385-395.

Ruchkin , D.S., Sutton, S., Mahafey, D., Glaser, J. (1986) Terminal CNV in the absence of motor response. Electroenceph. clin. Neurophysiol. 63, 445-463.

Sasaki, K., Gemba, H., Hashimoto, S., Mizuno, N. (1979) Influences of cerebellar hemispherectomy on slow potentials in the motor cortex preceding self-paced hand movements in the monkey. Neuroscience Letters 15, 23-28.

Shibasaki, H., Barrett, G., Halliday, A.M., Halliday, E. (1980) Components of the movement-related cortical potential and their scalp topography. Electroenceph. clin Neurophysiol. 49, 213-226.

Shibasaki, H., Barret, G., Halliday, E., Halliday, A.M. (1981) Cortical potentials associated with voluntary foot movement in man. Electroenceph. clin. Neurophysiol. 52, 507-516.

Schultz, W., Romo, R. (1992) Role of primate basal ganglia and frontal cortex in the internal generation of movements. Exp. Brain Res. 91, 363-384.

Singh, J., Knight, R.T., Rosenlicht, N., Korun, J.M., Beckley, D.J., Woods, D.L. (1992) Abnormal premovement brain potentials in schizophrenia. Schizophr. Res. 8 (1), 31-41.

Sochůrková, D., Rektor, I., Stančák, Jr., A. (2000) Intracranial recordings of readiness potential and event-related desynchronisation in hand and foot movements. Clin. Neurophysiol. 111 (1), 85.

Stančák, Jr., A., Riml, A., Pfurtscheller, G. (1996) The effects of external load on movement-related changes of the sensorimotor EEG rhythms. Electroenceph. clin. Neurophysiol. 102, 495-504.

Streitová, H., Rektor, I., Kubová, D., Bareš, M., Hortová, H. (1999) Activity of the gyrus cinguli in the movement preparation and performance and in the cognitive functions. A SEEG study of readiness potentials, movement accompanying potentials, CNV and P3. Suppl. Parkinsonism & Related Disord. 5, 122.

Vidailhet, M., Atchison, P., Stocchi, F., Thompson, P.D., Rothwell, J.C., Marsden, C.D. (1995) The Bereitschaftspotential preceding stepping in patients with isolated gait ignition failure. Movement Disord. 10, 18-21.

Walter, W.G., Cooper, R., Aldridge, V.J., McCallum, C., Cohen, J. (1964) The contingent negative variation: an electro-cortical sign of sensorimotor association in man. Electroenceph. clin. Neurophysiol. 17, 340-344.

Yazawa, S., Ikeda, A., Kunieda, T., Ohara, S., Mima, T., Nagamine, T., Taki, W., Kimura, J., Hori, T., Shibasaki, H. (2000) Human presupplementary motor area is active before voluntary movement: subdural recording of Bereitschaftspotential from medial frontal cortex. Exp. Brain Res. 131 (2), 165-177.

DISTRIBUTED SOURCE MODELING IN THE ANALYSIS OF MOVEMENT-RELATED ACTIVITY

Rumyana Kristeva-Feige

Neurological Clinic
Albert-Ludwigs-University
Breisacher Straße 64
79106 Freiburg, Germany

INTRODUCTION

This chapter deals with the generators of movement-related activity in the Bereitschaftspotential paradigm as revealed by distributed source modeling. The chapter starts with a short overview of the developments in recording and analysis techniques in the study of the Bereitschaftspotential. This is followed by a short tutorial on source reconstruction considering individual brain anatomy from MRI and particularly distributed source modeling. After that, the studies using this modeling in time and frequency domain are presented. I conclude with some critical comments and suggestions for future research.

During the last four decades, the analysis of movement-related cortical sources has been influenced by the dramatic developments in recording and analysis techniques and by the introduction of new concepts of brain function. The analysis of the movement-related potentials (MRPs), as well as magnetic fields, started with mere description of latencies and amplitudes of the potentials and fields recorded from a few scalp positions overlying the contralateral, midline and ipsilateral frontoparietal areas (Kornhuber and Deecke, 1965; Deecke et al., 1969, 1976; Kristeva et al., 1979; Kristeva and Kornhuber, 1980). At a further stage of investigation recordings from the whole scalp were made. Simultaneously with these advances of the recording techniques, new topographic analyses like Laplacian methods (Urbano et al., 1998; Babiloni et al., 1999; Cui and Deecke, 1999; Cui et al., 2000; Stancak, Jr. et al., 2000) and source reconstruction methods using moving dipole and spatiotemporal models were introduced (Cheyne and Weinberg, 1989; Cheyne et al., 1991; Kristeva et al., 1991; Lang et al., 1991a,b; Toro et al., 1993; Kristeva-Feige et al., 1994; Nagamine et al., 1994; Hallett and Toro, 1996; Gerloff et al., 1998). The contribution of the dipole model analysis for understanding the generators of the Bereitschaftspotential is reviewed in another chapter (Toma and Hallett, this volume). This chapter deals with the introduction of distributed source models in the analysis of movement-related sources. These models have the advantage that they do not use *a priori* knowledge about the number and location of cortical sources. Therefore, they are important not only for analysis of movement-related activity but also for analysis of the dynamic organization and reorganization of the information processing in all systems based on distributed networks.

SOURCE RECONSTRUCTION BASED ON THE INDIVIDUAL BRAIN ANATOMY

Before introducing the idea of distributed source modeling, I would like to briefly discuss the principles of the source reconstruction that consists of estimation of the underlying cerebral sources from the measured distribution of the electric potentials or of the magnetic fields on the scalp.

Interestingly, the development of electric source reconstruction was accelerated by the rapid developments of magnetic source reconstruction in the eighties (Kaufman and Williamson, 1982; Williamson and Kaufman, 1990). The reason for these quick developments were:

- The magnetic fields recorded from the scalp are almost not smeared by the different head compartments, i.e. the different head compartments are quasitransparent for the magnetic fields.
- The magnetic fields are sensitive only to tangential components of an active neuronal population (*i.e.* to sources in sulci cerebri such as MI).
- In most commercially available magnetoencephalography (MEG) systems gradiometers are used in order to suppress homogeneous or far-fields, but that also lead to a reduced sensitivity for deep sources. Thus there is an increased specificity for neo-cortical sources.

For these three reasons magnetic source reconstruction was easier to cope with. And last, but not least, concerted efforts of the scientists from the biomagnetic society were aimed at its quick development. However, the blindness of MEG towards radial sources (sources in gyri cerebri) has given an impetus to the electric source reconstruction and the researchers have started working on the problem how to realistically approximate the head morphology and how to compute the smearing effects of the electric fields recorded from the scalp through the different compartments (cerebrospinal fluid, skull, skin) taking into consideration their conductivities. Thus the source reconstruction based on high-resolution EEG taking into consideration the individual head anatomy from the three-dimensional (3D) MRI was developed. This advanced source reconstruction is based on:

- High-resolution (*i.e.* multichannel) EEG (most of the labs use 64-channels) to obtain a potential distribution with high spatial resolution,
- Coregistration of EEG and MRI data to transform the electrode positions and the MRI into a single coordinate system,
- Advanced volume conductor head models,
- Advanced source models,
- Advanced source space models, and
- Visualization of the results into the individual MRI.

The different elements of the source reconstruction are shown in Figure 1.

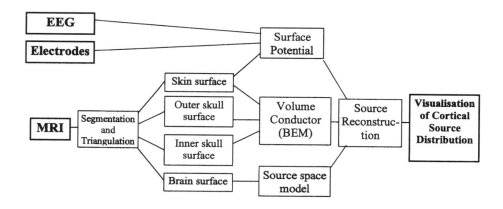

Figure 1. Flow chart of the advanced source reconstruction

During the experiment, multichannel EEG is recorded in a typical Bereitschaftspotential paradigm, *i.e.* the subject performs a voluntary movement every 12-15 seconds. To transform the electrode positions and the MRI into a single coordinate system after the experimental session, the electrode positions and the head contour are digitized by 3D position digitizers (e.g. ZEBRIS or Polemus; for more details see Huppertz et al. (1998)).

To estimate the underlying cerebral sources from the measured distribution of the electric potentials on the scalp an "inverse solution" has to be made. It is clear that the relation between the potential distribution on the scalp and the source configuration is not that simple because one and the same distribution can be due to completely different source configurations, since so called silent sources (e.g. ring-currents in the electric case or radial currents in the magnetic case) can be added to every generator configuration without changing the external field distribution. Therefore, the inverse problem does not have an unique solution.

For solving the inverse problem a head model, a source model and a source space model are needed. The volume conductor head model accounts for the shape of the three different compartments (brain with CSF, skull and skin and their conductivities) after segmenting them from MRI (segmentation means finding the borders between the different compartments on the basis of their different tissue densities). Instead of using a spherical head model the more realistically shaped three compartment volume conductor head model computed using the boundary element method (BEM) is used. Thus the shapes of the different compartments are considered and smearing effects are better accounted for than by using the oversimplifying spherical models as shown in Fuchs et al. (1998).

The source model is the mathematical representation of the intracerebral sources. While the equivalent current dipole models require *a priori* assumptions, the distributed source models (one of them is the cortical current density, CCD) do not require *a priori* assumptions about the number and location of underlying cortical sources. They need a less severe minimum norm constraint only. With current density reconstructions regularization plays an important role. If the signal to noise ratio (SNR) of the measured data can be determined a discrepancy measure can be used (Morozov, 1968; Fuchs et al., 1999).

Another important element of the source model is the depth normalization, which compensates for the lower sensitivity of the amplifier system to deep sources, particularly those in the depth of sulci or in the interhemispheric fissure.

The source space model represents the space in which sources have to be calculated. For cortical current density (CCD) analysis the sources are calculated on the segmented brain surface.

After the source reconstruction, the results (in the case of CCD the cortical current density maps) are superimposed (visualized) onto the individual MRI.

Image segmentation, volume conductor modeling, source reconstruction, and visualization can be performed using some software packages available on the market (e.g. the CURRY software (Neuroscan Labs, Virginia, USA)).

Used in this sophisticated way the electric source reconstruction is a powerful tool because it provides both high temporal and high spatial resolution. Moreover, the EEG is sensitive to both tangential and radial sources. And this is something very important for analysis of the cortical motor function because the MI is located in the depth of the central sulcus (White et al., 1997) and what we see in a top view of the brain is mostly the premotor area (Zilles et al., 1995).

DISTRIBUTED SOURCE MODELS

The distributed source models need no assumptions about the number, shape, or size of activated areas. They were introduced by Wang et al. (1992), Hämäläinen and Ilmoniemi (1994), Pascual-Marqui et al. (1994a), Fuchs et al. (1998) and Grave de Peralta Menendez et al. (2001). They all have in common a predetermined distribution of elementary current dipoles on given positions in the head, either on a regular grid (Pascual-Marqui et al., 1994b) or constrained to the cortical surface (Dale and Sereno, 1990; Fuchs et al., 1999) as in CCD analysis. While the positions of these dipoles are fixed, their orientations and strengths must be determined. The reconstructed CCD maps obtained in this way show the current distribution on the cortex that can account for the potentials measured on the head surface.

The number of electrodes is usually much smaller than the number of modeling dipoles. Therefore, a highly underdetermined problem has to be solved. This is possible only by introducing additional constraints (e.g., minimum norm). The constraining model term has to be weighted against the data term by a regularization parameter whose choice is crucial in order to avoid overfit or underfit of the data (Fuchs et al., 1999). For the minimum norm least square (MNLS) approaches (called also L_2-norm approaches), a linear system of equations is solved. When using other than the L_2-norm (e.g., the L_1-norm), a system of non-linear equations with the same number of unknowns has to be solved. Therefore, the L_1-norm needs much more computational efforts than the L_2-norm. However, the L_1-norm approaches give much more focal solutions and have a more robust behaviour against outliers in the measured data. For more details about distributed source models the reader is referred to the review article of Fuchs et al. (1999).

DISTRIBUTED SOURCE MODELS IN THE STUDY OF MOVEMENT-RELATED ACTIVITY

The distributed source reconstruction models were introduced in the analysis of movement-related potentials by Knösche et al. (1996). Using MNLS the authors have recognized the activation of the MI and of the frontal medial wall motor areas (SMA) during preparation of a voluntary movement. Due to the small number of electrodes, the suboptimal electrode layout, the simple spherical head and source space model used, the spatial resolution of this study was still limited. Another reason for the low spatial resolution was the MNLS approach used. Despite the limited spatial resolution achieved, this study represents a considerable step in the analysis of movement-related cortical sources because it demonstrated the value of the distributed source models for analysis of movement-related activity.

A further considerable step in the distributed source model analysis of movement-related potentials was made with the study of Ball et al. (1999) using high-resolution EEG (64 channels) combined with source reconstruction considering the individual brain morphology from MRI. The source model used in this study was L_1-norm and a realistically shaped BEM volume conductor head model was used. Due to this advanced source reconstruction a better spatial resolution was achieved and the role of the executive motor (MI and posterior SMA) and higher-order motor (anterior and intermediate SMA, cingulate motor area and inferior parietal lobe area) areas during voluntary movement preparation and execution was shown in detail. In this study the authors examined a simple finger flexion movement in six healthy right handed subjects, integrating high-resolution EEG with the individual structural and functional MRI. Highly converging results from EEG and fMRI were obtained for both executive and higher order motor areas.

All 6 subjects showed activation of the MI and of the frontal medial wall motor areas. Figure 2 shows the results for one of the subjects investigated. Note, that in Fig. 2 the lateral maximum of CCD corresponds to the fMRI activation of the hand area of the left primary sensorimotor cortex and the medial current density maximum corresponds to the fMRI activation of the frontal medial wall motor areas. Note also, that the currents shown in Fig. 2b point in an anterior direction, consistent with activation of MI in the anterior bank of the central sulcus, *i.e.* of the primary motor area, while 70 ms after EMG onset the currents point in a posterior direction, consistent with activation of the primary somatosensory area in the posterior bank of the central sulcus (Kristeva-Feige et al., 1994).

Another interesting result in the study of Ball et al. (1999) was that three out of the six subjects investigated showed activation in the inferior parietal lobe (IPL) starting during early movement preparation as seen from Figure 3 for one of the subjects.

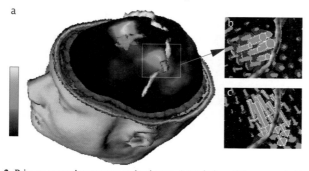

Figure 2. Primary sensorimotor area activation: **a.** A rendering of the segmented head of one of the subjects, view from left and above. The segmented left and right central sulci are shown in grey. Current density map (maximum in yellow, minimum in black, linear scale shown on the left) at EMG onset is shown embedded in the section of the segmented head. fMRI activation map is shown in turquoise. **b.** Magnification of the hand area of the left primary sensorimotor area as shown in (a). The individual current vectors of the CCD source are displayed, whereas in the current density map in (a) only the absolute values of the currents are visualised. In their sum they are equivalent to the current density source at EMG onset shown in (a). **c.** Current vectors in the same area as in (b) are shown 70 ms after EMG onset (from Ball et al. (1999) with kind permission of Neuroimage).

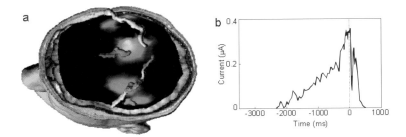

Figure 3. Inferior parietal lobe (IPL) activation: a. CCD map 100 ms before EMG onset and fMRI activation for one of the subjects. Both modalities show a medial source, a contralateral MI source and a contralateral IPL source. In the CCD map ipsilateral MI and IPL sources can be seen as well. b. Waveform of the contralateral IPL source shown in (a). Note the early onset of activity and the minimum of source strength occurring 70 ms after the EMG onset (from Ball et al. (1999) with kind permission of Neuroimage).

Due to the novel L_1-norm analysis used, the study also shed more light on the frontal medial wall activation prior to the voluntary movement. Interestingly, two different types of activation were observed: Four of the subjects showed an anterior type of activation and two of the subjects a posterior type of activation. In the former, activity started in the anterior cingulate motor area (CMA) and subsequently shifted its focus to the intermediate supplementary motor area (SMA). Approx. 120 ms before the movement started, the intermediate SMA showed a drop of source strength, and simultaneously MI showed an increase of source strength. In the posterior type, activation was restricted to the posterior SMA. Figure 4 shows the frontal medial wall activation for one of the subjects investigated having anterior type of frontal medial wall activation.

Fig. 4c shows the pre-movement shift of the location of the CCD source from the anterior CMA to the intermediate SMA. As illustrated in Figure 4d this occurs about 2200 ms prior to the movement onset.

As shown in Fig. 4d, there is a reciprocal relation between intermediate SMA and MI activity in terms of source strength around EMG-onset. Note that exactly when there is a drop in the source strength of the intermediate SMA, there is a positive deflection just before EMG onset in the BP, called pre-motion positivity, PMP, shown in Fig. 4e. This frontal medial wall activation pattern was found in four of the investigated subjects. The remaining two subjects did not show neither a drop in the strength of the intermediate SMA nor PMP. Therefore, the authors claimed that the generator site of the PMP is in the frontal medial wall motor areas. This is consistent with the fact that PMP is not seen in MEG which is silent for these areas (Kristeva et al., 1991; Kristeva-Feige et al., 1994). This assumption is also well in line with recent results based on surface Laplacian estimates (Urbano et al., 1996).

Let us summarise the results from this study:

- In all subjects showing activation of higher-order motor areas (anterior CMA, intermediate SMA, IPL) these areas were shown to become active before the executive motor areas (MI and posterior SMA). The SMA was active not only before but also during the MI activity.
- The early activation of the anterior CMA and the IPL may be related to attentional functions of these areas.
- The premotion positivity possibly has its generator site in the intermediate SMA. The drop in source strength of the intermediate SMA can disinhibit the primary MI and initiate the movement.

- The cortical motor system is not generally activated in a global way, but even during one motor task (simple finger flexion/extension) in different individuals different subsets of cortical motor areas team up to perform the task, while others remain silent.

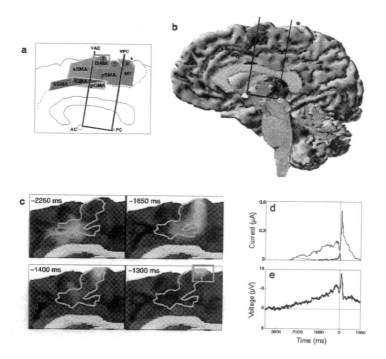

Figure 4. Anterior type frontal medial wall activation: a) Schematic illustration of the motor areas of the frontal medial wall, modified after Picard and Strick (1996) and Stephan et al. (1995). The ACPC (anterior commissure – posterior commissure)-line, VAC (vertical AC)-plane and VPC (vertical PC)-plane are marked in blue. The VAC-plane is a gross anatomical landmark for the border between the anterior and the posterior SMA (Zilles et al., 1996). The point, where the central sulcus meets the edge of the medial wall is marked by an asterisk. The cingulate sulcus is marked with a red dotted line. Both the SMA and the CMA are subdivided into an anterior, an intermediate and a posterior part. The anterior CMA (aCMA) corresponds to the anterior sector of the rostral cingulate zone of Picard and Strick (1996), the intermediate CMA (iCMA) and the posterior CMA (pCMA) to their posterior sector of the rostral cingulate zone and to their caudal cingulate zone. In the SMA, the anterior SMA (aSMA) was defined as the part of the SMA anterior to the VAC plane, our intermediate and posterior SMA (iSMA, pSMA) are equivalent to the rostral and caudal posterior SMA of Stephan et al. (1995), respectively. The precentral gyrus is marked with an encircled "p", the first and second gyrus anterior to the precentral gyrus with an encircled "1" and "2", respectively. The area of the intermediate SMA corresponds to the dorsal medial aspect of the second gyrus anterior to the precentral gyrus. This area is marked with a blue box. b) A median saggital section, showing the medial aspect of the right hemisphere for subject 1. fMRI activation (turquoise) of the CMA and of the SMA. For orientation the same anatomical landmarks as shown in (b) are displayed. c) Sequential CCD maps corresponding to the fMRI activation shown in (b). The outline of the fMRI activation shown in (a) is marked by a turquoise line. In the lower right panel the intermediate SMA is marked by a blue box as in (a). d) Waveforms of the time course of the CCD sources shown in (c): CMA (green), intermediate SMA (red). The time course of activation of the MI contralateral to the movement side of the same subject is shown in blue. e) Bereitschaftspotential of the same subject recorded at Cz (common average reference) (from Ball et al. (1999) with kind permission of Neuroimage).

It is worth mentioning that the multimodal imaging used in the study of Ball et al. (1999) revealed individual cortical brain activity with high temporal and spatial resolution, independent of *a priori* physiological assumptions. The fact, that the activation of the areas was shown with both modalities gave assurance to the authors in the results obtained. However, one has to have in mind that both techniques are in development. Let us consider what can be improved in the EEG method used: While the source reconstruction for the lateral (MI) sources was performed using the smoothed cortex as a source space model, the source reconstruction of the frontal medial wall sources was performed using the realistic one. Furthermore, the influence of the depth normalization on the reconstruction result was not systematically investigated. Altogether, the study has shown that a systematic evaluation of all the reconstruction parameters was needed.

THE BEREITSCHAFTSPOTENTIAL PARADIGM IN THE STUDY OF CORTICO-MOTONEURONAL SYNCHRONIZATION

Before describing how the distributed source models were applied to better understand the cortico-motoneuronal synchronization I would like to make a short introduction to the concept of binding that is closely related to the idea of "cell assemblies".

The idea of "cell assemblies", introduced by Hebb (1949) was well received in the Neurosciences and still remains a major conceptual force in theories of brain function. A "cell assembly" is the population of neurons that is activated by a single object. Based on the powerful idea of "cell assemblies" is the concept of binding: functions are determined by cell populations and these populations are assembled in varying constellations (the "binding"). Coherent 40 Hz oscillations recorded in the visual system were suggested to bind together distributed but functionally related neuronal populations to get a unified percept (Singer, 1993a,b). This distributed code represents a very economic and flexible strategy because a restricted number of neurons suffice to encode a wide range of stimuli. Therefore, a wide range of combinations is possible.

The popularity of the binding concept is reflected in the many works in different fields during the last two decades: Synchronization between distributed neuronal activity patterns on a fine temporal scale has been proposed as a candidate mechanism for integration in the visual system (Eckhorn et al., 1988; Gray and Singer, 1989; Singer et al., 1990; Engel et al., 1992; Singer and Gray, 1995a,b; Engel et al., 1997), in frontal areas (Prut et al., 1998; Abeles et al., 1993), and in the visuomotor areas (Roelfsema et al., 1997; Classen et al., 1998). The coherence in the beta frequency range observed in the human motor system between the electromyogram (EMG) and cortical activity measured by MEG (Conway et al., 1995; Salenius et al., 1997) or EEG (Halliday et al., 1998; Mima et al., 2000; Mima et al., 2001) during sustained voluntary muscle contraction was also interpreted as a binding in the human motor system, similar to the synchronization observed in monkeys (Murthy and Fetz, 1992; Sanes and Donoghue, 1993; Murthy and Fetz, 1996a,b; Riehle et al., 1997; Baker et al., 1997; Baker et al., 1997; Riehle et al., 1997).

Due to the fact that in most of the studies aimed at investigating the cortico-motoneuronal synchronization an isometric contraction was used, for a few years the consensus was that the beta range synchronization occurs mainly during such tasks. However, the maintained motor contraction task did not allow studying the dynamic properties of the cortico-muscular synchronization. The appropriate paradigm to assess the dynamic properties of this synchronization was again the Bereitschaftspotential paradigm as used by Feige et al. (2000).

Besides the dynamic properties of the cortico-motoneuronal synchronization this study approaches another open question: Is only the contralateral MI synchronized with the motoneuronal pool or are there also other motor areas participating on the cortico-motoneuronal network?. Due to the fact, that most of the previous studies were performed using MEG, which is sensitive to tangential sources mostly (*i.e.* sources in sulci like the MI), the opinion was that only the contralateral MI is synchronized with the motoneuronal pool. This was the motivation for Feige et al. (2000) to use EEG, which is sensitive not only to tangential sources in sulci, but also to sources in gyri (like the premotor areas). The authors simultaneously recorded cortical activity by high-resolution EEG (61 channels) and EMG activity from one of the prime mover muscles (right flexor digitorum muscle) during a phasic voluntary movement in seven healthy subjects. The synchronization between the EEG and the EMG was assessed by coherence analysis. Figure 5 shows the results for one of the seven subjects investigated.

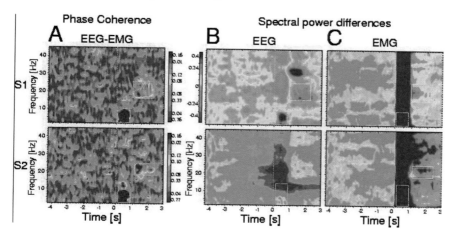

Figure 5. Frequency vs. time plots of EEG-EMG phase coherence (**A**) and movement-related EEG- (**B**) and EMG- (**C**) spectral power changes. Data, taken from repeated trials for two (out of seven) subjects investigated. Time window from 5 s before to 3 s after EMG onset (vertical white dotted line at 0 ms). Phase coherence values were coded according to the shown colour scales on the right (values above indicating the phase coherence value, and values below denoting the probability of observing this phase coherence by chance (Rayleigh test). The colour scale for the spectral power changes applies to both subjects. Frequency analysis was performed using a sliding time window of width 768 ms, in which two 512 ms FFTs overlapping by half were calculated (for this reason, the figure only shows values starting at 4.2 s prior to the movement). The data in each FFT window was demeaned and detrended prior to FFT analysis. The resulting frequency resolution was 1.9 Hz. The analysis window was shifted in steps of 42 ms across each epoch for each of the channels. All reported latencies refer to the delay between EMG onset and the end of the analysis time window.
A. Movement-related frequency vs. time distributions of the EEG-EMG phase coherence. The EEG channel shown is the one with the maximum 16-28 Hz post-movement coherence with the EMG, as measured against common average reference. Observe that subjects exhibited two patches of strong EEG-EMG coherence: (1) Low-frequency (2–14 Hz) coherence, starting immediately at EMG-onset. This coherence (demarcated by a dotted white frame) most likely reflects the movement-related EEG potentials, found by EMG-onset triggered averaging of the EEG; they show up in the phase coherence analysis because the EMG-burst contains components of practically all frequencies, synchronized to movement onset (cf. **C**). (2) High-frequency (23±3 Hz) coherence, starting after the movement (which lasted approx. 400 ms) and peaking at 964±283 ms after EMG onset. This high-frequency coherence (demarcated by a solid white frame) after a phasic voluntary movement was not previously described.
B. Frequency vs. time distributions of movement-related EEG spectral power changes, measured as the reliability of difference relative to a reference interval between 5 and 4 s before movement onset (relative gain G_r (Feige et al., 1996; Feige, 1999). G_r is –1 for fully reliable power decrease, 0 for no change and 1 for fully reliable power increase. Both subjects showed a clear increase in spectral power, covering a wide frequency range from 10-45 Hz (details depending on the subject), which started about 800 ms after EMG onset (*i.e.* beginning after movement termination). Note that the excess spectral power in the EEG, reflecting the well-known post-movement cortical beta-synchronization, includes the patch of beta-range EEG-EMG coherence, but covers a distinctly wider frequency range (cf. solid white frames, copied from A).
C. Movement-related frequency vs. time distributions of the EMG, computed in the same way as done for the EEG (B). The post-movement EEG/EMG synchronization (solid white frames, copied from (A) occurred after termination of the movement-related EMG burst (dark vertical strip between 0 and 1 s). Observe, however, that the EMG spectral power does not return to baseline levels immediately, but exhibits a residual enhancement that - like the EEG - includes, but clearly extends beyond the beta-band EEG-EMG coherence (from Feige et al. (2000) with kind permission of Journal of Neurophysiology).

Interestingly, as seen from Fig. 5, two different patches of coherence with different spectrotemporal properties were found: a low-frequency coherence (demarcated by a dotted white frame) during the movement and a high-frequency coherence (demarcated by solid white frame) at 23 Hz after the movement execution.

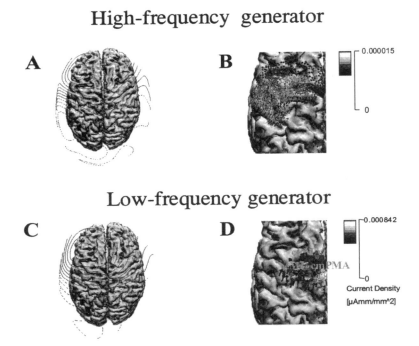

Figure 6. Sources of the high- and low-frequency EMG-coherent EEG activity for one of the subjects. Top view of the MRI reconstruction of the cerebral cortex for this subject (nose pointing upwards). cMI – contralateral primary motor area. cPMA – contralateral premotor area. cmPMA – contralateral mesial premotor area. (A) Potential distribution of the high-frequency EMG-coherent EEG activity superimposed onto MRI reconstruction. Complex potential distribution pattern, suggesting activation of multiple areas. (B) Magnification of the reconstructed CCD map underlying the potential distribution in A. The CCD map is displayed with individual current vectors (clipped at 75% of the maximum current). Scaling according to colour bar. Extended cortical current density source in cMI and cPMA. The tangential currents are consistent with activation of the cMI in the anterior bank of the central sulcus, the radial currents are consistent with activation of the cPMA. (C) Potential distribution of the low-frequency EMG-coherent EEG activity superimposed onto MRI reconstruction. Complex potential distribution pattern, suggesting again activation of multiple areas. (D) Magnification of the reconstructed CCD map underlying the potential distribution in C. The CCD map is displayed with individual current vectors. Scaling according to colour bar. Extended cortical current density source in cMI, cPMA and cmPMA. The tangential currents are consistent with activation of the cMI in the anterior bank of the central sulcus, the radial currents are consistent with additional activation of the medial cPMA (from Feige et al. (2000) with kind permission of Journal of Neurophysiology).

To localize the cortical sources of the EMG-coherent EEG activity, the authors employed a new variant of phase coherence analysis, called phase reference analysis (Feige, 1999). Using this analysis it is possible to extract the EMG-coherent EEG maps for both, low- and high-frequency beta range synchronization. This is possible because unlike the conventional coherence analysis the phase reference analysis preserves both amplitude and phase information of the EEG. Thus, the sources of the EMG-coherent EEG activity can be localized by applying source reconstruction to the extracted EMG-coherent electrical potentials. The electric source reconstruction of the extracted EMG-coherent EEG maps was performed with respect to the individual brain anatomy from MRI using a distributed source model (cortical current density analysis, minimum norm least square, MNLS) and a realistic head model. As seen from Fig. 6 the generators of the beta-range synchronization were not only located in the primary motor area, but also in premotor areas. The same was true for the low-frequency generator, which is directly related to movement execution. That the generators of the beta range synchronization are located not only in the MI but also in the premotor area was also suggested by using Laplacian method (Mima et al., 2000).

Surprisingly, the beta range synchronization occurs after the voluntary movement, when the subjects has to put the finger at exactly the same position from which the movement has started. This may reflect the transition of the collective motor network into a new equilibrium state when more attentive demands are needed. This favors the hypothesis that the beta range synchronization possibly reflects attention related to higher attentive demands.

As seen from Figure 5 only a part of the post-movement high-frequency synchronization (in the beta range) was coherent with the motoneuronal pool. This means that the post-movement synchronization is of composite nature and contains a part (in the beta range), which is coherent with EMG and another part (in the gamma range) which is not coherent. The finding that the postmovement beta synchronization is synchronized with the motoneuronal pool suggest that it plays an active role in motor control (Feige et al., 1996; Kilner et al., 1999) and does not reflect idling (Pfurtscheller, 1992) or inhibition (Salmelin and Hari, 1994).

Let us summarize the results from this study:

• The distributed source modelling of the EMG-coherent EEG has shown that multiple motor areas are dynamically synchronized with the motoneuronal pool.

• This synchronization occurs after the phasic voluntary movement when attentive demands are higher.

• The beta range synchronization is coherent with the motoneuronal pool. This finding supports the view that it plays an active role in motor control.

Although this study has shed more light on the synchronization in the motor system it is still an open question whether this beta range synchronization reflects binding in the motor system, attention, precision in force production or motor learning as noticed also in the few reviews in this field (Farmer, 1998; Baker et al., 1999; Hari and Salenius, 1999; Mima and Hallett, 1999; Brown, 2000; Schnitzler et al., 2000). Further studies are necessary to investigate the functional significance of the cortico-motoneuronal synchronization and to clarify what it means in terms of motor control and to exclude that it is an epiphenomenon.

CONCLUDING COMMENTS AND SUGGESTIONS FOR FUTURE RESEARCH

This chapter summarized the contribution of the distributed source models in analyzing the generators of the movement-related activity in the Bereitschaftspotential paradigm. In summary, the greatest advantage of the distributed source models is that they do not require *a priori* knowledge about the number, shape and size of the activated sources. Therefore, they can be successfully applied for investigation not only of the functional organization but also of the reorganization in the human motor system and in all other systems based on distributed information processing in the cortical macronetworks.

Where do I see possible pitfalls using distributed source models? Although the distributed source modeling does not require *a priori* knowledge, a lot of *post priori* knowledge is needed for understanding the results obtained. What does this mean? For example, when looking at the maps in Figure 6 an investigator who has previous experience using MEG in investigating the generators of the Bereitschaftspotential will immediately see on the left side of the Figure the complex potential distribution, suggesting the activation of multiple cortical areas. If an investigator is aware of the fact that MI is in the depth of the central sulcus and that on the top is mostly the premotor area the tangential currents are clear from the activation of MI and the radial mostly from the premotor area. Only by integrating a lot of knowledge coming from different approaches used in human and animal studies can the results be correctly interpreted.

A second danger of using distributed source models is to use them without evaluation: For the future it is very important to systematically evaluate on a theoretical basis the influence of the different parameters of source reconstruction based on cortical current density analysis. The reconstruction methods have to be evaluated with respect to volume conductor head model (spherical three shell or boundary element method head model), lead field normalization (depth normalization by full or downweighted powers of leadfield gain), and minimum norm formulation (minimum norm least squares or nonlinear Lp norm with different model term weightings). Simulated sources in settings similar to the experimental ones have to be used at the beginning. The advantages of the simulated sources are that the location, orientation and strength of the test-dipoles is known. In addition, the influence of factors playing a role under real conditions (changes of electrode impedance, attentional problems) is excluded. Such an evaluation of the influence of the different source reconstruction parameters with simulated sources similar to the movement-related cortical sources is under study in our laboratory (Sick et al., 2001). Such systematic evaluation has started also for epilepsy data (Fuchs et al., 1999). Only after such thorough evaluation conclusions can be drawn about parameters for source reconstruction. Only in this way can the currently available sophisticated software (e.g. CURRY) be competently and successfully used.

What is important for the future? To my mind, it is very important to use multimodal imaging with the same experimental procedure in order to combine the high spatial resolution of fMRI with the high temporal resolution of EEG. This was the case with the abovementioned work of Ball et al. (1999) which integrated EEG source reconstruction and fMRI activations, *i. e.* the high spatial resolution of fMRI was used to confirm the results from the electric source reconstruction. When the activation of a cortical area is shown by both techniques one can be more confident in the results obtained. This was the case with the activation of the inferior parietal lobe area and of the frontal medial wall motor areas in the study of Ball et al. (1999) where the activation shown using fMRI gave us confidence in the results from the electric source reconstruction. Furthermore, there is a trend in the biomagnetic and bioelectric society to develop fMRI-constrained EEG and MEG source localization, *i.e.* to use the fMRI activation maps as constraints for the source localization in order to improve it. However, several problematic constellations may occur from the fact that there are false positives (fMRI activations without a corresponding current source) and false negatives (current sources that do not show in the fMRI images). Another problem can arise from the fact that *a priori* EEG and fMRI reflect different processes and therefore one can not expect to find the fMRI activation maps and the source reconstruction results at one and the same spot. And of course, it should not be forgotten that both techniques are in development.

Generally, it is very important for the future to get more insight into the association between preparation and execution of a movement. As done in monkeys with single unit recordings, the association between preparation and movement parameters should be established in detail. In humans, one should operate with single trials. The repetitive very simple (not object or goal-related) movement would not be the appropriate task to get further insight into the problem of readiness.

In closing, I hope to have shown the power of the Bereitschaftspotential paradigm 35 years after its discovery. The paradigm still gives new insights into motor control when new recording and analysis techniques and new concepts of brain function are introduced. These insights help us to go into more sophisticated paradigms and to understand them.

ACKNOWLEDGEMENTS

It is a pleasure for me to thank my teachers Profs. Kornhuber and Deecke. I also thank my colleagues who were actively involved in parts of the reported research: Drs. B. Feige, A. Schreiber, Cand. Drs. med. C. Sick and T. Ball, Dr. M. Wagner and Profs. A. Aertsen and H. Burkhard. I am grateful to Profs. C.H. Lücking and M. Wiesendanger and to Dr. M. Fuchs and Dr. M. Wagner for many stimulating talks and for critical reading of an earlier version of this chapter. The financial support received by DFG (grant Kr 1392/7-1) and by the Research Fund of the Albert-Ludwig-University is gratefully acknowledged.

REFERENCES

Abeles, M., Bergman, H., Margalit, E., and Vaadia, E. (1993) Spatiotemporal firing patterns in the frontal cortex of behaving monkeys. *Journal of Neurophysiology* 70, 1629-1638.

Babiloni, C., Carducci, F., Cincotti, F., Rossini, P. M., Neuper, C., Pfurtscheller, G., and Babiloni, F. (1999) Human movement-related potentials vs desynchronization of EEG alpha rhythm: a high-resolution EEG study. *Neuroimage* 10, 658-665.

Baker, S. N., Kilner, J. M., Pinches, E. M., and Lemon, R. N. (1999) The role of synchrony and oscillations in the motor output. *Experimental Brain Research* 128, 109-117.

Baker, S. N., Olivier, E., and Lemon, R. N. (1997) Coherent oscillations in monkey motor cortex and hand muscle EMG show task-dependent modulation. *Journal of Physiology* 501, 225-241.

Ball, T., Schreiber, A., Feige, B., Wagner, M., Lucking, C. H., and Kristeva-Feige, R. (1999) The role of higher-order motor areas in voluntary movement as revealed by high-resolution EEG and fMRI. *Neuroimage* 10, 682-694.

Brown, P. (2000) Cortical drives to human muscle: the Piper and related rhythms. *Progress in Neurobiology* 60, 97-108.

Cheyne, D., Kriseva, R., and Deecke, L. (1991) Homuncular organization of human motor cortex as indicated by neuromagnetic recordings. *Neuroscience Letters* 122, 17-20.

Cheyne, D. and Weinberg, H. (1989) Neuromagnetic fields accompanying unilateral finger movements: pre-movement and movement-evoked fields. *Experimental Brain Research* 78, 604-612.

Classen, J., Gerloff, C., Honda, M., and Hallett, M. (1998) Integrative visuomotor behavior is associated with interregionally coherent oscillations in the human brain. *Journal of Neurophysiology* 79(3), 1567-73.

Conway, B. A., Halliday, D. M., Farmer, S. F., Shahani, U., Maas, P., Weir, A. I., and Rosenberg, J. R. (1995) Synchronization between motor cortex and spinal motoneuronal pool during the performance of a maintained motor task in man. *Journal of Physiology* 489, 917-924.

Cui, R. Q. and Deecke, L. (1999) High resolution DC-EEG analysis of the Bereitschaftspotential and post movement onset potentials accompanying uni- or bilateral voluntary finger movements. *Brain Topography* 11, 233-249.

Cui, R. Q., Huter, D., Egkher, A., Lang, W., Lindinger, G., and Deecke, L. (2000) High resolution DC-EEG mapping of the Bereitschaftspotential preceding simple or complex bimanual sequential finger movement. *Experimental Brain Research* 134, 49-57.

Dale, A.M. and Sereno, M. (1990) Improved localization of cortical activity by combining EEG and MEG with MRI cortical surface reconstruction: a linear approach. *Journal of Cognitive Neurosciences* 18, 192-205.

Deecke, L., Grozinger, B., and Kornhuber, H. H. (1976) Voluntary Finger Movement in Man: Cerebral Potentials and Theory. *Biological Cybernetics* 23, 99-119.

Deecke, L., Scheid, P., and Kornhuber, H. H. (1969) Distribution of readiness potential, pre-motion positivity, and motor potential of the human cerebral cortex preceding voluntary finger movements. *Experimental Brain Research* 7, 158-168.

Eckhorn, R., Bauer, R., Jordan, W., Brosch, M., Kruse, W., Munk, M., and Reitboeck, H. J. (1988) Coherent oscillations: a mechanism of feature linking in the visual cortex? Multiple electrode and correlation analyses in the cat. *Biological Cybernetics* 60, 121-130.

Engel, A. K., Konig, P., Kreiter, A. K., Schillen, T. B., and Singer, W. (1992) Temporal coding in the visual cortex: new vistas on integration in the nervous system. [Review] *Trends in Neurosciences* 15, 218-226.

Engel, A. K., Roelfsema, P. R., Fries, P., Brecht, M., and Singer, W. (1997) Role of the temporal domain for response selection and perceptual binding. *Cerebral Cortex* 7, 571-582.

Farmer, S. F. (1998) Rhythmicity, synchronization and binding in human and primate motor systems. *Journal of Physiology* 509, 3-14.

Feige, B. (1999). *Oscillatory brain activity and its analysis on the basis of MEG and EEG*, pp. 1-144. Waxmann, Münster/NewYork.

Feige, B., Aertsen, A., and Kristeva-Feige, R. (2000) Dynamic synchronization between multiple cortical motor areas and muscle activity in phasic voluntary movements. *Journal of Neurophysiology* 84 , 2622-2629.

Feige, B., Kristeva-Feige, R., Rossi, S., Pizzella, V., and Rossini, P. M. (1996) Neuromagnetic study of movement-related changes in rhythmic brain activity. *Brain Research* 734, 252-260.

Fuchs, M., Drenckhahn, R., Wischmann, H. A., and Wagner, M. (1998) An improved boundary element method for realistic volume-conductor modeling. *IEEE Transections in Biomedical Engineering* 45, 980-997.

Fuchs, M., Wagner, M., Kohler, T., and Wischmann, H. A. (1999) Linear and nonlinear current density reconstructions. *Journal of Clinical Neurophysiology* 16, 267-295.

Gerloff, C., Uenishi, N., and Hallett, M. (1998) Cortical activation during fast repetitive finger movements in humans: dipole sources of steady-state movement-related cortical potentials. *Journal of Clinical Neurophysiology* 15, 502-

513.

Grave de Peralta Menendez, R., Hauk, O., Andini S.G., Vogt, H, and Mosher, J. C. (2001) Linear inverse solutions with optimal resolution kernels applied to electromagnetic tomography. *Human Brain Mapping* 5, 454-467.

Gray, C. M. and Singer, W. (1989) Stimulus-specific neuronal oscillations in orientation columns of cat visual cortex. *Proceedings of National Academy of Sciences U.S.A* 86, 1698-1702.

Hallett, M. and Toro, C. (1996) Generation of movement-related potentials in the supplementary sensorimotor area. *Advances in Neurology* 70, 147-152.

Halliday, D. M., Conway, B. A., Farmer, S. F., and Rosenberg, J. R. (1998) Using electroencephalography to study functional coupling between cortical activity and electromyograms during voluntary contractions in humans. *Neuroscience Letters* 241, 5-8.

Hamalainen, M. S. and Ilmoniemi, R. J. (1994) Interpreting magnetic fields of the brain: minimum norm estimates. *Med. Biol. Eng Computation* 32, 35-42.

Hari, R. and Salenius, S. (1999) Rhythmical corticomotor communication. *Neuroreport* 10, 1-10.

Hebb, D. O. (1949). *Organization of behavior: A neuropsychological theory.* Wiley, New York.

Huppertz, H. J., Otte, M., Grimm, C., Kristeva-Feige, R., Mergner, T., and Lücking, C. H. (1998) Estimation of the accuracy of a surface matching technique for registration of EEG and MRI data. *Electroencephalography and Clinical Neurophysiology* 106, 409-415.

Kaufman, L. and Williamson, S. J. (1982) Magnetic location of cortical activity. *Annals of the New York Academy of Sciences* 388, 197-213.

Kilner, J. M., Baker, S. N., Salenius, S., Jousmaki, V., Hari, R., and Lemon, R. N. (1999) Task-dependent modulation of 15-30 Hz coherence between rectified EMGs from human hand and forearm muscles. *Journal of Physiology* 516, 559-570.

Knosche, T., Praamstra, P., Stegeman, D., and Peters, M. (1996) Linear estimation discriminates midline sources and a motor cortex contribution to the readiness potential. *Electroencephalography and Clinical Neurophysiology* 99, 183-190.

Kornhuber, H. and Deecke, L. (1965) Hirnpotentialènderungen bei Willkürbewegungen und Passiven Bewegungen des Menschen: Bereitschaftspotential und reafferente Potentiale. *Pflügers Archiv* 284, 1-17.

Kristeva-Feige, R., Walter, H., Lutkenhoner, B., Hampson, S., Ross, B., Knorr, U., Steinmetz, H., and Cheyne, D. (1994) A neuromagnetic study of the functional organization of the sensorimotor cortex. *European Journal of Neurosciences* 6, 632-639.

Kristeva, R., Cheyne, D., and Deecke, L. (1991a) Neuromagnetic fields accompanying unilateral and bilateral voluntary movements: topography and analysis of cortical sources. *Electroencephalography and Clinical Neurophysiology* 81, 284-298.

Kristeva, R., Keller, E., Deecke, L., and Kornhuber, H. H. (1979) Cerebral potentials preceding unilateral and simultaneous bilateral finger movements. *Electroencephalography and Clinical Neurophysiology* 47, 229-238.

Kristeva, R. and Kornhuber, H. H. (1980) Cerebral potentials related to the smallest human finger movement. *Progress in Brain Research* 54, 178-182.

Lang, W., Cheyne, D., Kristeva, R., Beisteiner, R., Lindinger, G., and Deecke, L. (1991a) Three-dimensional localization of SMA activity preceding voluntary movement. A study of electric and magnetic fields in a patient with infarction of the right supplementary motor area. *Experimental Brain Research* 87, 688-695.

Lang, W., Cheyne, D., Kristeva, R., Lindinger, G., and Deecke, L. (1991b) Functional localisation of motor processes in the human cortex. *Electroencephalography and Clinical Neurophysiology Supplement* 42, 97-115.

Mima, T. and Hallett, M. (1999) Corticomuscular coherence: a review. *Journal of Clinical Neurophysiology* 16, 501-511.

Mima, T., Matsuoka, T., and Hallett, M. (2001) Information flow from the sensorimotor cortex to muscle in humans. *Clinical Neurophysiology* 112, 122-126.

Mima, T., Steger, J., Schulman, A. E., Gerloff, C., and Hallett, M. (2000) Electroencephalographic measurement of motor cortex control of muscle activity in humans. *Clinical Neurophysiology* 111, 326-337.

Morozov VA. The error principle in the solution of operator equations by the regularization method. (1968) *USSR Comp Math Phys* 28, 69-80. 1968.

Murthy, V. N. and Fetz, E. E. (1992) Coherent 25- to 35-Hz oscillations in the sensorimotor cortex of awake behaving monkeys. *Proceedings of the National Academy of Sciences U.S.A* 89, 5670-5674.

Murthy, V. N. and Fetz, E. E. (1996a) Oscillatory activity in sensorimotor cortex of awake monkeys:Synchronization of local field potentials and relation to behavior. *Journal of Neurophysiology* 76, 3949-3967.

Murthy, V. N. and Fetz, E. E. (1996b) Synchronization of neurons during local field potential oscillations in sensorimotor cortex of awake monkeys. *Journal of Neurophysiology* 76, 3968-3982.

Nagamine, T., Toro, C., Balish, M., Deuschl, G., Wang, B., Sato, S., Shibasaki, H., and Hallett, M. (1994) Cortical magnetic and electric fields associated with voluntary finger movements. *Brain Topography* 6, 175-183.

Pascual-Marqui, R. D., Michel, C. M., and Lehmann, D. (1994b) Low resolution electromagnetic tomography: a new method for localizing electrical activity in the brain. *International Journal of Psychophysiology* 18, 49-65.

Pascual-Marqui, R. D., Michel, C. M., and Lehmann, D. (1994a) Low resolution electromagnetic tomography: a new method for localizing electrical activity in the brain. *International Journal of Psychophysiology* 18, 49-65.

Pfurtscheller, G. (1992) Event-related synchronization (ERS): an electrophysiological correlate of cortical areas at rest. *Electroencephalography and Clinical Neurophysiology* 83, 62-69.

Picard, N. and Strick, P. L. (1996) Motor areas of the medial wall: a review of their location and functional activation. *Cerebral Cortex* 6, 342-353.

Praamstra, P., Stegeman, D. F., Horstink, M. W., and Cools, A. R. (1996) Dipole source analysis suggests selective modulation of the supplementary motor area contribution to the readiness potential. *Electroencephalography and Clinical Neurophysiology* 98, 468-477.

Prut, Y., Vaadia, E., Bergman, H., Haalman, I., Slovin, H., and Abeles, M. (1998) Spatiotemporal structure of cortical activity: properties and behavioral relevance. *Journal of Neurophysiology* **79**, 2857-2874.

Riehle, A., Grun, S., Diesmann, M., and Aertsen, A. (1997) Spike synchronization and rate modulation differentially involved in motor cortical function. *Science* **278**, 1950-1953.

Roelfsema, P. R., Engel, A. K., Konig, P., and Singer, W. (1997) Visuomotor integration is associated with zero time-lag synchronization among cortical areas. *Nature* **385**, 157-161.

Salenius, S., Portin, K., Kajola, M., Salmelin, R., and Hari, R. (1997a) Cortical control of human motoneuron firing during isometric contraction. *Journal of Neurophysiology* **77**, 3401-3405.

Salenius, S., Portin, K., Kajola, M., Salmelin, R., and Hari, R. (1997b) Cortical Control of Human Motoneuron Firing During Isometric Contraction. *Journal of Neurophysiology* **77**, 3401-3405.

Salmelin, R. and Hari, R. (1994) Spatiotemporal characteristics of sensorimotor neuromagnetic rhythms related to thumb movement. *Neuroscience* **60**, 537-550.

Sanes, J. N. and Donoghue, J. P. (1993) Oscillations in local field potentials of the primate motor cortex during voluntary movement. *Proceedings of the National Academy of Sciences of the United States of America* **90**, 4470-4474.

Schnitzler, A., Gross, J., and Timmermann, L. (2000) Synchronised oscillations of the human sensorimotor cortex. *Acta Neurobiologica Exp.(Warsz.)* **60**, 271-287.

Sick, Ch., Huppertz, H. J., and Kristeva-Feige, R. (2001) Comparison of different electric source reconstruction methods based on cortical current density analysis for investigating movement-related cortical activity, in preparation

Singer, W. (1993a) Neuronal representations, assemblies and temporal coherence. *Progress in Brain Research* **95**, 461-474.

Singer, W. (1993b) Synchronization of cortical activity and its putative role in information processing and learning. *Annual Reviews in Physiology* **55**, 349-374.

Singer, W., Gray, C., Engel, A., Konig, P., Artola, A., and Brocher, S. (1990) Formation of cortical cell assemblies. *Cold Spring Harb. Symp. Quant. Biol.* **55**, 939-952.

Singer, W. and Gray, C. M. (1995) Visual feature integration and the temporal correlation hypothesis. *Annual Reviews in Neurosciences* **18**, 555-586.

Singer, W. and Gray, C. M. (1995b) Visual feature integration and the temporal correlation hypothesis. [Review]. *Annual Review of Neuroscience* **18**, 555-586.

Stancak, A., Jr., Feige, B., Lucking, C. H., and Kristeva-Feige, R. (2000) Oscillatory cortical activity and movement-related potentials in proximal and distal movements. *Clinical Neurophysiology* **111**, 636-650.

Stephan, K. M., Fink, G. R., Passingham, R. E., Silbersweig, D., Ceballos-Baumann, A. O., Frith, C. D., and Frackowiak, R. S. (1995) Functional anatomy of the mental representation of upper extremity movements in healthy subjects. *Journal of Neurophysiology* **73**, 373-386.

Toro, C., Matsumoto, J., Deuschl, G., Roth, B. J., and Hallett, M. (1993) Source analysis of scalp-recorded movement-related electrical potentials. *Electroencephalography and Clinical Neurophysiology* **86**, 167-175.

Urbano, A., Babiloni, C., Onorati, P., and Babiloni, F. (1996) Human cortical activity related to unilateral movements. A high resolution EEG study. *Neuroreport* **20:8**, 203-206.

Urbano, A., Babiloni, C., Onorati, P., Carducci, F., Ambrosini, A., Fattorini, L., and Babiloni, F. (1998) Responses of human primary sensorimotor and supplementary motor areas to internally triggered unilateral and simultaneous bilateral one-digit movements. A high-resolution EEG study. *European Journal of Neurosciences* **10**, 765-770.

Wang, J. Z., Williamson, S. J., and Kaufman, L. (1992) Magnetic source images determined by a lead-field analysis: the unique minimum-norm least-squares estimation. *IEEE Transaction in Biomedical Engineering* **39**, 665-675.

White, L. E., Andrews, T. J., Hulette, C., Richards, A., Groelle, M., Paydarfar, J., and Purves, D. (1997) Structure of the human sensorimotor system. I: Morphology and cytoarchitecture of the central sulcus. *Cerebral Cortex* **7**, 18-30.

Williamson, S. J. and Kaufman, L. (1990) Evolution of neuromagnetic topographic mapping. *Brain Topography* **3**, 113-127.

Zilles, K., Schlaug, G., Geyer, S., Luppino, G., Matelli, M., Qu, M., Schleicher, A., and Schormann, T. (1996) Anatomy and transmitter receptors of the supplementary motor areas in the human and nonhuman primate brain. *Advances in Neurology* **70**, 29-43.

Zilles, K., Schlaug, G., Matelli, M., Luppino, G., Schleicher, A., Qu, M., Dabringhaus, A., Seitz, R., and Roland, P. E. (1995) Mapping of human and macaque sensorimotor areas by integrating architectonic, transmitter receptor, MRI and PET data. *Journal of Anatomy* **187**, 515-537.

RECORDINGS OF MOVEMENT-RELATED POTENTIALS COMBINED WITH PET, fMRI OR MEG

Colum D. MacKinnon PhD

Department of Neurology
Mount Sinai School of Medicine
New York, New York 10029
USA

INTRODUCTION

The seminal recordings of the cerebral potentials preceding voluntary movement by Kornhuber and Deecke (1964, 1965) were the first to clearly show that the brain was active long before the initiation of movement. Since the premovement activity, termed the Bereitschaftspotential (BP), was time-locked to the onset of movement, it was thought to reflect synaptic activity associated with the planning, preparation and initiation of movement. Yet, despite considerable investigation over the past three decades, three fundamental questions about the nature of the BP remain unanswered: What are the neural structures contributing to the generation of the BP? What is the time course of their activity? What is the function of this activity?

The first two questions reflect the fact that movement preparation is a dynamic process involving the activity of multiple cortical areas with overlapping time courses of activity. Hence, electromagnetic fields recorded at an electrode can reflect the summation of potentials from multiple sources, and peaks in the scalp surface topography do not necessarily correspond to activity in immediately underlying regions of the brain. Recent advances in the use of high-resolution electroencephalography (EEG) and magnetoencephalography (MEG), combined with linear and non-linear source reconstruction techniques using MRI-derived models of the brain (Simpson et al., 1995; Fuchs et al., 1999), offer the possibility to decompose the scalp recorded signal into its underlying source generators (the inverse problem). Yet, due to the indeterminant nature of the inverse problem, the validity of source generator models remains in question. Consequently, multimodal integration techniques have been developed that combine electrophysiological recordings with temporal resolution sufficient to capture changes in synaptic activity (in milliseconds) such as EEG and MEG, with imaging technologies that have high spatial resolution (less than 10 mm) such as functional magnetic resonance imaging (fMRI) and positron emission tomography (PET). Co-registration of electrophysiological recordings with functional neuroimaging serves two purposes. Firstly, the location of source generators derived from source reconstructions such as dipole source analysis can be directly compared to the regions of significantly increased regional cerebral blood flow (rCBF) or blood oxygen level detection (BOLD) signal derived from PET or fMRI respectively. Since the changes in rCBF and BOLD signal are both thought to reflect changes in the metabolic demands induced by increases in local neuronal activity, dipole locations should correspond closely to the regions of PET or fMRI activation. Secondly, regions of activation can be used to place physiologically valid constraints on source reconstruction models.

In this chapter I will briefly review experiments that have recorded movement-related potentials (MRPs) in conjunction with PET, fMRI or MEG in the same subjects, either simultaneously or sequentially using identical (or near identical) experimental paradigms, in order to examine the source generators of the BP. The results of these experiments will be discussed with reference to the three principal hypotheses proposed to explain the cortical generators of the BP. The findings of these experiments provide compelling evidence that the mesial frontal cortex, including the supplementary motor area (SMA), pre-SMA, and cingulate motor areas (CMAs), and contralateral primary motor cortex (MI) are the principal generators of the BP. A contribution from the ipsilateral MI is questioned.

BACKGROUND

The timing and topography of the scalp surface potentials preceding voluntary movement have been described earlier in this book. Briefly, the cerebral potentials preceding self-initiated movements are comprised of two primary components: an initial slow-rising potential beginning as much as 3 s prior to movement onset that is bilaterally symmetric with maximal negativity over the vertex, termed the BP1 component (NS1 or BP are other terms), and a later steeper-sloped potential beginning approximately 400-500 ms prior to movement onset with a preponderance of negativity over the contralateral sensorimotor cortex, termed the BP2 component (NS' or NS2 are other terms) component. Two additional premovement components include an abrupt decrease in negativity occurring immediately prior to EMG onset, termed the premovement positivity (PMP) (not present in all subjects), followed by a sharp increase in negativity that peaks over the frontal cortex approximately 90 ms after EMG onset, termed the motor potential (MP).

Kornhuber and Deecke (1964, 1965) originally considered that the increased negativity over the region of the contralateral sensorimotor cortex (BP2 component) was generated by activity within the precentral gyrus. Later, they proposed that activity within regions of the mesial frontal cortex, including the SMA and anterior cingulate, contributed to the large negative potentials recorded over the vertex (Deecke et al., 1976; Deecke and Kornhuber, 1978). Synaptic activity within the SMA was thought to reflect aspects of motor planning and preparation, whereas activity within MI was proposed to mediate movement initiation.

The idea that the SMA participates in higher-order aspects of motor function was strongly supported by early rCBF studies showing that the SMA was preferentially activated during the performance of a complex series of finger movements or during the mental rehearsal of the same movements, but not during simple repetitive finger tapping (Orgogozo and Larsen, 1979; Roland et al., 1980). In contrast, the contralateral MI was activated during both simple and complex movements, but not during the mental rehearsal of complex movements. Roland et al. (1980) interpreted the data to mean that the SMA was responsible for the internal programming of motor subroutines and that "...these areas form a queue of time-ordered motor commands before voluntary movements that are executed by way of the primary motor area". The results of a variety of MRP studies seemed consistent with Roland's hypothesis by showing that the BP magnitude increased commensurately with the requirement for movement preparation, such as complex, sequential or simultaneous movements (Kitamura et al., 1993; Lang et al., 1988, 1990; Simonetta et al., 1991). It has subsequently been shown that the caudal region of the SMA is active in association with the execution of simple repetitive movements and shows frequency-, velocity- and force-dependent changes in rCBF similar to MI suggesting a role for this area in motor execution (Dettmers et al., 1995; Sadato et al., 1996; Turner et al.,

1998). Yet Roland's interpretation continues to contribute to the view that the SMA is involved in higher-order motor processes, such as generating the intention to act or movement planning and preparation, and exerts its influence on motor output via MI.

Rethinking the Source Generators of the BP

Two developments have prompted a rethinking of the neural generators of premovement activity. Firstly, a variety of neuroanatomical, electrophysiological, and functional neuroimaging studies in human and non-human primates have shown that the frontal cortex contains at least seven distinct motor regions. In the mesial frontal cortex alone, at least four separate motor regions have been identified including the anterior SMA (pre-SMA), posterior SMA (SMA-proper), and rostral and caudal cingulate motor areas (CMAr and CMAc) (Picard and Strick, 1996). Each of these mesial regions processes a large proportion of movement-related neurons that demonstrate long-lead activity (> 1 s) prior to movement onset (e.g. Matsuzaka et al., 1992; Romo and Schultz, 1992; Shima and Tanji, 1991). With the exception of the pre-SMA, each of these regions projects to the corticospinal tract and has extensive terminations within the intermediate zone and ventral grey matter (He et al., 1995; Dum and Strick, 1996). In this manner, motor regions of the mesial frontal cortex possess the capacity to directly influence spinal motor network activity. Thus, the division of the primary generators of the BP into SMA and MI contributions and the idea of a hierarchically organized sequential order of activation of these regions is likely overly simplistic.

Secondly, a role for the SMA in the generation of the BP has been questioned based upon recordings from subdural and epidural electrodes implanted in patients with epilepsy (Neshige et al., 1988), MEG recordings of premovement magnetic fields (Cheyne and Weinberg, 1989; Kristeva et al., 1991; Cheyne et al., 1995; Salmelin et al., 1995), and dipole source analysis of MRPs (Bötzel et al., 1993; Toro et al., 1993; Böcker et al., 1994). Neshige et al. (1988) were the first to show that BPs could be recorded from intracranial electrode grids placed over the region of the sensorimotor cortex during both contralateral and ipsilateral hand movements. These results suggested that the large surface negative potentials recorded over the surface might be the result of summation of bilateral potentials originating from MI and not the SMA. This idea was supported by MEG studies showing that movement-related fields (MRFs) beginning approximately 500 ms prior to unilateral finger extension movements were best explained by dipole models with source generators located bilaterally near MI (Kristeva et al., 1991; Salmelin et al., 1995; Hoshiyama et al., 1997). A number of spatiotemporal dipole source analysis studies have also reported that models incorporating sources located bilaterally within MI provide the best fit to the scalp recorded BP (Bötzel et al., 1993; Toro et al., 1993, 1994; Böcker et al., 1994). These studies concluded that the SMA contributed little to the generation of the BP. It is important to note that the iterative best-fit solutions incorporated by these models placed bilateral hemispheric symmetry constraints on the locations of the dipoles. These constraints were rationalized based upon PET studies that had shown bilateral MI activation during unilateral motor tasks (e.g. Colebatch et al., 1991) and the MEG and intracranial recordings cited above.

Based on the findings cited above, three principal hypotheses have emerged to explain the source generators of the BP: (1) the BP1 originates from bilateral SMA activity whereas the BP2 is generated by the addition of activity within the contralateral MI; 2) both the BP1 and BP2 are generated by bilateral MI activity with minimal contribution from the SMA; and (3) the BP1 and BP2 are generated by bilateral activity within the SMA and MI. The discussion that follows will review studies that have used combined MRP recordings with PET, fMRI or MEG to assess the validity of these hypotheses.

COMBINED EEG AND MEG

One method of evaluating the possible source generators proposed to mediate MRPs and MRFs is to record both EEG and MEG in the same subjects using the same task. EEG recordings obtained from the surface of the scalp are greatly influenced by inhomogeneities in the conductivity of the different tissue layers interposed between the synaptically induced current and the scalp surface, but can detect potentials that have been volume conducted from distant regions of the brain irrespective of dipole orientation. In contrast, MEG signals recorded on the surface are relatively unaffected by changes in tissue conductivity and principally reflect the primary current source, with relatively little contribution from secondary or volume conducted currents. However, MEG sensors are only able to detect dipole currents that are oriented tangentially with respect to the scalp surface and are "blind" to radial dipoles. For these reasons, combined EEG and MEG can provide complementary information about underlying source generators.

To date, only two studies have combined MRP and MRF recordings in the same subjects (Nagamine et al., 1994, 1996). MRPs and MRFs were collected in separate sessions in both studies. In those studies, a readiness field (RF) began approximately 600 ms prior to EMG onset whereas the BP began up to 750 ms earlier. In contrast to the bilateral distribution of the BP1 and BP2, the RF was predominant over the contralateral MI in all subjects. A RF over the region of the ipsilateral MI was observed in only one of the seven subjects tested. These findings were interpreted to show that the BP1 is generated by bilateral sources that are oriented radially to the surface of the scalp, such as the crown of the precentral sulcus or SMA, whereas activity within the banks of the contralateral central sulcus contributes to the BP2. Alternatively, the absence of an early RF might be explained by the use of inappropriate high-pass filter settings (0.3 Hz) or noise introduced by the use of oligochannel recordings. This former idea is supported by a recent study with whole-scalp MEG (143 channel) showing RFs beginning as much as 1.9s prior to the onset of complex unilateral finger movements (Erdler et al., 2000).

COMBINED EEG AND PET OR fMRI

Initial findings from a pilot study using combined spatiotemporal dipole source analysis of MRP recordings and $H_2^{15}O$ PET by Toro et al. (1994) appeared to support the bilateral MI hypothesis. PET activations associated with a simple finger movement task were reported to be within the sensorimotor cortex (SI/MI) bilaterally, although changes in rCBF were much greater within the contralateral SI/MI. Only one of the four subjects studied showed activation of the mesial frontal cortex. Bilateral MI dipole sources derived using the hemispheric symmetry fitting strategy described above (Toro et al., 1993) appeared to correspond to the regions of PET activation. The contralateral MI dipole was reported to be within 4 to 20 mm of the location of peak rCBF in the contralateral SI/MI. Data pertaining to the correspondence between the ipsilateral MI dipole and SI/MI activation was not presented due to the weak and variable ipsilateral PET activation across subjects. The absence of mesial frontal cortex activation in this study was conspicuous considering that subsequent PET and fMRI studies using movement rates comparable to those used in MRP studies (< 1 Hz) have invariably reported activity within the mesial frontal cortex, particularly rostral regions of the SMA and CMA (e.g. Deiber et al., 1991, 1996, 1999; Sadato et al., 1996), and not within the ipsilateral MI. The reason for the discrepancies between the PET activations reported by Toro et al. (1994) and other studies is unclear.

In contrast to the above findings, three studies have presented evidence supporting a prominent role for the mesial frontal cortex in the generation of the BP (Jahanshahi et al., 1995; MacKinnon et al., 1996; Ball et al., 1999). Jahanshahi et al. (1995) used combined MRP recordings and $H_2^{15}O$ PET to explore the functional anatomy of self-initiated and externally triggered movements in both normal subjects and patients with Parkinson's disease. In this study MRPs and PET images were collected in separate sessions. EEG was recorded from a montage of 12 scalp surface electrodes during four conditions: (1) self-initiated extension of the right index finger at a rate of once every 3 s; (2) the same movements externally triggered by a tone presented at a rate identical to the self-initiated movements (regularly-timed external trigger condition); (3) the same movements triggered by a tone that varied in onset time between 3 and 8 s (variably-timed external trigger condition); and (4) a rest condition.

The MRPs recorded for each of these conditions in the normal subjects are shown in Figure 1. Both the self-initiated and regularly-timed external trigger conditions were associated with premovement potentials beginning up to 1.5 s prior to EMG onset with maximal negativity over the vertex. In contrast, premovement potentials were absent prior to movements triggered by a variably-timed tone. These MRPs were consistent with other studies that have compared premovement activity preceding movements cued by stimuli of predictable (regularly-timed) and unpredictable (variably-timed) timing (Papa et al., 1991; Cunnington et al., 1995).

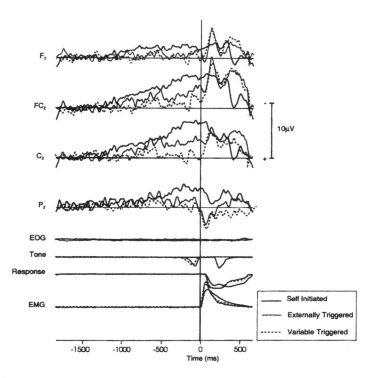

Figure 1. Grand averages of MRPs preceding movements that were self-initiated (thick traces) or externally triggered by regularly-timed (thin traces) or variably-timed (dotted lines) tones in normal control subjects. (From Jahanshahi et al., 1995. Reproduced by permission of Oxford University Press).

The PET activations associated with these tasks are shown in Figures 2B, 2D and 2F. Self-initiated movements were associated with activity in the contralateral SI/MI, thalamus and rostral CMA, bilaterally in the SMA (peak activations were near the rostral SMA),

Variable

Regular

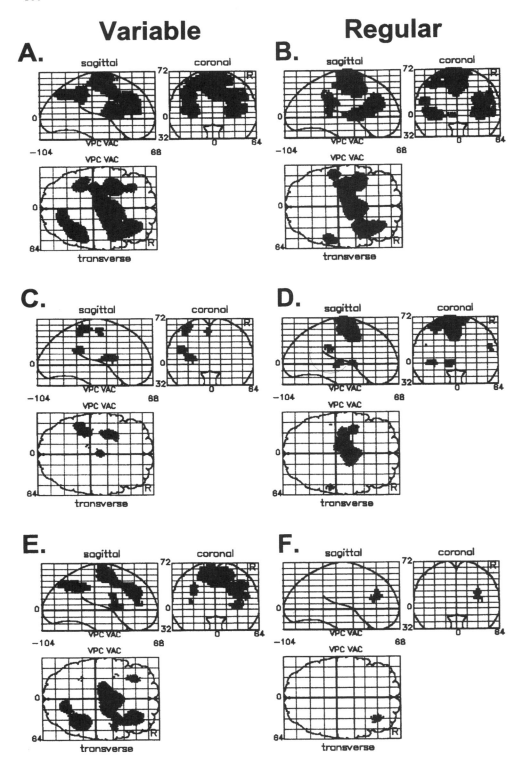

Figure 2. SPM projections of the sites of significantly increased rCBF (threshold at P < 0.05 corrected for multiple comparisons) for variably paced (left) and regularly paced (right) movements. The activated areas are shown projected onto single sagittal, coronal and transverse planes conforming to the stereotactic atlas of Talairach and Tournoux (Talairach and Tournoux, 1988). **A.** Self-initiated movements performed at a variable rate of once every 3 to 8 s compared to rest. **B.** Self-initiated movements performed at a regular rate of approximately once every 3 s (from Jahanshahi et al., 1995). **C.** Movements externally triggered at a variable rate of once every 3 to 8 s compared to rest. **D.** Movements externally triggered at a regular rate of approximately once every 3 s compared to rest (from Jahanshahi et al., 1995). **E.** Self-initiated compared to externally triggered movements performed at a variable rate of once every 3 to 8 s. **F.** Self-initiated compared to externally triggered movements performed at a regular rate of approximately once every 3 s (from Jahanshahi et al., 1995). (Figures from Jenkins et al., 2000. Reproduced by permission of Oxford University Press).

premotor cortex, insula, and parietal area 40, and the ipsilateral dorsolateral prefrontal cortex (DLPFC) (Figure 2B). No activation was observed in the ipsilateral SI/MI. The areas of activation for the regularly-timed external trigger condition were the same as those for the self-initiated movements (Figure 2D) with the exception of the absence of activity in the DLPFC (Figure 2F). The marked similarity of the premovement potentials and PET activations obtained for the self-initiated and regularly-timed external trigger conditions suggested that the areas of the brain mediating movement preparation in both tasks were essentially the same. Patients with Parkinson's disease showed impairment in the activation of rostral regions of the cingulate motor areas and SMA in conjunction with a significant attenuation of the BP for the self-initiated movements.

In a follow-up $H_2^{15}O$ PET study, Jenkins et al. (2000) recognized that similarities in the MRPs and PET activations associated with the self-initiated and regularly-timed external trigger conditions reported by Jahanshahi et al. (1995) might have been due to the predictability of the timing of movement onset. This idea was supported by the absence of premovement potentials preceding movements triggered by a tone of variably-timed onset (Figure 1). Unfortunately, the variably-timed external trigger condition was not examined with PET in the initial study. Consequently, Jenkins et al. (2000) re-examined the PET activations associated with self-initiated right finger extension movements performed at a variable rate of once every 2 to 7 s, and the same movements externally triggered by tones yoked to the same rate as the self-initiated movements (comparable to the paradigm used to record MRPs associated with variably-timed tones by Jahanshahi et al., 1995). The results of these PET studies are shown in Figures 2A, 2C and 2E. The activations for the self-initiated movements were essentially the same as those obtained in the initial experiments (compare Figures 2A and 2B). In contrast, movements cued by variably-timed tones were associated with activity in the contralateral SI/MI, caudal SMA, and putamen, but not the rostral SMA and CMA (Figure 2C). Direct comparisons between the self-initiated and variably-timed external trigger conditions (Figure 2E) demonstrated that regions of the rostral SMA and CMA and bilateral DLPFC were selectively activated in association with self-initiated movements. Comparable results have also been obtained with fMRI using a similar movement cueing paradigm (Deiber et al., 1999). Taken together with the MRP recordings of Jahanshahi et al. (1995), these results show that activity within the rostral SMA and CMA is correlated with the presence of premovement potentials and strongly supports a role for these areas in the preparation for predictably timed voluntary movements, irrespective of whether the movement is internally or externally triggered.

Similar findings were reported in a combined MRP and $H_2^{15}O$ PET study by MacKinnon et al. (1996). In that study 32-channel EEG recordings and PET were acquired simultaneously during the performance of self-initiated right thumb-to-index finger opposition movements. Two movement frequencies were compared: (1) movements performed at an irregular rate with a mean inter-movement interval = 8.6 s, and (2) movements performed at a fixed rate of 2 Hz. The latter paradigm was chosen to assess if there were frequency-dependent changes in the location and extent of precentral and mesial frontal activity. In addition, the voxels of significant change in rCBF obtained with movements at the two frequencies were used to place *a priori* constraints on the location of dipoles used to model the scalp surface potentials.

The grand average MRPs recorded during the PET sessions are shown in Figure 3. All components of the MRPs were evident in the grand average waveform. The corresponding PET activations for this task are shown in Figure 4B. Movements performed at a variable rate of once every 5 to 12 s were associated with significant increases in rCBF in the regions of the contralateral rostral SMA and CMA, ipsilateral cerebellum, and superior temporal gyrus (area 22), but not in SI/MI. Additional scans and EEG recordings with movements performed at a rate of 0.2 Hz in a subset of three subjects were associated with increased rCBF in the regions of the rostral SMA and CMA and contralateral SI/MI (data not shown). The same movements performed at 2 Hz (Figure 4A) were associated with activity in the contralateral SI/MI, thalamus and putamen, bilateral caudal SMA and ipsilateral cerebellum. Similar frequency-dependent changes in the location of mesial frontal activation, and the presence or absence of contralateral SI/MI activation, have been reported in other studies (Sabatini et al., 1993; Sadato et al., 1996).

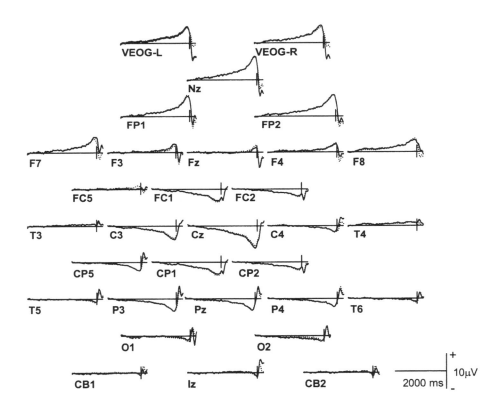

Figure 3. Grand average MRPs (solid lines) recorded from 31 scalp-surface electrodes in 8 subjects during the acquisition of $H_2^{15}O$ PET (from MacKinnon et al., 1996). Note the maximal negativity at Cz and reversal in polarity over the midline anterior frontal cortex. Dotted lines represent the surface potentials modeled by dipoles in the region of the mesial frontal cortex, near the rostral SMA and CMA, and contralateral MI.

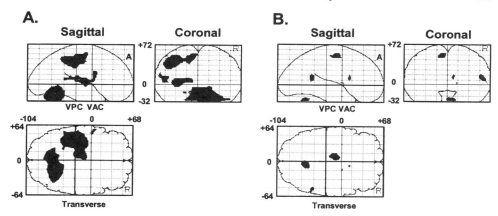

Figure 4. SPM projections of the sites of significantly increased rCBF (P < 0.05) for the comparison of self-initiated right finger movements with rest. **A.** Self-initiated movements performed at a constant rate of 2 Hz. **B.** Self-initiated movements performed at a variable rate of once every 3 to 12 s. (from MacKinnon et al., Contributions of the mesial frontal cortex to the premovement potentials associated with intermittent hand movements in humans. Human Brain Mapping 4(1): 1-22. Copyright © 1996 Wiley-Liss. Reprinted by permission of Wiley-Liss, Inc. a subsidiary of John Wiley & Sons, Inc.)

These data were used to derive four spatiotemporal dipole source analysis models of the generators of the BP (Figure 5). The first model consisted by a single unconstrained dipole fitted over the BP time interval (Figure 5A). This dipole invariably localized to the region of the contralateral mesial frontal cortex, within or just outside the rostral SMA/CMA region of PET activation for the 0.1 Hz movements, and explained over 93% of the BP. The second model was composed of symmetrical bilateral generators within the region of MI. The location of the contralateral MI dipole was constrained to remain within a 1 cm^3 volume surrounding the location of peak SI/MI activation obtained from the PET study with movements at 2 Hz. Note that the fit of the bilateral MI model to the BP (> 93%) was comparable to the single dipole solution. The third model directly compared the possible contributions of models 1 and 2 by placing both into the same solution. The dipole moment generated by the mesial frontal dipole captured the majority of the BP signal. The bilateral MI dipoles appeared to contribute to the BP2 component, although with considerably less magnitude than the mesial frontal cortex. The fourth model used one unconstrained dipole and a second dipole constrained to the contralateral MI to model the BP and MP components. The BP component was modeled by a dipole within the region of rostal SMA/CMA PET activation, whereas the MI dipole explained the MP. The fit of this model to the scalp-recorded MRP is shown in Figure 3.

The findings of the combined MRP and PET experiments by MacKinnon et al. (1996) were interpreted to reflect a major contribution from the mesial frontal cortex, particularly the rostral CMA and SMA, to the generation of the BP. Furthermore, contributions from the contralateral MI were small and confined to the BP2 time interval, but became prevalent just prior to and during the onset of EMG activity. A contribution from the ipsilateral MI was not supported.

Ball et al. (1999) used improved source reconstruction techniques based upon MRI-derived models of the brain and skull in conjunction with fMRI to re-examine the source generators of MRPs. In that study, MRPs were recorded from a montage of 61 scalp surface electrodes during self-paced right index finger flexion movements performed once every 12 to 24 s. A similar paradigm was used during fMRI experiments with the exception that movements were performed approximately once every 3 seconds. Source

Model 1.

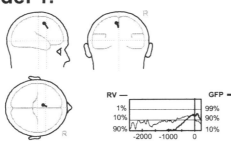

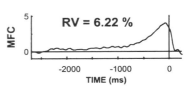

Model 2.

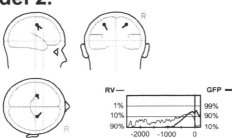

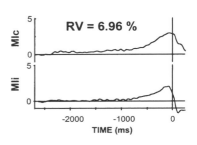

Model 3.

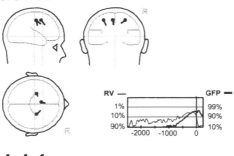

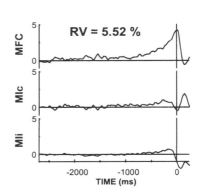

Model 4.

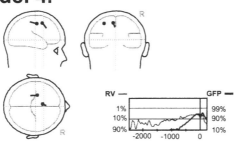

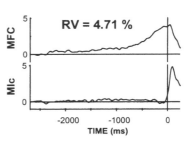

Figure 5. Four spatiotemporal dipole source analysis models of the neural generators of the BP. A. Single unconstrained dipole fitted over the BP time interval. B. Bilateral MI model with contralateral MI dipole constrained to the region of SI/MI PET activation and ipsilateral dipole to hemispheric symmetry. C. Models A and B placed into the same solution. D. One unconstrained dipole and a second dipole constrained to the region of SI/MI PET activation fitted over the BP and MP time intervals. Abbreviations: MFC, mesial frontal cortex; MIc, and MIi, contralateral and ipsilateral primary motor cortex respectively; RV, residual variance; GFP, global field power. (from MacKinnon et al., Contributions of the mesial frontal cortex to the premovement potentials associated with intermittent hand movements in humans. Human Brain Mapping 4(1): 1-22. Copyright © 1996 Wiley-Liss. Reprinted by permission of Wiley-Liss, Inc. a subsidiary of John Wiley & Sons, Inc.)

generators of the MRPs were reconstructed *within-subjects* by co-registering EEG and anatomical MRI, constructing source space and volume conductor models from the MRI, and using a current density reconstruction technique termed the L1-norm approach (Fuchs et al., 1999). The inverse procedure was constrained by restricting source space to the cortical surface and only dipoles oriented perpendicular to this surface were considered. The principal advantage of current density reconstruction techniques such as this, as opposed to spatiotemporal dipole source analysis, is that no *a priori* assumptions need to be made about the number or location of the source generators.

In these experiments, all subjects (n = 6) showed overlapping fMRI activation and current source density peaks in the motor regions of the mesial frontal cortex, corresponding to the pre-SMA, SMA-proper, and cingulate motor areas, and contralateral MI. Mean distances between the centre of mass of the current sources and fMRI ranged from 5.7 to 16.3 mm. Two different patterns of medial wall activation were identified. In four of six subjects, the BP was initiated within anterior regions of the CMA and pre-SMA (anterior to VAC), then activity progressed dorso-caudally toward the SMA-proper as movement onset approached. A decrease in SMA activity occurred approximately 120 ms before EMG onset followed by a marked increase in activity at EMG onset. In contrast, activity within the contralateral MI was relatively low throughout the majority of the BP, and did not show a clear increase in source activity until approximately 100 ms before EMG onset, peaking between 30 and 50 ms later. The decrease in SMA activity and reciprocal increase in MI activity just prior to EMG onset corresponded to the PMP waveform measured on the scalp. In two subjects, the BP appeared to be initiated later (onset time near 1800 ms) and activity first originated within the SMA proper (15 mm posterior of the VAC). Subjects with this "posterior-type" pattern of activation did not show a premovement decrease in SMA activity and corresponding PMP on the scalp.

A unique finding reported in this study was the presence of a region of significant BOLD signal and overlapping current source density in the region of the contralateral inferior parietal lobe (area 44) in three subjects. Current sources were also observed in the ipsilateral IPL, but without corresponding fMRI activation. The activity within the contralateral inferior parietal lobe was a major contributor to the BP in all three subjects. It is interesting to note that significant PET activation in area 40 has been reported in many of the PET studies examining motor preparatory activity (Jahanshahi et al., 1995; Jenkins et al., 2000) but was not discussed. The prevalence and function of premovement activity in the inferior parietal lobe is a topic that requires further investigation.

COMBINED MEG AND PET OR fMRI

MEG studies have consistently reported RFs over time periods equivalent to the BP2 interval preceding unilateral upper arm movements. The field pattern of these RFs appeared to be consistent with a source generator in the region of the contralateral MI. Several studies using combined MEG and fMRI or PET have shown an apparent correspondence between the contralateral MI dipoles used to models RFs and the regions of increased rCBF or BOLD signal (Walter et al., 1992; Beisteiner et al., 1995; Gerloff et al., 1996). Walter et al. (1992) restricted the recording of MRFs to the contralateral hemisphere and were able to show that dipole sources that best modeled the RFs preceding foot, thumb, index finger and mouth movements showed a somatotopic organization similar to that observed with PET. However, in each of the studies, cited above co-registration errors were considerable with mean differences in the location of dipoles and PET or fMRI activation of greater than 16 mm. These differences could have been due to errors in the co-registration of the modalities, the combined spatial inaccuracies associated with these modalities, or errors in the source reconstruction models. Like most MRF studies, a single moving dipole was used to model the field within a hemisphere. Models such as these do not account for contributions from additional sources with overlapping time courses, which can lead to inaccuracies in dipole localization. Nonetheless, the demonstration of a medial-lateral somatotopic organization of RFs and corresponding PET activation is in keeping with previous MEG studies (Cheyne et al., 1991) and provides compelling evidence that the contralateral MI contributes to the BP2.

In contrast to a single MI source, Pederson et al. (1998) found a sequential pattern of cortical activation associated with finger extension movements. A Multiple Signal Classification (MUSIC) algorithm was implemented to estimate the location and time course of source activity. This source modeling approach showed that premovement activity was initiated in the region of the contralateral DLPFC (-900 to -250 ms), followed by activity near the rostal SMA (-300 to –100 ms), premotor cortex (-100 to 0 ms, where 0 corresponded to movement onset), then the contralateral MI (0 to +100 ms). Sources within each of these fields appeared to correspond to regions of peak PET activation with co-registration errors ranging from 8 to 19 mm. This rostral-caudal-lateral sequence of activation was similar to that reported by Ball et al. (1999) using combined EEG and fMRI. However, these results should be interpreted with caution since only two subjects were reported.

SUMMARY

The principal goal of the studies reviewed in this chapter was to evaluate the spatial and temporal dynamics of cortical activity mediating MRPs or MRFs by combining electrophysiological recordings of scalp surface activity with functional neuroimaging. A variety of decomposition techniques have been used in these studies to model the underlying source generators of the scalp-recorded electromagnetic fields. In many cases, the co-registration errors between dipole locations and regions of PET or fMRI activation were considerable (commonly greater than 15 mm). Only one of the studies cited above used source reconstruction techniques using realistic MRI-derived models of the skull and brain (Ball et al., 1999), yet mean co-registration errors in that study were still in the vicinity of 9 mm. Inaccuracies in this range might be expected given the summative effects of errors inherent to source reconstruction and neuroimaging techniques. The fact that rCBF and BOLD signals are measures of changes in hemodynamic responses, and do not directly reflect the synaptic activity modeled by dipole source analysis, may also contribute of localization error bias. Nonetheless, the marked similarities in the spatial and temporal profiles of cortical activity reported in many of these studies have helped to clarify a variety of questions regarding the neural generators of the BP.

What are the neural structures contributing to the generation of the BP?

Taken together, experiments that have combined MRP recordings with functional neuroimaging have provided compelling evidence that bilateral regions of the mesial frontal cortex, including the pre-SMA, CMAs, and SMA-proper are the principal generators of the BP1 component of the BP. This finding is consistent with studies that have used high-resolution EEG recordings and current source density reconstruction methods (e.g. Cui et al., 1999, 2000; Knösche et al., 1996), whole-scalp MEG recordings of RFs (Erdler et al., 2000), and intracranial recordings in patients with intractable epilepsy (Ikeda et al., 1992; Yazawa et al., 2000), without the addition of functional neuroimaging correlation. Both the

BP1 and earliest RF begin around the same time (Lang et al., 1991; Erdler et al., 2000) and gradually increase until approximately 500 ms before EMG onset. Source reconstructions over this time interval have isolated source generators in the region of the mesial frontal cortex that overlap with PET and fMRI activations, with little or no contribution from MI (MacKinnon et al., 1996; Pedersen et al., 1998; Ball et al., 1999).

The balance of evidence, principally from combined MEG studies, supports a role for the contralateral MI in the generation of the BP2. The emergence of a prominent RF equivalent in time course and slope to the BP2, the medial-lateral somatotopic organization of the fields and the resultant dipoles (Cheyne et al., 1991; Walter et al., 1992), and the correspondence of regions of PET or fMRI activation, suggests that this activity is generated by synaptic input to neurons deep within the anterior bank of the central sulcus. The EEG-derived source models of the BP discussed in this chapter have tended to show that the contralateral MI activity begins late (near 100 ms before EMG onset), whereas MEG studies have clearly shown MI activity beginning up to 500 ms prior to EMG onset. Discrepancies between these techniques could be due to differences in the source reconstruction methods used (e.g. single vs. multiple dipole solutions). Alternatively, MEG may be better able to resolve the MI activity due to its inherent sensitivity to sources within a fissure.

Is there a contribution from the ipsilateral MI?
The one issue that is likely to remain contentious is the role of the ipsilateral MI in the generation of the BP2. The PET and fMRI studies discussed above consistently did not find significant activation of the ipsilateral MI. Bilateral MI source models have typically shown substantial contributions from both the contralateral and ipsilateral MI (Bötzel et al., 1993; Toro et al., 1993; MacKinnon et al., 1996, see Figure 5, Model 2). If this were the case, then the ipsilateral MI activity would be expected to generate sufficient signal to be detected with PET or fMRI (Jahanshahi et al., 1995; Jenkins et al., 2000). Furthermore, source reconstructions derived from combined EEG and PET or fMRI have identified contributions from the mesial frontal cortex and contralateral MI, but not from the ipsilateral MI (MacKinnon et al., 1996; Ball et al., 1999). Taken together, these findings suggest that the ipsilateral MI does not contribute to the BP.

Yet, many MEG studies have reported bilateral RFs that were best explained by bilateral MI dipoles (Cheyne and Weinberg, 1989; Kristeva et al., 1991; Salmelin et al., 1995; Hoshiyama et al., 1997; Erdler et al., 2000). Two factors may contribute to these divergent views. Firstly, it has been shown that ipsilateral RFs are not present in the majority of subjects (Nagamine et al., 1994). Secondly, ipsilateral RFs may be accompanied by bilateral EMG activity (Kristeva et al., 1991). If this is the case, then the presumed ipsilateral MI activity may simply reflect its role in contralateral movement. Nonetheless, the role of the ipsilateral MI requires further investigation and combined high-resolution MEG, EEG, and functional neuroimaging studies may help to resolve this issue.

What is the time course of their activity?
It has been proposed that the BP is mediated by serial activation of the mesial frontal cortex, then contralateral MI, eventually culminating in the generation of descending volleys via the corticospinal tract to initiate movement (e.g. Roland et al., 1980; Cui et al., 1999). Alternatively, activation of the mesial frontal cortex may be sustained until just before or after the onset of EMG activity, and thereby is active in parallel with the contralateral MI during the BP2 interval. The former hypothesis implies that the BP1 and BP2 have distinctly separate generators and is supported by studies of the topography of scalp current density maps (Cui et al., 1999) and the differential effects of neurological disorders on the BP1 or BP2 components (e.g. Dick et al., 1989).

The source models reviewed in this chapter clearly support the view that the mesial frontal cortex and contralateral MI are active in parallel during the BP2. In all subjects, activity within the mesial frontal cortex, modeled by dipole source analysis or current density reconstruction, was maintained until after EMG onset, or was associated with a marked drop in activity immediately preceding EMG onset (MacKinnon et al., 1996; Ball et al., 1999). These data are in accord with single unit recordings from the SMA-proper, pre-SMA, and CMA in behaving non-human primates that have shown that neurons with long-lead activity maintain or increase firing rates until the onset of EMG activity (e.g. Matsuzaka et al., 1992; Romo and Schultz, 1992; Shima and Tanji, 1991, 1998, 2000). The excitability of corticomotoneurons within the contralateral MI does not increase until less than 100 ms before EMG onset (Chen et al., 1998), therefore activity within MI during this period is unlikely to directly influence spinal motoneuron excitability.

How does the mesial frontal cortex generate the scalp surface negativity?

Since the region of hand representation within the SMA is thought to reside deep within the medial wall, there is likely to be some degree of cancellation of electromagnetic fields between adjacent banks of the SMA (Lang et al., 1991). However, asymmetries in synaptic activity and convolutions in adjacent banks of the SMA (Ikeda et al., 1992) will result in electromagnetic fields that are detectable by either EEG or MEG. This idea is supported by whole-scalp MEG recordings (143 channel) showing that RFs with onset comparable to the BP1 were best explained by a dipole in region of the SMA (Erdler et al., 2000). Furthermore, extracellular currents generated by synaptic input to neurons within the banks of the cingulate sulcus or the additional enfolding of the SMA (Ikeda et al., 1992) would create dipoles that are principally oriented radially to the surface of the scalp. These currents probably account for the large surface negative potentials over the vertex recorded with EEG, but cannot be detected with MEG.

What is the function of the premovement activity?

The studies reviewed in this chapter have not specifically addressed this question. However, the combined MRP and PET studies of Jahanshahi et al. (1995) and Jenkins et al. (2000) in conjunction with other MRP studies (Papa et al., 1991; Cunnington et al., 1995) provide compelling evidence that the premovement activity within mesial frontal cortex mediates the production of movements with a predictable onset timing (whether cued or uncued). Deecke and colleagues have postulated that the SMA subserves the temporal organization of sequential and bilateral coordination (Cui et al., 1999). This hypothesis is supported by recent studies in monkeys showing that neurons within the pre-SMA and SMA proper are preferentially involved in the temporal organization of movement sequences (Shima and Tanji, 1998, 2000). Yet, the presence of activity preceding simple unilateral movements further suggests that regions of the mesial frontal cortex mediate the temporal organization of voluntary movements irrespective of movement complexity. The marked attenuation or ablation of the BP with lesions of the cerebellum (Shibasaki et al., 1986), and presence of cerebellar activation associated with self-initiated movements imaged with PET (MacKinnon et al., 1996), implies that a motor network comprising the pre-SMA, rostral CMA, SMA-proper and ipsilateral cerebellum may mediate this timing function.

CONCLUSION

These studies recapitulate and substantiate the original hypotheses put forward by Kornhuber and Deecke that the BP principally reflects activity originating bilaterally from the SMA, anterior cingulate, and contralateral MI (Deecke and Kornhuber, 1978). Kornhuber and Deecke were also the first to identify individual differences in the timing and topography of the MRPs (Deecke and Kornhuber, 1969; Deecke et al., 1976). In keeping with this idea, source reconstruction techniques combined with functional neuroimaging have identified different patterns of mesial frontal activation, such as a possible contribution from the IPL (Ball et al., 1999). Experiments are now required that use these techniques to capture the individual variations in the spatial and temporal patterns of cortical activity mediating simple and complex voluntary movements in order to examine the influence of this activity on motor output in both health and disease.

REFERENCES

Ball, T., Schreiber, A., Feige, B., Wagner, M., Lucking, C.H. and Kristeva-Feige, R. (1999) The role of higher-order motor areas in voluntary movement as revealed by high-resolution EEG and fMRI. Neuroimage 10, 682-94.

Beisteiner, R., Gomiscek, G., Erdler, M., Teichtmeister, C., Moser, E. and Deecke, L. (1995) Comparing localization of conventional functional magnetic resonance imaging and magnetoencephalography. Eur J Neurosci 7, 1121-4.

Böcker, K.B.E., Brunia, C.H.M. and Cluitmans, P.J.M. (1994) A spatio-temporal dipole model of the readiness potential in humans. I. Finger movement. Electroenceph clin Neurophysiol 91, 275-285.

Bötzel, K., Plendl, H., Paulus, W. and Scherg, M. (1993) Bereitschaftspotential: is there a contribution of the supplementary motor area? Electroenceph clin Neurophysiol 89, 187-196.

Chen, R., Yaseen, Z., Cohen, L.G. and Hallett, M. (1998) Time course of corticospinal excitability in reaction time and self-paced movements. Ann Neurol 44, 317-325.

Cheyne, D., Kristeva, R. and Deecke, L. (1991) Homuncular organization of human motor cortex as indicated by neuromagnetic recordings. Neurosci Lett 122, 17-20.

Cheyne, D. and Weinberg, H. (1989) Neuromagnetic fields accompanying unilateral finger movements: pre-movement and movement-evoked fields. Exp Brain Res 78, 604-612.

Cheyne, D., Weinberg, H., Gaetz, W. and Jantzen, K.J. (1995) Motor cortex activity and predicting side of movement: neural network and dipole analysis of pre-movement magnetic fields. Neurosci Lett 188, 81-84.

Colebatch, J.G., Deiber, M.-P., Passingham, R.E., Friston, K.J. and Frackowiak, R.S.J. (1991) Regional cerebral blood flow during voluntary arm and hand movements in human subjects. J Neurophysiol 65, 1392-1401.

Cui, R.Q. and Deecke, L. (1999) High resolution DC-EEG analysis of the Bereitschaftspotential and post movement onset potentials accompanying uni- or bilateral voluntary finger movements. Brain Topogr 11, 233-49.

Cui, R.Q., Huter, D., Egkher, A., Lang, W., Lindinger, G. and Deecke, L. (2000) High resolution DC-EEG mapping of the Bereitschaftspotential preceding simple or complex bimanual sequential finger movement. Exp Brain Res 134, 49-57.

Cui, R.Q., Huter, D., Lang, W. and Deecke, L. (1999) Neuroimage of voluntary movement: topography of the Bereitschaftspotential, a 64-channel DC current source density study. Neuroimage 9, 124-34.

Cunnington, R., Iansek, R., Bradshaw, J.L. and Phillips, J.G. (1995) Movement-related potentials in Parkinson's disease. Brain 118, 935-950.

Deecke, L., Grözinger, B. and Kornhuber, H.H. (1976) Voluntary finger movements in man: cerebral potentials and theory. Biol Cybern 23, 99-119.

Deecke, L. and Kornhuber, H.H. (1978) An electrical sign of participation of the mesial supplementary motor cortex in human voluntary finger movement. Brain Res 159, 473-476.

Deecke, L., Scheid, P. and Kornhuber, H.H. (1969) Distribution of readiness potential, pre-motion positivity, and motor potential of the human cerebral cortex preceding voluntary finger movements. Exp Brain Res 7, 158-168.

Deiber, M.-P., Honda, M., Ibanez, V., Sadato, N. and Hallett, M. (1999) Mesial motor areas in self-initiated versus externally triggered movements examined with fMRI: effect of movement type and rate. J Neurophysiol 81, 3065-77.

Deiber, M.-P., Ibañez, V., Sadato, N. and Hallett, M. (1996) Cerebral structures participating in motor preparation in humans: a positron emission tomography study. J Neurophysiol 75, 233-247.

Deiber, M.-P., Passingham, R.E., Colebatch, J.G., Friston, K.J., Nixon, P.D. and Frackowiak, R.S.J. (1991) Cortical areas and the selection of movement: a study with positron emission tomography. Exp Brain Res 84, 393-402.

Dettmers, C., Fink, G.R., Lemon, R.N., Stephan, K.M., Passingham, R.E., Silbersweig, D., Holmes, A., Ridding, M. C., Brooks, D.J. and Frackowiak, R.S.J. (1995) Relation between cerebral activity and force in the motor areas of the human brain. J Neurophysiol 74, 802-815.

Dick, J., P.R., Rothwell, J.C., Day, B.L., Cantello, R., Buruma, O., Gioux, M., Benecke, R., Berardelli, A., Thompson, P.D. and Marsden, C.D. (1989) The Bereitschaftspotential is abnormal in Parkinson's disease. Brain 112, 233-244.

Dum, R.P. and Strick, P.L. (1996) Spinal cord terminations of the medial wall motor areas in macaque monkeys. J Neurosci 16, 6513-25.

Erdler, M., Beisteiner, R., Mayer, D., Kaindl, T., Edward, V., Windischberger, C., Lindinger, G. and Deecke, L. (2000) Supplementary motor area activation preceding voluntary movement is detectable with a whole-scalp magnetoencephalography system. Neuroimage 11, 697-707.

Fuchs, M., Wagner, M., Köhler, T. and Wischmann, H.-A. (1999) Linear and non-linear current density reconstructions. J Clin Neurophysiol 16, 267-295.

Gerloff, C., Grodd, W., Altenmuller, E., Kolb, R., Naegele, T., Klose, U., Voigt, K. and Dichgans, J. (1996) Coregistration of EEG and fMRI in a simple motor task. Hum Brain Mapp 4, 199-209.

He, S.Q., Dum, R.P. and Strick, P.L. (1995) Topographic organization of corticospinal projections from the frontal lobe: motor areas on the medial surface of the hemisphere. J Neurosci 15, 3284-306.

Hoshiyama, M., Kakigi, R., Berg, P., Koyama, S., Kitamura, Y., Shimojo, M., Watanabe, S. and Nakamura, A. (1997) Identification of motor and sensory brain activities during unilateral finger movement: spatiotemporal source analysis of movement-associated magnetic fields. Exp Brain Res 115, 6-14.

Ikeda, A., Lüders, H.O., Burgess, R.C. and Shibasaki, H. (1992) Movement-related potentials recorded from supplementary motor area and primary motor area: role of supplementary motor area in voluntary movements. Brain 115, 1017-1043.

Jahanshahi, M., Jenkins, I.H., Brown, R.G., Marsden, C.D., Passingham, R.E. and Brooks, D.J. (1995) Self-initiated versus externally triggered movements. I. An investigation using measurement of regional cerebral blood flow with PET and movement-related potentials in normal and Parkinson's disease subjects. Brain 118, 913-933.

Jenkins, I.H., Jahanshahi, M., Jueptner, M., Passingham, R.E. and Brooks, D.J. (2000) Self-initiated versus externally triggered movements. II. The effect of movement predictability on regional cerebral blood flow. Brain 123, 1216-28.

Kitamura, J., Shibasaki, H., Takagi, A., Nabeshshima, H. and Yamagushi, A. (1993) Enhanced negative slope of cortical potentials before sequential as compared with simultaneous extensions of two fingers. Electroenceph clin Neurophysiol 86, 176-182.

Knösche, T., Praamstra, P., Stegeman, D. and Peters, M. (1996) Linear estimation discriminates midline sources and a motor cortex contribution to the readiness potential. Electroenceph clin Neurophysiol 99, 183-90.

Kornhuber, H.H. and Deecke, L. (1964) Hirnpotentialänderungen beim Menschen vor und nach Willkürbewegungen, dargestellt mit Magnetbandspeicherung und Rückwärtsanalyse. Pflugers Arch 281, 52.

Kornhuber, H.H. and Deecke, L. (1965) Hirnpotentialänderungen bei Willkürbewegungen und passiven bewegungen des menschen: bereitschaftspotential und reafferente potential. Pflugers Arch 284, 1-17.

Kristeva, R., Cheyne, D. and Deecke, L. (1991) Neuromagnetic fields accompanying unilateral and bilateral voluntary movements: topography and analysis of cortical sources. Electroenceph clin Neurophysiol 81, 284-298.

Lang, W., Cheyne, D., Kristeva, R., Beisteiner, R., Lindinger, G. and Deecke, L. (1991) Three-dimensional localization of SMA activity preceding voluntary movement. Exp Brain Res 87, 688-695.

Lang, W., Lang, M., Uhl, F., Koska, C., Kornhuber, A. and Deecke, L. (1988) Negative cortical DC shifts preceding and accompanying simultaneous and sequential finger movements. Exp Brain Res 71, 579-587.

Lang, W., Obrig, H., Lindinger, G. and Deecke, L. (1990) Supplementary motor area activation while tapping bimanually different rhythms in musicians. Exp Brain Res 79, 504-514.

MacKinnon, C.D., Kapur, S., Hussey, D., Verrier, M.C., Houle, S. and Tatton, W.G. (1996) Contributions of the mesial frontal cortex to the premovement potentials associated with intermittent hand movements in humans. Hum Brain Mapp 4, 1-20.

Matsuzaka, Y., Aizawa, H. and Tanji, J. (1992) A motor area rostral to the supplementary motor area (presupplementary motor area) in the monkey: neuronal activity during a learned motor task. Neurophysiol 68, 653-662.

Nagamine, T., Kajola, M., Salmelin, R., Shibasaki, H. and Hari, R. (1996) Movement-related slow cortical magnetic fields and changes of spontaneous MEG- and EEG-brain rhythms. Electroencephal clin Neurophysiol 99, 274-86.

Nagamine, T., Toro, C., Balish, M., Deuschl, G., Wang, B., Sato, S., Shibasaki, H. and Hallett, M. (1994) Cortical magnetic and electric fields associated with voluntary finger movements. Brain Topogr 6, 175-83.

Neshige, R., Lüders, H. and Shibasaki, H. (1988) Recording of movement-related potentials from scalp and cortex in man. Brain 111, 719-736.

Orgogozo, J.M. and Larsen, B. (1979) Activation of the supplementary motor area during voluntary movements in man suggests it works as a supramotor area. Science 206, 847-850.

Papa, S.M., Artieda, J. and Obeso, J.A. (1991) Cortical activity preceding self-initiated and externally triggered voluntary movement. Movt Dis 6, 217-224.

Pedersen, J.R., Johannsen, P., Bak, C.K., Kofoed, B., Saermark, K. and Gjedde, A. (1998) Origin of human motor readiness field linked to left middle frontal gyrus by MEG and PET. Neuroimage, 8, 214-20.

Picard, N. and Strick, P.L. (1996) Motor areas of the medial wall: a review of their location and functional activation. Cereb Cortex 6, 342-353.

Roland, P.E., Larsen, B., Lassen, N.A. and Skinhoj, E. (1980a) Supplementary motor area and other cortical areas in organization of voluntary movements in man. J Neurophysiol 43, 118-136.

Romo, R. and Schultz, W. (1992) Role of primate basal ganglia and frontal cortex in the internal generation of movements. III. Neuronal activity in the supplementary motor area. Exp Brain Res 91, 396-407.

Sabatini, U., Chollet, F., Rascol, O., Celsis, P., Rascol, A., Lenzi, G.L. and Marc-Vergnes, J.-P. (1993) Effect of side and rate of stimulation on cerebral blood flow changes in motor areas during finger movements in humans. J Cereb Blood Flow Metab 13, 639-645.

Sadato, N., Ibanez, V., Deiber, M.-P., Campbell, G., Leonardo, M. and Hallett, M. (1996) Frequency-dependent changes of regional cerebral blood flow during finger movements. J Cereb Blood Flow Metab 16, 23-33.

Salmelin, R., Forss, N., Knuutile, J. and Hari, R. (1995) Bilateral activation of the human somatomotor cortex by distal hand movements. Electroenceph clin Neurophysiol 95, 444-452.

Shibasaki, H., Barrett, G., Neshige, R., Hirata, I. and Tomoda, H. (1986) Components of the movement-related cortical potential and their scalp topography. Brain Res 368, 361-365.

Shima, K., Aya, K., Mushiake, H., Inase, M., Aizawa, H. and Tanji, J. (1991) Two movement-related foci in the primate cingulate cortex observed in signal-triggered and self-paced forelimb movements. J Neurophysiol 65, 188-202.

Shima, K. and Tanji, J. (1998) Both supplementary and presupplementary motor areas are crucial for the temporal organization of multiple movements. J Neurophysiol 80, 3247-60.

Shima, K. and Tanji, J. (2000) Neuronal activity in the supplementary and presupplementary motor areas for temporal organization of multiple movements. J Neurophysiol 84, 2148-60.

Simonetta, M., Clanet, M. and Rascol, O. (1991) Bereitschaftspotential in a simple movement or in a motor sequence starting with the same simple movement. Electroenceph clin Neurophysiol 81, 129-134.

Simpson, G.V., Pflieger, M.E., Foxe, J.J., Ahlfors, S.P., Vaughan, H.G., Hrabe, J., Ilmoniemi, R.J. and Lantos, G. (1995) Dynamic neuroimaging of brain function. J Clin Neurophysiol 12, 432-49.

Toro, C., Matsumoto, J., Deuschl, G., Roth, B.J. and Hallett, M. (1993) Source analysis of scalp-recorded movement-related electrical potentials. Electroenceph clin Neurophysiol 86, 167-175.

Toro, C., Wang, B., Zeffiro, T., Thatcher, R.W. and Hallett, M. (1994) Movement-related cortical potentials: source analysis and PET/MRI correlation. In: Thatcher, R.W., Hallett, M., Zeffiro, T., John, E.R. and Huerta, M., (Eds.) Functional Neuroimaging: Technical Foundations, pp. 259-267. Orlando, Fl: Academic Press.

Turner, R.S., Grafton, S.T., Votaw, J.R., Delong, M.R. and Hoffman, J.M. (1998) Motor subcircuits mediating the control of movement velocity: a PET study. J Neurophysiol 80, 2162-76.

Walter, H., Kristeva, R., Knorr, U., Schlaug, G., Huang, Y., Steinmetz, H., Nebeling, B., Herzog, H. and Seitz, R.J. (1992) Individual somatotopy of primary sensorimotor cortex revealed by intermodal matching of MEG, PET, and MRI. Brain Topogr 5, 183-7.

Yazawa, S., Ikeda, A., Kunieda, T., Ohara, S., Mima, T., Nagamine, T., Taki, W., Kimura, J., Hori, T. and Shibasaki, H. (2000) Human presupplementary motor area is active before voluntary movement: subdural recording of Bereitschaftspotential from medial frontal cortex. Exp Brain Res 131, 165-77.

GENERATORS OF THE MOVEMENT-RELATED CORTICAL POTENTIALS AND DIPOLE SOURCE ANALYSIS

Keiichiro Toma, M.D. and Mark Hallett, M.D.

Human Motor Control Section, Medical Neurology Branch
National Institute of Neurological Disorders and Stroke
National Institutes of Health
Bethesda, MD 20892-1428, U.S.A.

INTRODUCTION

The precise location and timing of cortical activation in voluntary movements have been major issues in motor control physiology. For example, during movement preparation, whether the supplementary motor area (SMA) is activated in a sequential (Deecke and Kornhuber, 1978; Orgogozo and Lasen, 1979; Deecke, 1987; Deecke and Lang, 1996; Deecke et al., 1999) or parallel (Hyland et al., 1989; Ikeda et al., 1992) manner has been intensively debated with regard to the primary motor area (M1). During movement execution, contribution of the gyrus and sulcus part of motor cortex has also been an important issue (Strick and Preston, 1982a, 1982b; Kawashima et al., 1995; Geyer et al., 1996). Intracranial recordings with subdural (Neshige et al., 1988; Ikeda et al., 1992) as well as depth (Rektor et al., 1994) electrodes are useful approaches to gain information about the precise cortical regions and time course of the activation. However, these techniques invasively record pathological brains, and the limitations of electrode placement over distributed cortical regions prevent simultaneous collection of data from various brain areas. In addition, subdural recording from an electrode placed over the gyrus provides little information about activity from the sulcus. Recent development and improvement of electromagnetic or neuroimaging measurements enable us to investigate human brain function noninvasively. At the present time, however, no non-invasive methodology with sufficient temporal and spatial resolution has been developed. Electromagnetic measurements such as the electroencephalogram (EEG) and magnetoencephalogram (MEG) provide information about cortical activation with millisecond resolution, but with limited spatial localization. By contrast, neuroimaging techniques such as positron emission tomography (PET) (Herscovitch, 1994) and functional magnetic resonance imaging (fMRI) (Friston et al., 1995; Rosen et al., 1998) yield good spatial information with millimeter resolution, although they are devoid of good timing information.

The Bereitschaftspotential
Edited by Jahanshahi and Hallett, Kluwer Academic/Plenum Publishers, New York, 2003

The movement-related cortical potential (MRCP) is the electrophysiological evidence of motor cortical involvement during movement preparation and execution (Kornhuber and Deecke, 1965). Although components of the MRCP are well established (Kornhuber and Deecke, 1965; Vaughan et al., 1968; Shibasaki et al., 1980; Tarkka and Hallett, 1991), their generator sources within the brain are incompletely understood. The MRCP reflects synchronous postsynaptic potentials generated by the large number of cortical projection neurons that are arrayed perpendicular to the cortical surface, thereby producing dipolar current sources orthogonal to the cortical mantle (Niedermeyer and Lopes da Silva, 1993). Dipole source modeling is a technique to estimate the location and temporal behavior of the current sources within the brain from the scalp EEG or MEG signals (Fender, 1987).

In this chapter, we review dipole source analyses for the MRCP described in previous studies and discuss the physiological relevance and problems of dipole source modeling. Since EEG and MEG provide complementary information to each other, we also review studies on the movement-related magnetic field (MRMF). Likely generator sources of the MRCP are then discussed from anatomical and physiological points of view. The major problem for dipole source modeling arises from the ambiguities in determining various parameters including the number of dipoles, dipole location, and dipole orientation. We finally propose using coregistration of electromagnetic data with neuroimaging techniques such as PET or fMRI to constrain these multiple parameters during dipole source modeling.

DIPOLE SOURCE ANALYSIS OF MRCP

Estimating the location and distribution of current sources within the brain, based on voltage potential measurements at the scalp, requires a solution of the "inverse problem" (Sarvas, 1987). Given a subset of potentials measured at a finite number of scalp electrode positions, calculation of current sources and potential fields within the brain does not have a unique solution. In other words, an infinite number of different combinations of generator sources can account for a given set of data (Van Oosterom, 1991). Due to this "inverse problem", the dipole source modeling in previous studies commonly tried to use several constraints with *a priori* anatomical and physiological hypothesis. These hypothesis-based analyses, however, contain arbitrary processes, and thus the findings have been inconsistent. To minimize residual variance (%RV) between activity derived from the dipole model and the measured data, location and direction of the dipoles are computed by an iterative least square fit. A volume conductor head model ("forward model") is used for this computation, in which different conductivities are assigned to different tissues surrounding the cortical surface. Volume conductors are grouped into two major categories, i.e., spherical (Nunetz, 1981; Fender, 1991; Van Oosterom, 1991) and realistic-head (Roth et al., 1993, 1997) models. Although, so far, the three-layer spherical volume conductor has commonly been used, a realistic-head model that is more analogous to the shape of each individual head is expected to provide more accurate fitting results.

EEG study

There are several different nomenclatures for the MRCP among different researchers (Kornhuber and Deecke, 1965; Vaughn et al., 1968; Shibasaki et al., 1980; Tarkka and

Hallett, 1991), where components of the MRCP are defined based on their latencies as well as spatial distributions over the scalp. In figure 1A, the distinguishable MRCP components are presented together with their scalp topography. Consistently identified components of the MRCP are the bereitschaftspotential (BP), negative slope (NS') and motor potential (MP). The BP is a gently rising negativity beginning a few seconds before movement with symmetrical scalp distribution. The BP is followed by a steeper component called NS' over the contralateral central areas, starting about several hundred

MRCP and MRMF Components

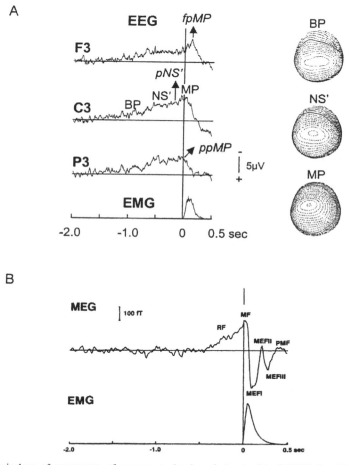

Figure 1. A. Terminology of components of movement-related cortical potentials (MRCP) for right index finger movement. Waveforms are from the F3, C3 and P3 electrodes. The bereitschaftspotential (BP) is a gentle slope with symmetrical, widespread distribution over the scalp, beginning about 1.5 s before EMG onset. The following steeper component is termed negative slope (NS'), occurring at about several hundred milliseconds prior to EMG onset with more focal topography over the centroparietal region. The highest amplitude of MRCP, labeled motor potential (MP), is observed at the time of the movement itself. B. Waveform of movement-related magnetic fields (MRMF) from a lateral position over the left hemisphere during right index finger movement. The readiness field (RF) begins approximately 500 ms prior to EMG onset. The motor field (MF) forms a peak from –50 ms to 50 ms, and movement-evoked magnetic field I (MEFI) follows at 100 ms after EMG onset. It is noteworthy that the NS' and RF as well as fpMP and MEFI share similar latency. From Kristeva et al. (1991) with permission of Elsevier.

milliseconds prior to movement. After the peak of the NS' (pNS') is the MP that was subdivided into three components based on latency and scalp topography by Tarkka and Hallett (1991). The initial slope of the MP (isMP) is identified immediately before the electromyogram (EMG) onset, showing focal topography at the primary sensorimotor area (SM1). This contralateral negativity persists for 30 to 50 ms after onset of the EMG discharge, and then suddenly drops off in the centroparietal areas, forming a parietal peak of the MP (ppMP). Subsequently, the peak negativity shifts toward the anterior part of the scalp, termed the frontal peak of the MP (fpMP). The isMP and ppMP likely represent activation of the motor cortex to execute movement because of their focal scalp topography at the contralateral central area. Despite its anterior distribution, it is assumed that the fpMP is generated by afferent feedback from the movement to the sensory cortical regions (Arezzo and Vaughan, 1980).

Toro et al. (1993, 1994) chose the pNS' and fpMP for dipole source modeling because they are the most distinct peak events immediately before and after movement onset, possibly representing maximal activation of the underlying cortices (figure 2A). Toro and colleagues conducted spatiotemporal dipole analysis (Scherg, 1989; Scherg et al., 1989; Scherg and Berg, 1991) using the brain electric source analysis (BESA) package with a three-layer spherical head model. In the spatiotemporal dipole source analysis, the sources modeled for the potentials at single time instance (e.g., pNS' or fpMP) are assumed to have a fixed location and orientation, while source strength varies over time. In figure 2A, they presumed one source (source 1) in the right hemisphere and one source (source 2) in the left hemisphere at the instant of the pNS'. Positions of these two dipoles were constrained to be mirror images of each other, but each dipole of the pair was free to orient in any direction. The fpMP was fitted by two sources (3a and 3b). Source 3a was constrained to the location of source 2 but free of orientation constraint, whereas source 3b was unconstrained in both location and orientation. After fixation of the location and direction of these four dipoles, source strength of each dipole was calculated as a function of time for an expanded time range (e.g., –1500 to 500 ms). Four of 10 subjects were well modeled by the three-dipole model without a 3b source, while such a source improved %RV in the remaining 6 subjects. In all 6 subjects, source 3b was in the frontal midline. Sources at the bilateral SM1 (1 and 2) showed a similar time course and strength, which started rising at about –1000 ms and reached maximal strength at EMG onset. In contrast, sources 3a and 3b demonstrated a sharp peak at 100 ms after EMG onset. The most remarkable finding of the study was a change in direction of the dipoles in the contralateral SM1 before (source 2) and after (source 3a) the movement (figure 2A).

Tarkka (1994) conducted an unconstrained dipole modeling using BESA and produced a result with no ipsilateral source (figure 2B). The analysis included a very short period around movement (250 ms preceding and 200 ms following EMG onset). An iterative optimization of location, orientation and strength of three unconstrained dipoles for the whole time range (-250 ms to 200 ms) gave a satisfactory small %RV (about 5%). Tarkka found two dipoles (source 2 and 3) in the contralateral SM1 and one dipole (source 1) in the midline area, but an additional fourth dipole which fitted in the occipital area did not significantly reduce %RV. The midline dipole (source 1), presumably reflecting the SMA activation, was most active during the preparatory period before movement onset, while the two dipoles (source 2 and 3) located contralaterally to the moving finger were most active during and after the motor act. The findings suggested that the SMA is engaged in movement preparation, the contralateral M1 during movement execution, and the contralateral primary sensory area (S1) following the movement, which agrees with the idea of serial activation between the SMA and M1 (Deecke and Kornhuber, 1978;Orgogozo and Lasen, 1979; Deecke, 1987).

EEG dipole source modeling for right finger movement

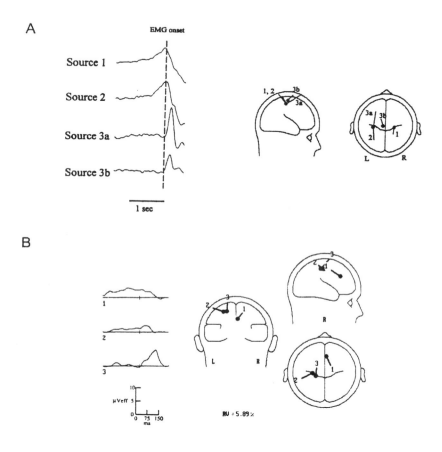

Figure 2. Two different types of dipole modeling for the MRCP. Strength of each source as a function of time (left) and location and orientation of the sources (right) are presented. Tail of the dipoles indicates an intracellular current with positive pole. A. Spatiotemporal dipole modeling for the grand averaged data of 10 subjects performing self-paced, right index finger movement once every 10 s. Location and orientation of source 1 and 2 are calculated at the instant of pNS' and those of the source 3a and 3b are at the fpMP. Bilateral primary sensorimotor areas (SM1) are active during premovement period. The midline source contributes only to postmovement period. From Toro et al. (1994) with permission of Academic Press. B. Unconstrained dipole modeling over the whole analysis time for the grand averaged data of 14 subjects performing self-paced, right index finger movement once every 3-7 s. No source is present on the ipsilateral side. Instead, a midline source is active during the premovement period. The short vertical line in the baseline indicates EMG onset. From Tarkka et al. (1994) with permission of Elsevier.

With regard to the contribution of activities from the SMA and ipsilateral SM1 to the MRCP, the findings are discrepant in the above studies (Toro et al., 1993, 1994; Tarkka, 1994), although both studies successfully achieved more than 90% of goodness of fit (%RV under 10%). To separate the SMA and SM1 source during finger flexion movement, Praamstra et al. (1996) applied a two-step analysis. In the first step, the symmetrical part of the potential was eliminated in homologous electrodes over the left and right hemispheres. Then, dipole sources responsible for the lateralized readiness potential were computed and identified in the SM1. In the next step, the original waves were analyzed with two dipoles in the contralateral M1 and S1 and one homologous dipole in the ipsilateral M1 which were fixed in a location and orientation as determined in the first step. Another source was introduced to significantly improve %RV and, as a result, it was consistently fitted in the medial frontocentral area (MFC). Praamstra and colleagues found that the sources located in the bilateral M1 and MFC explained the premovement activity of the MRCP. The source modeling for the lateralized readiness potential, however, does not necessarily guarantee accurate location and orientation of the dipoles. Although subdural recording showed that both SM1 and SMA generate the MP (Neshige et al., 1988; Ikeda et al., 1992), MP activity in the study by Praamstra et al. is not distributed between the M1 and SMA sources but mostly explained by the SMA source. It seems difficult to properly separate the potentials over the scalp into the SM1 and SMA sources.

MEG study

MEG, which is the magnetic counterpart of EEG, is a useful tool to localize generators of the event-related brain activity with high spatial accuracy (Williamson and Kaufman, 1981; Romani and Rossini, 1988; Cohen et al., 1990; Cheyne et al., 1991). MEG picks up magnetic fields orthogonal to the electric current source because magnetic fields are generated as surrounding the electric current. Whereas EEG is sensitive to both tangential and radial sources, the MEG selectively records the tangential activity with respect to the scalp surface. Because of the insensitivity of the magnetic flux to different conductivities of the tissues surrounding the cortical surface, MEG signals are not distorted as a result of the volume conduction effect that is unavoidable in EEG recordings (Kristeva et al., 1997). To explore involvement of the anterior (Brodmann's area 4) and posterior (area 3) banks of the central sulcus during movement, the selective sensitivity of MEG to the tangential sources can be taken advantage of, excluding contamination of the activity from the crown of pre-and postcentral gyrus. Thus, knowledge obtained from MEG studies provides useful complementary information for MRCP source modeling.

Magnetic fields associated with voluntary movements are referred to as the MRMF, which has a series of waveform components (Deecke et al., 1982; Hari et al., 1983; Cheyne and Weinberg, 1989; Kristeva et al., 1991). In figure 1B, the first detectable component is called the readiness field (RF) starting about 0.5 s before the EMG onset. The RF is regarded as a counterpart of the NS' of MRCP because of their similar latencies. The magnetic field then slightly increases in amplitude between 50 ms prior to and 30-50 ms after the EMG onset; this is called the motor field (MF). The RF and MF show bilateral topography, suggesting bilateral generator sources. Subsequently, the movement-evoked field I (MEFI) is detected about 100 ms after EMG onset, which is thought to correspond to the fpMP of MRCP. Figure 3 shows MEG dipole sources associated with unilateral and bilateral finger movement. Dipoles for the MF were modeled bilaterally where the ipsilateral dipole was localized about 1 cm deeper than the contralateral dipole and had significantly less strength (Kristeva et al., 1991) (figure 3). Cheyne and Weinberg (1989) also showed that the ipsilateral dipole fitted inconsistently at deeper locations or outside the head. The MEFI source for unilateral movement is present only on the contralateral side with the opposite polarity to the MF source.

MEG sources associated with unilateral (left: LH, right: RH) and bilateral (BH) finger movements

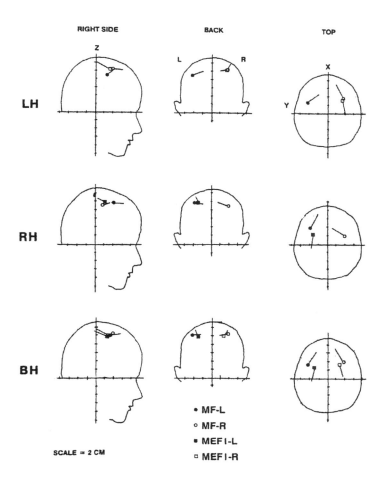

SCALE = 2 CM

- MF-L
- ○ MF-R
- MEFI-L
- □ MEFI-R

Figure 3. Location and orientation of the MEG dipoles during left (LH), right (RH) and bilateral (BH) finger movement. Dipoles for the MF over the left (MF-L) and right (MF-R) hemispheres and those for MEFI over the left (MEFI-L) and right (MEFI-R) hemispheres are shown. For the LH and RH, dipole is fitted bilaterally for the MF, but unilaterally on the contralateral side for the MEFI. For the BH, MEFI sources are present bilaterally. The dipole polarity of the MF and MEFI is opposite to each other. The ipsilateral MF source is localized deeper than the contralateral MF source for the LH and RH, whereas the bilateral MF sources are at the same depth for the BH. Tails of the dipoles indicate strength and direction of current flow with each plane of projection. Note that tail of the dipoles indicates negative pole, opposite from that in figure 2 and 3. From Kristeva et al. (1991) with permission of Elsevier.

In the early MEG studies, only small brain regions were investigated simultaneously due to limitation of available sensors. Recently developed multichannel neuromagnetometers allow simultaneous recording of magnetic fields from multiple brain areas. Multichannel MEG recording further confirmed the bilateral RF from 1 to 0.5 s prior to EMG onset and the contralateral MEFI within 200 ms after EMG onset (Nagamine et al., 1994, 1996; Salmelin et al., 1995). Moreover, simultaneous recording of MEG and EEG clearly demonstrated the contralateral preponderance of the RF, in contrast to the symmetrical distribution of the BP (Nagamine et al., 1994, 1996).

Taking EEG and MEG findings together, it is likely that the crown of the precentral gyrus (area 6) is active during the BP period, and subsequently the anterior bank of the central sulcus (area 4) becomes active bilaterally from -0.5 s (NS'/RF) through movement execution. After maximal activation of the anterior bank of the central sulcus at movement onset (MF), afferent feedback from the movement causes activation of the posterior bank of the central sulcus (area 3) on the contralateral side around 100 ms (fpMP/MEFI).

Steady-state MRCP: a variant of MRCP

Several movement factors including movement rate (interval between two consecutive movements) affect latency, amplitude or slope of the MRCP. MRCP associated with fast repetitive movements (e.g., 2 movements per one second) is termed steady-state MRCP (ssMRCP) (Gerloff et al., 1997, 1998a, 1998b). Dipole analysis for such fast repetitive movements showed different generators from the slowly performed movement that has commonly been used in MRCP studies. Three components are detectable for ssMRCP. The first component (pre-MP) peaks about 60 ms before EMG onset. Dipole source modeling revealed that sources for the pre-MP are largely radial and located over central regions bilaterally, consistent with activation of crown parts of bilateral precentral gyri. The second component called the motor peak (MP) is at about 10 ms after the EMG onset, and its dipole source is located at the contralateral central region, oriented with the positive end posteriorly. The following peak in the postmovement phase, labeled post-MP, is seen at about 95 ms. The post-MP is located over the contralateral central region and shows an anteriorly oriented dipole. Thus, two tangential dipoles with opposite direction before and after EMG onset appear to indicate sequential activation from the anterior (at the MP) to the posterior (at the post-MP) bank of the central sulcus. With regard to the contribution of the midline source to the ssMRCP, an additional source in the MFC increased the goodness of dipole fitting in 8 subjects out of 12 (Gerloff et al., 1998a).

THE LIKELY GENERATORS OF THE MRCP

Here, we discuss likely cortical generators of the MRCP and the timing relationship between these multiple cortical generators, based on findings from dipole source modeling in both EEG and MEG. We refer to knowledge obtained from direct cortical recording with subdural electrodes to clarify the problems with dipole modeling. As is evident from subdural recordings (Neshige et al., 1988; Ikeda et al., 1992), the MFC and bilateral SM1 share a similar temporal activation pattern, beginning to rise a few seconds preceding movement and peaking around movement onset. In the scalp EEG, moreover, the currents within the brain are propagated, distorted and attenuated by the geometric properties and conductivities of various tissues between the cortical surface and scalp electrodes such as cerebrospinal fluid, skull and scalp. Thus, activities with similar temporal behaviors overlap over the scalp, resulting in a difficulty to separate multiple sources within the brain.

Contralateral SM1

M1 is both precentral bank and gyrus in primates, but only precentral bank in humans; the crown of precentral gyrus of humans is area 6 (premotor area). Simultaneous recording of the MRCP and MRMF (Nagamine et al., 1994, 1996) showed different temporal evolution of the activity from these two areas: crown of precentral gyrus and anterior bank of central sulcus.

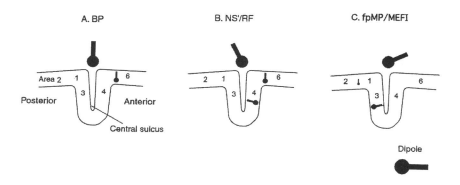

Figure 4. Simplified schematic diagram of sequential activation within SM1 contralateral to the moving hand. Summed vector of dipoles at each area (small black arrows) is presented as a large black arrow. A. For the BP, only crown of precentral gyrus (area 6) seems to be active. No counterpart of the BP is detectable with MEG, indicating no activation of the sulcus at this period. B. From about 0.5 s preceding EMG onset, both the gyrus (area 6) and sulcus (area 4) likely become active, generating the NS'. Only the tangential component from area 4 is detectable with MEG as the RF. C. After the movement, activity from sensory areas including area 3 may produce an anteriorly oriented dipole, which can be recorded as fpMP in EEG and MEFI in MEG. Tail of the dipoles indicates the postive pole of intracellular current.

In the SM1, the magnetic field equivalent to the BP of the MRCP cannot be detected by MEG, which indicates that the BP is largely generated from the crown of the precentral gyrus (area 6) as illustrated in the schematic diagram in figure 4A. A change in the dipole orientation from posterior to anterior (Cheyne and Weinberg, 1989; Kristeva et al., 1991; Toro et al., 1993; Tarkka, 1994; Gerloff et al., 1998a) suggests sequential activation from precentral bank (area 4) to postcentral bank (area 3) immediately before and after movement onset. During the NS' period, both the crown of the precentral gyrus and precentral bank are likely active; vectorial summation of these dipoles introduces a predominantly radial and slightly posteriorly-oriented single dipole (figure 4B). After movement onset, this vectorial summed dipole changes its direction anteriorly, probably as a result of activation of the postcentral bank (figure 4C). Thus, there is sequential activation of area 6, 4 and 3 in the contralateral central region during movement.

MFC

The MFC sources may generate relatively large radial potentials with spread over the bilateral central regions causing a symmetrical contribution to the BP (Deecke and Kornhuber, 1978). Praamstra et al. (1996) and Tarkka (1994) demonstrated possible modeling for activation of the MFC during the premovement period with or without the ipsilateral M1 activation, respectively. However, several EEG modeling studies did not find a generator source in the MFC (Bötzel et al., 1993; Böcker et al., 1994). Bötzel et al. (1993) calculated the M1 source close to the midsagittal plane. In the model of Böcker et al. (1993), premovement activity was mostly explained by a deep source at an ambiguous location. Although Toro et al. (1993, 1994) found a dipole source at the MFC in a half of the examined subjects, it was active only during the postmovement period without contributing to the premovement period (figure 2A).

MEG studies have failed to demonstrate evidence of involvement of the midline sources including the SMA and cingulate motor area (CM), at least in normal subjects performing simple movement task (Cheyne and Weinberg, 1989; Kristeva et al., 1991). This fact is indicative of no tangential activity in the MFC, because MEG is sensitive to tangential sources with respect to the scalp surface. One plausible reason for absence of tangential activity in the MFC is canceling out of two dipoles with opposite directions from left and right medial frontal walls (Lang et al., 1991). In addition, the MFC sources located deeply from the scalp may cause only a weak magnetic field at the scalp. In this regard, the tangential MFC dipole (source 1 in figure 2B) in the EEG study by Tarkka (1994) without constraint of dipole orientation is not compatible with the MEG studies. Orientation constraint radially to the scalp surface may be required for the MFC source.

To assume a contribution of the activity from the MFC to radial potentials in the scalp EEG, likely generator mechanisms in the MFC should be postulated. One possible candidate is a bank of the cingulate sulcus (Shima et al., 1991; Ikeda et al., 1992). The activity from banks of the medial frontal walls might produce the radial scalp potentials, as illustrated in figure 5. The SMA proper and caudal part of the CM share similar anatomical and functional properties (Luppino et al., 1991, 1994; Shima et al., 1991; Dum and Strick, 1992), and motor tasks often involve the CM as well as the SMA (Picard and Strick, 1996). Another explanation is that the finger representation of area 6 in the medial frontal walls could reach the edge of the medial fissure (Gerloff et al., 1998a; Talairach and Tournoux, 1988) (figure 5). Lim et al. (1994) demonstrated that predominantly tonic, proximal motor responses that are typical for SMA stimulation are elicited not only from the medial portion of the superior frontal gyrus but also from its dorsal convexity. The human SMA, therefore, is not always confined to the medial portion of the superior frontal gyrus. It is conceivable that anatomical diversity among different subjects may have yielded different results about the existence of MFC sources in previous studies. The bilateral medial frontal walls could have asymmetrical anatomy, which may result in relatively large residual activity from partial cancellation of activity. In previous studies using simple movement task, about a half of the examined subjects had an additional midline source (Toro et al., 1993, 6 of 10 subjects; Gerloff et al., 1996, 2 of 4; Gerloff et al., 1998a, 8 of 12). When complex movement is used, activity from the contralateral SMA becomes predominant during the BP period and relatively large residual activity in the MFC can be recorded with MEG (Erdler et al., 2000). It is still uncertain how much activity the MFC contributes to the scalp EEG.

Possible Generators of MRCP in Medial Frontal Regions

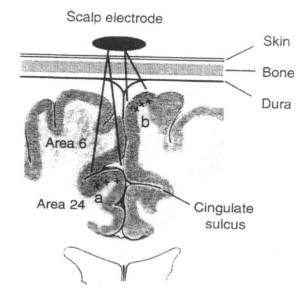

Coronal view of the medial frontal walls

Figure 5. Schematic illustration of possible mechanisms to generate scalp-recorded negative potentials in the medial frontocentral structures. Banks of the cingulate sulcus (a) and edges of the medial frontal wall (b) could produce negative potentials over the scalp. The scalp electrodes only record summed potentials arising from several generators that may have opposite polarity. Asymmetrical anatomical configuration of two medial walls juxtaposed at both sides of the interhemispheric fissure may result in partial cancellation and cause residual activity with various orientations.

Ipsilateral SM1

Symmetrical distribution of the scalp-recorded BP could be explained by the results of a summation of the potentials arising from the SM1 bilaterally even without significant contribution of activity from the MFC. In the subdural recordings, the early component of the MRCP, the BP, was recorded from the bilateral hand sensorimotor areas with similar amplitudes (Neshige et al., 1988) (figure 6). In studies with PET (Colebatch et al., 1991; Shibasaki et al., 1993) and fMRI (Rao et al., 1993), activation was detected in the SM1 bilaterally for complex movements, whereas simple movements only involved the contralateral SM1.

This discrepancy between electromagnetic and neuroimaging findings might be explained by an inhibitory role of the ipsilateral SM1 during movements. Waldvogel et al. (2000) discovered that cortical inhibition produces less increase in regional cerebral blood flow (rCBF) than facilitation, and activation in neuroimaging studies is probably caused mainly by excitatory postsynaptic potentials. Inhibitory synapses are present near the soma of pyramidal cells and excitatory synapses are at an apex of the dendrites (Peters et al., 1991); one inhibitory synapse is present for every five to six excitatory synapses (Beaulieu and Colonnier, 1985). This fact suggests that inhibitory synapses work more efficiently with less energy demand than excitatory synapses. Hence, the ipsilateral SM1 that may inhibit the hand of the opposite side (not engaged in the motor task) escapes detection with rCBF measurement, although EEG or MEG can detect the inhibitory activity similar to excitatory activity. Since deeply located inhibitory synapses produce surface-negative field potentials similar to superficially located excitatory synapses (Niedermeyer and Lopes da Silva, 1993), a surface-negative field potential can be produced by excitatory or inhibitory synapses.

MRCP associated with contralateral and ipsilateral finger movement from subdural electrodes

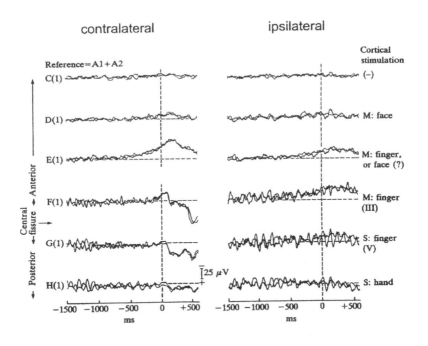

Figure 6. MRCP during voluntary middle finger extension is recorded from subdural electrodes placed over the left pre- and post-central gyri at the level of the hand sensorimotor area. The BP is present similarly for movement of either side. In contrast, NS' and MP are absent for ipsilateral movement. Left: contralateral recording to the movement. Right: ipsilateral recording to the movement. Electrodes are placed anteroposterior row crossing the left central fissure, as illustrated on the left side of the figure. Motor (M) and sensory (S) response, and no response (-) by cortical simulation are labeled on the right side of the figure. From Neshige et al. (1988) with permission of Oxford University Press.

The study by Tarkka (1994) showed no ipsilateral source; several studies demonstrated indeterminate source location (Cheyne and Weinberg, 1989) or weak source strength (Kristeva et al., 1991; Nagamine et al., 1994, 1996) on the ipsilateral side. Compared with the contralateral side, late components of the MRCP, NS' and MP are absent or of much smaller amplitude from the ipsilateral SM1 in subdural recording (Neshige et al., 1988) (figure 6). It can be speculated that failure in modeling the ipsilateral dipole is a result of weak and unclear dipolar voltage potentials of the MRCP on the ipsilateral side (Cheyne and Weinberg, 1989).

MULTIMODAL APPROACH: COMBINATION OF ELECTROMAGNETIC AND NEUROIMAGING TECHNIQUES

Comparison of the localization accuracy between different methodologies (EEG, MEG, PET and fMRI) is of great interest for functional imaging to validate the localization reliability of each technique. Once electromagnetic (EEG, MEG) data are coregistered with neuroimaging (PET, fMRI) data, the relation in space between current sources and PET or fMRI activation areas can be estimated. Walter et al. (1992) conducted a pioneering study to investigate intermodal matching of electromagnetic and neuroimaging measurements; PET and MEG were recorded during right foot, finger and mouth movements in a single subject. The results showed that location of the MEG sources follows the somatotopic organization of the PET activation. So far, however, very few investigations have been carried out in which EEG/MEG and PET/fMRI measurements are combined. It was reported that dipole sources for the MRCP differed less than 20 mm in location from the hand area showing an increase in rCBF measured by H_2O 15 PET (Wang at al., 1994) or fMRI (Beisteiner et al., 1995). Anatomical congruence of electromagnetic and metabolic activation signals was also demonstrated by a combined study of MEG and FDG-PET (Joliot et al., 1998). Their results showed that the distance between the center of the PET-activated mass and the MF and MEFI sources for the right finger movement was about 1 to 1.5 cm, within range of the error caused by the instruments, coregistration and modeling
(figure 7).

Combination of EEG/MEG and PET/fMRI information is useful to constrain solution for the "inverse problem" with anatomical and physiological knowledge (Sherg and Berg, 1991; Heinze et al., 1994; MacKinnon et al., 1996). Since a localization difference of about 1 cm has been shown between electromagnetic and neuroimaging techniques, it is reasonable to constrain dipole location at the activated spots revealed by neuroimaging measurements. In addition to the number and location of dipole sources, dipole orientation may also be constrained perpendicular to the activated cortical mantles as evident in PET or fMRI. However, the actual activated region is not a tiny spot as modeled in dipole source analysis but has some size. Spreading of activation influences the dipole localization (Hari et al., 1991). Also, dipole orientation can be affected by the spread of activation as well as complex configuration of the activated cortices.

Conventional neuroimaging techniques such as PET or block-designed fMRI presuppose a steady-state change in rCBF during repetitive task execution (e.g., two movements per one second) over substantially long period of time (e.g., 30 s). With these techniques, it is impossible to observe an increase in rCBF caused by single movement. In contrast, the recently developed fMRI technique with echoplanar imaging enables us to map the second-by-second time course of the hemodynamic response in association with an event of short duration; this is termed event-related fMRI (Buckner et al., 1996; Rosen et al., 1998). Event-related fMRI allows us to see a change in rCBF associated with a single movement (Toma et al., 1999) and employ the same tasks as used in the electromagnetic measurements; for instance, self-paced voluntary movement about every 15-20 s can be used for both the event-related fMRI and EEG/MEG acquisition. In addition, since location of the activated areas and time course of activation are variable for different subjects, coregistration of electromagnetic data with neuroimaging data on an individual basis should be required to obtain accurate dipole localization.

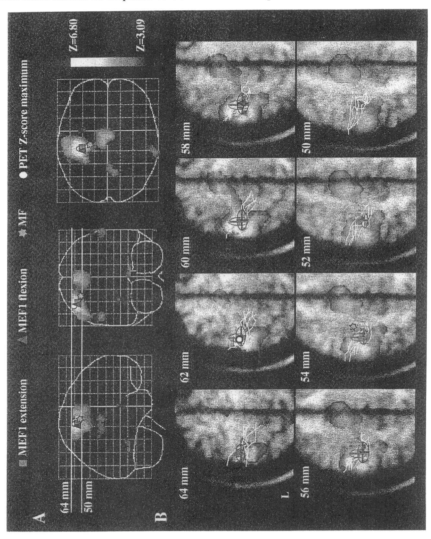

Figure 7. Coregistration of intersubject averaged PET, MEG and anatomical MRI. MEG dipoles for the MF and MEFI are localized within PET-activated regions in the SM1. A. SPM projection display of the PET activated areas with MEFI and MF locations associated with tapping of the right index finger. B. Overlay of the PET Z-score map and MEG components onto the averaged anatomic MRI across 5 subjects. Maximal PET activation is observed at 62 mm of the axial slice, MF source at 54 mm and MEFI extension/flexion sources at 58 mm. Cross surrounded by an ellipsoid indicates two times the source location standard error. From Joliot et al. (1998) with permission of Academic Press.

CONCLUSION

It is difficult to identify multiple generators within the brain from the MRCP measured at the scalp. However, attempts to localize generator sources using dipole-modeling techniques have revealed that the bilateral SM1 and MFC, including the SMA, are likely the major contributors to the MRCP. During the BP, the crown of the precentral gyrus (area 6) seems to be mainly active; the MFC may also participate to some extent. It is likely that the anterior bank (area 4) of the central sulcus generates the NS', while the posterior bank (area 3) generates the fpMP. To avoid as much as possible arbitrary processes inherent to the "inverse problem" during dipole modeling, constraint of multiple parameters by neuroimaging techniques seems to be useful. The number of dipoles and their precise location may be determined from the activated spots revealed by neuroimaging study. In addition, orientation constraint of the dipoles orthogonal to the activated cortical mantle may be reasonable, since cortical projection neurons are arrayed perpendicular to the cortex. Event-related fMRI will allow us to employ the same or similar tasks as those used in EEG recording on an individual basis. It is expected that precise knowledge of the location and timing of the motor cortical activation in healthy humans will be a valuable foundation to elucidate of the pathophysiology of various movement disorders.

REFERENCES

Arezzo, J. and Vaughan, H.G. (1980) Intracortical sources and surface topography of the motor potential and somatosensory evoked potential in the monkey. In: Kornhuber, H.H. and Deecke, L., (Eds.) Motivation, Motor and Sensory Processes of the Brain: Electrical Potentials, Behavior and Clinical Use. Progr. Brain Res., Vol. 54, p.p. 77-83, Elsevier, Amsterdam.

Beaulieu, C. and Colonnier, M. (1985) A laminar analysis of the number of round-asymmetrical and flat-symmetrical synapses on spines, dendritic trunks, and cell bodies in area 17 of the cat. J. Comp. Neurol. 23, 180-189.

Beisteiner, R., Gomiscek, G., Erdler, M., Teichtmeister, C., Moser, E. and Deecke, L. (1995) Comparing localization of conventional functional magnetic resonance imaging and magnetoencephalography. Eur. J. Neurosci. 7, 1221-1224.

Böcker, K.B., Brunia, C.H. and Cluitmans, P.J. (1994) A spatio-temporal dipole model of the readiness potential in humans. I. Finger movement. Electroenceph. Clin. Neurophysiol. 91, 275-285.

Bötzel, K., Plendl, H., Paulus, W. and Scherg, M. (1993) Bereitschaftspotential: is there a contribution of the supplementary motor area? Electroenceph. Clin. Neurophysiol. 89, 187-196.

Buckner, R.L., Bandettini, P.A., O'Craven, K.M., Savoy, R.L., Petersen, S.E., Raichle, M.E. and Rosen, B. (1996) Detection of cortical activation during averaged single trials of a cognitive task using functional magnetic resonance imaging. Proc. Natl. Acad. Sci. U.S.A. 93, 14878-14883.

Cheyne, D. and Weinberg, H. (1989) Neuromagnetic fields accompanying unilateral finger movements: pre-movement and movement-evoked fields. Exp. Brain Res. 78, 604-612.

Cheyne, D., Kristeva, R. and Deecke, L. (1991) Homuncular organization of human motor cortex as indicated by neuromagnetic recordings. Neurosci. Lett. 122, 17-20.

Cohen, D., Cuffin, B.N., Yunokuchi, K., Maniewski, R., Purcell, C., Cosgrove, G.R., Ives, J., Kennedy, J.G. and Schomer, D.L. (1990) MEG versus EEG localization test using implanted sources in the human brain. Ann. Neurol. 28, 811-817.

Colebatch, J.G., Deiber, M.P., Passingham, R.E., Friston, K.J. and Frackowiak, R.S. (1991) Regional cerebral blood flow during voluntary arm and hand movements in human subjects. J. Neurophysiol. 65, 1392-1401.

Deecke, L. and Kornhuber, H.H. (1978) An electrical sign of participation of the mesial 'supplementary' motor cortex in human voluntary finger movement. Brain Res. 159, 473-476.

Deecke, L., Weinberg, H. and Brickett, P. (1982) Magnetic fields of the human brain accompanying voluntary movement: Bereitschaftsmagnetfeld. Exp. Brain Res. 48, 144-148.

Deecke, L. (1987) Bereitschaftspotential as an indicator of movement preparation in supplementary motor area and motor cortex. In: Porter, R., (Ed.) Motor Areas of the Cerebral Cortex. p.p. 231-245. Wiley, Chichester.

Deecke, L. and Lang, W. (1996) Generation of movement-related potentials and fields in the supplementary sensorimotor area and the primary motor area. In: Lüders, H.O., (Ed.) Supplementary Sensorimotor Area. Advances in Neurology, Vol. 70, p.p. 127-146, Lippincott-Raven, New York.

Deecke, L., Lang, W., Uhl, F., Beisteiner, R., Lindinger, G. and Cui, R.Q. (1999) Movement-related potentials and magnetic fields: new evidence for SMA activation leading MI activation prior to voluntary movement. In: Comi, G., Lücking, C.H., Kimura, J. and Rossini, P.M., (Eds.) Clinical Neurophysiology: From Receptors to Perception. Electroenceph. Clin. Neurophysiol. Suppl. 50, 386-401.

Dum, R.P. and Strick, P.L. (1992) Medial wall motor areas and skeletomotor control. Curr. Opin. Neurobiol. 2, 836-839.

Erdler, M., Beisteiner, R., Mayer, D., Kaindl, T., Edward, V., Windischberger, C., Lindinger, G. and Deecke, L. (2000) Supplementary motor area activation preceding voluntary movement is detectable with a whole scalp magnetoencephalography system. NeuroImage 11, 697-707.

Fender, D.H. (1987) Source localization of brain electrical activity. In: Gevins, A.S. and Remond A., (Eds.) Methods of Analysis of Brain Electrical and Magnetic Signals. EEG Handbook, Revised Ser. Vol.1, p.p. 355-403, Elsevier, Amsterdam.

Fender, D.H. (1991) Models of the human brain and the surrounding media: their influence on the reliability of source localization. J. Clin. Neurophysiol. 8, 381-390.

Friston, K.J., Holmes, A.P., Worsley, K.J., Poline, J-P., Frith, C.D. and Frackowiak, R.S.J. (1995) Statistical parametric maps in functional imaging: a general linear approach. Hum. Brain Mapp. 2, 189-210.

Gerloff, C., Grodd, W., Altenmüller, E., Kolb, R., Nägele, T., Klose, U., Voigt, K. and Dichgans, J. (1996) Coregistration of EEG and fMRI in a simple motor task. Hum. Brain Mapp. 4,199-209.

Gerloff, C., Toro, C., Uenishi, N., Cohen, L.G., Leocani, L. and Hallett, M. (1997) Steady-state movement-related cortical potentials: a new approach to assessing cortical activity associated with fast repetitive finger movements. Electroenceph. Clin. Neurophysiol. 102, 106-113.

Gerloff, C., Uenishi, N. and Hallett, M. (1998a) Cortical activation during fast repetitive finger movements in humans: dipole sources of steady-state movement-related cortical potentials. J. Clin. Neurophysiol. 15, 502-513.

Gerloff, C., Uenishi, N., Nagamine, T., Kunieda, T., Hallett, M. and Shibasaki, H. (1998b) Cortical activation during fast repetitive finger movements in humans: steady-state movement-related magnetic fields and their cortical generators. Electroenceph. Clin. Neurophysiol. 109, 444-453.

Geyer, S., Ledberg, A., Schleicher, A., Kinomura, S., Schormann, T., Burgel, U., Klingberg, T., Larsson, J., Zilles, K. and Roland, P.E. (1996) Two different areas within the primary motor cortex of man. Nature 382, 805-807.

Hari, R., Antervo, A., Katila T., Poutanen T., Seppänen M., Tuomisto, T. and Varpula, T. (1983) Cerebral magnetic fields associated with voluntary limb movements. Nuova Cimento 2D, 484-494.

Hari, R. (1991) On brain's magnetic responses to sensory stimuli. J. Clin. Neurophysiol. 8, 157-169.

Heinze, H.J., Mangun, G.R., Burchert, W., Hinrichs, H., Scholz, M., Munte, T.F., Gos, A., Scherg, M., Johannes, S., Hundeshagen, H., Gazzaniga, M.S. and Hillyard, S.A. (1994) Combined spatial and temporal imaging of brain activity during visual selective attention in humans. Nature 372, 543-546.

Herscovitch, P. (1994) Radiotracer techniques for functional neuroimaging with positron emission tomography. In: Thatcher, R.W., Hallett, M., Zeffiro T., John, E.R. and Huerta, M., (Eds.) Functional Neuroimaging: Technical Foundations. p.p. 29-46, Academic Press, Orlando.

Hyland, B., Chen, D.F., Maier, V., Palmeri, A. and Wiesendanger, M. (1989) What is the role of the supplementary motor area in movement initiation? Prog. Brain Res. 80, 431-436.

Ikeda, A., Lüders, H.O., Burgess, R.C. and Shibasaki, H. (1992) Movement-related potentials recorded from supplementary motor area and primary motor area. Role of supplementary motor area in voluntary movements. Brain 115, 1017-1043.

Joliot, M., Crivello, F., Badier, J.M., Diallo, B., Tzourio, N. and Mazoyer, B. (1998) Anatomical congruence of metabolic and electromagnetic activation signals during a self-paced motor task: a combined PET-MEG study. NeuroImage 7, 337-351.

Kawashima, R., Itoh, H., Ono, S., Satoh, K., Furumoto, S., Gotoh, R., Koyama, M., Yoshioka, S., Takahashi, T., Yanagisawa, T. and Fukuda, H. (1995) Activity in the human primary motor cortex related to arm and finger movements. NeuroReport 6, 238-240.

Kornhuber, H.H. and Deecke, L. (1965) Hirnpotentialänderungen bei willkürbewegungen und passiven bewegungen des menschen: bereitschaftspotential und reafferente potentiale. Pflügers. Arch. Ges. Physiol. 284, 1-17.

Kristeva, R., Cheyne, D. and Deecke L. (1991) Neuromagnetic fields accompanying unilateral and bilateral voluntary movements: topography and analysis of cortical sources. Electroenceph. Clin. Neurophysiol. 81, 284-298.

Kristeva, R., Rossi, S., Feige, B., Mergner, T., Lucking, C.H. and Rossini, P.M. (1997) The bereitschaftspotential paradigm in investigating voluntary movement organization in humans using magnetoencephalography (MEG). Brain Res. Brain Res. Protoc. 1, 13-22.

Lang, W., Cheyne, D., Kristeva, R., Beisteiner, R., Lindinger, G. and Deecke, L. (1991) Three-dimensional localization of SMA activity preceding voluntary movement. A study of electric and magnetic fields in a patient with infarction of the right supplementary motor area. Exp. Brain Res. 87, 688-695.

Lim, S.H., Dinner, D.S., Pillay, P.K., Lüders, H., Morris, H.H., Klem, G., Wyllie, E. and Awad, I.A. (1994) Functional anatomy of the human supplementary sensorimotor area: results of extraoperative electrical stimulation. Electroenceph. Clin. Neurophysiol. 91, 179-193.

Luppino, G., Matelli, M., Camarda, R.M., Gallese, V. and Rizzolatti, G. (1991) Multiple representations of body movements in mesial area 6 and the adjacent cingulate cortex: an intracortical microstimulation study in the macaque monkey. J. Comp. Neurol. 311, 463-482.

Luppino, G., Matelli, M., Camarda, R. and Rizzolatti, G. (1994) Corticospinal projections from mesial frontal and cingulate areas in the monkey. Neuroreport 5, 2545-2548.

MacKinnon, C.D., Kapur, S., Hussey, D., Verrier, M.C., Houle, S. and Tatton, W.G. (1996) Contributions of the mesial frontal cortex to the premovement potentials associated with intermittent hand movements in humans. Hum. Brain Mapp. 4, 1-22.

Nagamine, T., Toro, C., Balish, M., Deuschl, G., Wang, B., Sato, S., Shibasaki, H. and Hallett, M. (1994) Cortical magnetic and electric fields associated with voluntary finger movements. Brain Topogr. 6, 175-183.

Nagamine, T., Kajola, M., Salmelin, R., Shibasaki, H. and Hari, R. (1996) Movement-related slow cortical magnetic fields and changes of spontaneous MEG- and EEG-brain rhythms. Electroenceph. Clin. Neurophysiol. 99, 274-286.

Neshige, R., Lüders, H. and Shibasaki, H. (1988) Recording of movement-related potentials from scalp and cortex in man. Brain 111, 719-736.

Niedermeyer, E. and Lopes da Silva, F. (1993) Electroencephalography. Basic Principles, Clinical Applications, and Related Fields. William and Wilkins, Baltimore.

Nunez, P. (1981) Electric Fields of the Brain. Oxford University Press, New York.

Orgogozo, J.M. and Larsen, B. (1979) Activation of the supplementary motor area during voluntary movement in man suggests it works as a supramotor area. Science 206, 847-850.

Peters, A., Oalay, S. and Webster, H. (1991) The fine structure of the nervous system. Neurons and their supporting cells. Oxford University Press, New York.

Picard, N. and Strick, P.L. (1996) Motor areas of the medial wall: a review of their location and functional activation. Cereb. Cortex 6, 342-353.

Praamstra, P., Stegeman, D.F., Horstink, M.W. and Cools, A.R. (1996) Dipole source analysis suggests selective modulation of the supplementary motor area contribution to the readiness potential. Electroenceph. Clin. Neurophysiol. 98, 468-477.

Rao, S.M., Binder, J.R., Bandettini, P.A., Hammeke, T.A., Yetkin, F.Z., Jesmanowicz, A., Lisk, L.M.,

Morris, G.L., Mueller, W.M. and Estkowski, L.D. (1993) Functional magnetic resonance imaging of complex human movements. Neurology 43, 2311-2318.

Rektor, I., Feve, A., Buser, P., Bathien, N. and Lamarche, M. (1994) Intracerebral recording of movement related readiness potentials: an exploration in epileptic patients. Electroenceph. Clin. Neurophysiol. 90, 273-283.

Romani, G.L. and Rossini, P. (1988) Neuromagnetic functional localization: principles, state of the art, and perspectives. Brain Topogr. 1, 5-21.

Rosen, B.R., Buckner, R.L. and Dale, A.M. (1998) Event-related functional MRI: past, present, and future. Proc. Natl. Acad. Sci. U.S.A. 95: 773-80.

Roth, B.J., Balish, M., Gorbach, A. and Sato, S. (1993) How well does a three-sphere model predict positions of dipoles in a realistically shaped head? Electroenceph. Clin. Neurophysiol. 87, 175-184.

Roth, B.J., Ko, D., Von Albertini-Carletti, I.R., Scaffidi, D. and Sato, S. (1997) Dipole localization in patients with epilepsy using the realistically shaped head model. Electroenceph. Clin. Neurophysiol. 102, 159-166.

Salmelin, R., Forss, N., Knuutila, J. and Hari, R. (1995) Bilateral activation of the human somatomotor cortex by distal hand movements. Electroenceph. Clin. Neurophysiol. 95, 444-452.

Sarvas, J. (1987) Basic mathematical and electromagnetic concepts of the biomagnetic inverse problem. Phys. Med. Biol. 32, 11-22.

Scherg, M. (1989) Fundamentals of dipole source potential analysis. In: Grandori, F., Hoke, M. and

Romani G.L., (Eds.) Auditory evoked Magnetic Fields and Potentials. Advances in Audiology, Vol.6, p.p. 40-69, Karger, Basel.

Scherg, M., Vajsar, J.J. and Picton, T.W. (1989) A source analysis of the human auditory evoked potentials. J. Cog. Neurosci. 1, 336-355.

Scherg, M. and Berg, P. (1991) Use of prior knowledge in brain electromagnetic source analysis. Brain Topogr. 4, 143-150.

Shibasaki, H., Barrett, G., Halliday, E. and Halliday, A.M. (1980) Components of the movement-related cortical potential and their scalp topography. Electroenceph. Clin. Neurophysiol. 49, 213-226.

Shibasaki, H., Sadato, N., Lyshkow, H., Yonekura, Y., Honda, M., Nagamine, T., Suwazono, S., Magata, Y., Ikeda, A., Miyazaki, M., Fukuyama, H., Asato, R. and Konishi, J. (1993) Both primary motor cortex and supplementary motor area play an important role in complex finger movement. Brain 116, 1387-1398.

Shima, K., Aya, K., Mushiake, H., Inase, M., Aizawa, H. and Tanji, J. (1991) Two movement-related foci 18.in the primate cingulate cortex observed in signal-triggered and self-paced forelimb movements. J. Neurophysiol. 65, 188-202.

Strick, P.L. and Preston, J.B. (1982a) Two representations of the hand in area 4 of a primate. I. Motor output organization. J. Neurophysiol. 48, 139-149.

Strick, P.L. and Preston, J.B. (1982b) Two representations of the hand in area 4 of a primate. II. Somatosensory input organization. J. Neurophysiol. 48, 150-159.

Tarkka, I.M. and Hallett, M. (1991) Topography of scalp-recorded motor potentials in human finger movements. J. Clin. Neurophysiol. 8, 331-341.

Tarkka, I.M. (1994) Electrical source localization of human movement-related cortical potentials. Int. J. Psychophysiol. 16, 81-88.

Talairach, J. and Tournoux, P. (1988) Co-planar Stereotaxic Atlas of the Human Brain. Thieme, Stuttgart.

Toma, K., Honda, M., Hanakawa, T., Okada, T., Fukuyama, H., Ikeda, A., Nishizawa, S., Konishi, J. and Shibasaki, H. (1999) Activities of the primary and supplementary motor areas increase in preparation and execution of voluntary muscle relaxation: an event-related fMRI study. J. Neurosci. 19, 3527-3534.

Toro, C., Matsumoto, J., Deuschl, G., Roth, B.J. and Hallett, M. (1993) Source analysis of scalp-recorded movement-related electrical potentials. Electroenceph. Clin. Neurophysiol. 86, 167-175.

Toro, C., Wang, B., Zeffiro, T., Thatcher, R.W. and Hallett, M. (1994) Movement-related cortical potentials: source analysis and PET/MRI correlation. In: Thatcher, R.W., Hallett, M., Zeffiro, T., John, E.R. and Huerta, M., (Eds.) Functional Neuroimaging: Technical Foundations. pp. 259-267. Academic Press, Orlando.

Van Oosterom, A. (1991) History and evolution of methods for solving the inverse problem. J. Clin. Neurophysiol. 8, 371-380.

Vaughan, H.G., Costa, L.D. and Ritter, W. (1968) Topography of the human motor potential. Electroenceph. Clin. Neurophysiol. 25, 1-10.

Waldvogel, D., van Gelderen, P., Muellbacher, W., Ziemann, U., Immisch, I. and Hallett, M. (2000) The relative metabolic demand of inhibition and excitation. Nature 406, 995-998.

Walter, H., Kristeva, R., Knorr, U., Schlaug, G., Huang, Y., Steinmetz, H., Nebeling, B., Herzog, H. and Seitz, R.J. (1992) Individual somatotopy of primary sensorimotor cortex revealed by intermodal matching of MEG, PET, and MRI. Brain Topogr. 5, 183-187.

Wang, B., Toro, C., Wassermann, E.M., Zeffiro T.A., Thatcher, R.W. and Hallett, M. (1994) Multimodal integration of electrophysiological data and brain images: EEG, MEG, TMS, MRI and PET. In: Thatcher, R.W., Hallett, M., Zeffiro T., John, E.R. and Huerta, M., (Eds.) Functional Neuroimaging: Technical Foundations. p.p.251-257, Academic Press, Orlando.

Williamson, S.J. and Kaufman, L. (1981) Magnetic fields of the cerebral cortex. In: Erne, S.N., Hahlbohm, H.D. and Lübbig, H., (Eds.) Biomagnetism. p.p.353-402, Walter de Gruyter, Berlin.

SURFACE RECORDINGS IN PATIENTS WITH MOVEMENT DISORDERS AND THE IMPACT OF SUBCORTICAL SURGERY

P. Praamstra[1], M. Jahanshahi[2], J.C. Rothwell[3]

[1] Behavioral Brain Sciences Centre and Queen Elizabeth Hospital, University of Birmingham, Edgbaston, Birmingham B15 2TT
[2] Department of Clinical Neurology, Institute of Neurology, The National Hospital for Neurology & Neurosurgery, Queen square, London WC1N 3BG
[3] Sobell Department of Neurophysiology, Institute of Neurology, The National Hospital for Neurology & Neurosurgery, Queen square, London WC1N 3BG

INTRODUCTION

Studies of the Bereitschaftspotential (BP) in patients with movement disorders have been motivated as often by curiosity about the behaviour of this potential in pathological conditions as by targeted questions regarding particular aspects of a disorder. It is fair to say that this curiosity has been amply rewarded. The aim of this chapter is to provide an overview and critical evaluation of the work carried out in this area.

The BP is typically recorded during the performance of self-paced voluntary movements, often consisting of the rotation of a single joint over a few degrees of arc, repeated at a rate of just once every few seconds. Such movements, by no means, represent a serious challenge to the motor system. Similarly, the movements used by clinical investigators to elicit the BP in patients with a movement disorder are not often the movements that truly characterise the movement problems. Nevertheless, investigators have found inventive ways to accommodate research approaches to the imposed limitations, as will be discussed below.

Among the neurological movement disorders that have been investigated with the BP, Parkinson's disease is by far the most frequently studied condition. This is not only due to the fact that Parkinson's disease is a common neurological movement disorder. Several factors working together have created the circumstances that made the BP fill a role in the investigation of the parkinsonian movement disorder. Of some importance in this respect has been an influential paper on the supplementary motor area (SMA) by Goldberg (1985). On the basis of an extensive review of anatomical, physiological, and clinical research, Goldberg elaborated a distinction between lateral and medial premotor systems. Consistent with lesioning experiments and intracortical recordings in monkeys (for review, see Passingham, 1993) the distinction capitalised on different modes of movement selection, contrasting internally generated movements, supported by the medial system, with movements elicited by an external cue, organised through the lateral premotor system. Given the prominence of basal ganglia-thalamocortical projections onto the SMA and

Parkinson patients' difficulties with self-initiated movements, Parkinson's disease was referred to as the prototypical human disease affecting the operation of the medial system. Moreover, the review appropriately discussed the BP as a means of studying SMA/medial premotor function during internally generated movements.

While accumulating knowledge on the division of labour between medial and lateral premotor structures appears to have stimulated the investigation of parkinsonian motor function, focusing on the SMA, a similar direction of research was suggested by investigations on corticospinal function in Parkinson's disease, using transcranial electric stimulation of the motor cortex. Movements elicited by electrical stimulation of the motor cortex, in contrast to those initiated voluntarily, are not influenced by whether Parkinson patients are *on* or *off* dopaminergic medication (Dick et al., 1984). From this observation, Dick and co-workers inferred that the parkinsonian movement disorder probably arises upstream of the motor cortex, i.e., in areas that elaborate the movement commands that are sent to the motor cortex. Assuming that the motor cortex is primarily responsible for the execution of movement and that premotor areas have a role in preparation, the electrical stimulation results thus implicated premotor areas of the cortex in the parkinsonian movement disorder. Likewise, from a functional perspective they suggested that the parkinsonian movement disorder should be understood as arising at preparatory stages of movement organisation.

The notion that bradykinesia may be caused at the stage of movement (pre) programming developed, at the same time, from reaction time and behavioral studies of Parkinson patients. Early studies using reaction time measurements reported that while Parkinson patients show slower response times than healthy control subjects, the deficit is more pronounced in simple reaction time tasks than in choice reaction tasks (for review and discussion, see Jahanshahi et al., 1992). This specific pattern, with delayed responses in circumstances where the movement can be prepared in advance, is found in some but not all studies, and has been taken as evidence that Parkinson patients fail to preprogram the required motor response, indicating deficient motor preparation.

Together, the research into medial and lateral premotor cortex function and the emerging view that parkinsonian bradykinesia arises at a functional stage before motor execution and at an anatomical level upstream of the motor cortex created the niche for an instrument that actually provided information on the cerebral events preceding movement. A variety of studies using the BP, not limited to Parkinson's disease, have drawn upon this theoretical background.

In the selection of studies reviewed in this report we have not confined ourselves to investigations that used the BP 'proper'. That is, we also included studies that recorded movement-related brain electrical activity as expressed in the contingent negative variation (CNV) and the lateralised readiness potential (LRP).

PARKINSON'S DISEASE

Early studies

Early reports on the BP in Parkinson's disease described the investigations in terms of "lesion" experiments, asking what effect a "lesion" has on the BP (Deecke et al., 1977; Shibasaki et al., 1978a). Assuming that motor cortical areas are largely responsible for the BP, research addressed how dysfunction of subcortical motor structures such as the basal ganglia affects the preparatory state of the motor cortex, as measured with the BP (Shibasaki et al., 1978a). As suggested by Shibasaki et al., combining results from different disease conditions might clarify the physiogenesis of the BP. At the same time, this early work is characterised by a vivid notion that alterations of the BP might be related to particular features of the movement disorder. As an example, the akinesia of Parkinson's disease was suggested to be reflected in a longer delay between the onset and the peak latency of the BP, i.e., in an earlier onset relative to movement (Shibasaki et al., 1978a). Whereas this change in onset cannot be regarded as a consistent feature of BP recordings in Parkinson's disease, the amplitude reduction of the BP, noted in the same study, has since been repeatedly replicated (e.g., Simpson and Khuraibet, 1987; Dick et al., 1989; Fève et al., 1992; Jahanshahi et al., 1995).

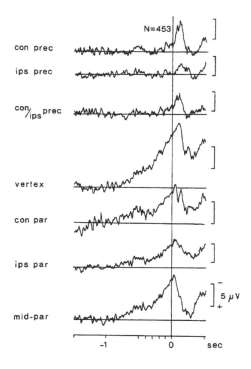

Figure 1. Bereitschaftspotential preceding voluntary finger movements in a patient with bilateral parkinsonism. Recordings from contralateral and ipsilateral precentral and parietal electrodes and the vertex, referred to linked earlobes. Note the high amplitude BP over the vertex contrasting with the low amplitude at precentral sites. (From Deecke and Kornhuber, 1978. Reproduced with permission).

Amplitude changes of the BP in Parkinson's disease have been extensively described by Deecke, Kornhuber, and co-workers (Deecke et al., 1977; Deecke, 1985), using multi-electrode recordings to document the distribution of amplitude reductions and help interpret the changes. In bilateral Parkinson's disease they found a significantly reduced amplitude of the BP with the slow negative shift of the BP being replaced with a BP of positive polarity in a number of patients (Deecke et al., 1977). In a group of 22 patients with pronounced asymmetry of motor symptoms they found a unilateral reduction of the BP over the affected hemisphere and a bilateral reduction when movements of the akinetic side were compared with the less affected side. Similar changes were found in the analyses of the Motor Potential in the same group of patients. Deecke and Kornhuber (1978) reported a relatively preserved BP at midline electrodes compared to attenuated BP amplitudes at electrodes overlying the motor cortex, suggesting that motor cortex function in Parkinson's disease is more severely affected than supplementary motor and cingulate areas (see Figure 1). In view of the later emphasis on deficient SMA function in Parkinson's disease (e.g., Dick et al., 1989; Jahanshahi et al., 1995), the results of this report are difficult to explain.

Recording of the BP while continuing dopaminergic medication and presence of dyskinesia in the selected patients are factors to consider in this respect (cf. Rascol et al., 1998).

Barrett, Shibasaki, and Neshige (1986) cautioned that previously reported abnormalities of the BP in Parkinson's disease might be due to inadequate methodology, in particular a faulty time-locking of selected EEG epochs to movement onsets. Applying a trial-by-trial visual onset detection of EMG to recordings obtained from 6 patients, they concluded that cortical potential shifts preceding voluntary movement are normal in Parkinson's disease. In an attempt to explain the discrepancy with earlier work, they also pointed out that age differences between control subjects and patients may have biased previous reports. With hindsight, it seems not implausible that latency effects on the BP, i.e., the earlier onset of the BP reported by Shibasaki et al (1978a), are indeed related to sub-optimal methodology. Amplitude reductions of the BP in Parkinson's disease, however, are not likely explained by inadequate timelocking of EEG to movement or EMG onset, as demonstrated in simulations by Dick et al. (1987). A reduced amplitude of the BP in Parkinson's disease is now well-established (e.g., Simpson and Khuraibet, 1987; Dick et al., 1989; Fève et al., 1992; Jahanshahi et al., 1995), although there are rare observations of an enhanced amplitude (Fattapposta et al., 2000).

Medication effects and disease severity

Investigations of drug effects on the BP in Parkinson's disease have partly been motivated by the large inter-individual variability of the BP. Comparing the same patients *on* and *off* medication has provided another approach to establishing a relation between motor deficits, dopaminergic status, and motor cortex physiology (Dick et al., 1987). Thus, Dick and co-workers found that patients had a significantly higher amplitude of the early BP when *on* medication compared with recordings made *off* medication. They also found that L-dopa and dopaminergic antagonists, given to healthy volunteers, specifically affected the early BP. These findings further supported the differences that they observed between Parkinson's disease patients and control subjects, which were similarly confined to the early BP phase.

Fève et al. (1992) investigated the effect of L-dopa administration on the BP in *de novo* patients, with measurements taken before the start of medication and after 3 months of treatment. Importantly, the early BP was absent in 4 out of 9 patients before the start of treatment, and reappeared in 3 after treatment. In the group statistics, however, the early BP component showed no significant change. In contrast, the slope of the late component increased significantly. In the same study, a group of patients with response fluctuations were measured whilst they were in an *off* state and while in an *on* state. This comparison showed dramatic amplitude differences, as documented in illustrations of what appear very careful recordings (see Figure 2). Similar to the effect of L-dopa on *de novo* patients, the BP increase in *on* compared to *off* states was limited to the late phase of the BP. In spite of this similarity, the authors proposed that the data suggest two kinds of BP modification. In *de novo* patients, L-dopa partially restores nigrostriatal function with consequent improvement of SMA function, as reflected in the early BP. In fluctuating patients, L-dopa possibly repletes dopamine at extrastriatal sites, such as limbic and cortical structures, reflected in a change of the lateralised late BP component.

A phenomenon that has not received much attention, is the alteration of movement-related activity occurring after movement. Dick et al. (1987) noted a consistently smaller post-movement positivity in healthy volunteers after L-dopa administration, referring to a sustained negativity in an epoch of 1000 ms after movement. A similar effect, i.e., a delayed return of the BP trace to baseline after the MP peak, was found by these authors in the comparison of Parkinson's disease patients with age-matched controls. Particularly robust examples of the phenomenon can be found illustrated in the study by Fève and co-workers (see Figure 2). The sustained negativity bears a resemblance to the post-imperative negative variation (PINV) observed as an effect of dopaminergic treatment of Parkinson's disease patients evaluated in a CNV paradigm (Amabile et al., 1986). These authors tentatively explained the PINV as the result of dopamine receptor hypersensitivity. Although the exact nature of the BP variant of this phenomenon is not yet clear, it is very likely related to L-dopa administration.

Whereas most of the BP studies in Parkinson's disease involved groups of not more than 10-15 patients, Simpson and Khuraibet (1987) reported on recordings performed on 55 patients. Even in this large group, however, they could not find evidence that abnormalities of the BP are related to the duration of clinical manifestations. By contrast, disease severity, as graded by the Webster Scale, correlated with a delayed onset of the BP and a reduced amplitude.

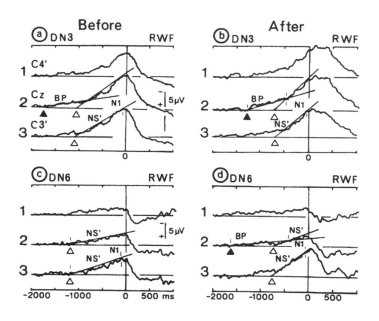

Figure 2. Bereitschaftspotential recordings before and after three month of L-dopa treatment in two *de novo* Parkinson patients. Changes in the slope of early BP and late BP (NS') are emphasised by the regression lines drawn in the figure. (From Fève et al., 1992. Reproduced with permission).

Basal ganglia-thalamocortical circuits and the SMA

While some studies recording the BP in Parkinson's disease have been of a largely explorative nature, others have been based on explicit assumptions regarding the neural generators of the BP. Dick et al. (1989) emphasised evidence supporting that a major portion of pallidal output is directed to nonprimary motor areas of the frontal lobe, in particular the SMA (Schell and Strick, 1984). Noting similarities between the effects of SMA lesions and the difficulties of Parkinson's disease patients in carrying out sequential movements, their study was undertaken to test the hypothesis of SMA dysfunction in Parkinson's disease, considered to be expressed as a reduced amplitude of the early BP. Consistent with their earlier report (Dick et al. 1987), measurements in a group of 14 Parkinson's disease patients, investigated after overnight withdrawal from medication, demonstrated a lower amplitude of the early BP, especially at electrode sites over the midline and the ipsilateral motor cortex. In contrast, at anterior electrode sites the late BP was of slightly higher amplitude in patients. Together, these results were interpreted as suggesting an impairment of SMA function in Parkinson's disease and, more speculatively, compensatory overactivity of the lateral premotor system (cf. Goldberg 1985).

Two different studies by Touge et al. (1995) and by Praamstra et al. (1996a) investigated SMA function in Parkinson's disease using an approach borrowed from earlier PET studies (Deiber et al.,1991; Playford et al., 1992). The approach contrasted different modes of movement selection using joy-stick movements. In one condition subjects moved a joy-stick in directions that were freely selected by the subject in every trial. This condition was compared with joy-stick movements for which the direction was fixed. In the imaging studies, a difference between Parkinson's disease patients and controls was especially apparent for the free selection condition, with the patients showing impaired activation of putamen, anterior cingulate, dorsolateral prefrontal cortex and SMA (Playford et al., 1992). The studies of Touge et al. (1995) and Praamstra et al. (1996a) adapted the task for BP recordings in more or less similar ways and both found higher BP amplitudes for freely selected than for fixed movements in healthy controls, but not in Parkinson's disease patients. It is noteworthy that in both studies the BP for freely selected movements diverged from the trace for fixed movements at a relatively late point in the time course of the BP. Praamstra et al. (1996a) acknowledged that, if reflecting SMA activity, one might have expected the "selection effect" to have occurred in the early phase of the BP rather than the late phase. Intracranial recordings of the BP, however, have found that the SMA is active during the entire time course of the BP (Ikeda et al., 1992). Hence, a late "selection effect" is not incompatible with it being generated by the SMA. Indeed, a dipole source analysis of the "selection effect" provided support for this effect having an origin in frontal midline structures (Praamstra et al., 1996b), consistent with the evidence from PET studies.

Adopting a comparable approach, Vidailhet et al. (1993) investigated the BP prior to gait initiation in patients with Parkinson's disease. Comparing simple foot movements, performed while sitting, with a stepping movement while standing, they found a much higher amplitude for the latter condition in control subjects. In Parkinson's disease patients this modulation was strongly attenuated, and was attributed to deficient SMA activation (see Figure 3). In isolated gait initiation failure, on the other hand, a normal increase of BP amplitude was found with stepping (Vidailhet et al., 1995).

As noted in the Introduction, scientific interest in the investigation of movement preparation and SMA function in Parkinson's disease has been stimulated by a specific pattern of findings in reaction time experiments. In some studies, Parkinson patients have demonstrated a more impaired performance in simple than in choice reaction time tasks, suggesting that they fail to take advantage of the possibility to preprogram responses in a simple reaction time task (cf. Jahanshahi et al., 1992). Filipovic et al. (1997) calculated the reaction time difference between simple and choice tasks (index of pre-programming in simple reaction time task) and found a correlation with the amplitude of the BP, which was recorded in a separate session. Thus, in patients where the choice vs. simple reaction time difference was small, the BP amplitude was generally lower than in patients who showed a faster reaction time in simple compared to choice tasks. Bötzel et al. (1995) also tested Parkinson patients in simple and choice tasks, but combined these with CNV recordings. Consistent with the pattern referred to above, Parkinson patients differed from elderly controls only in the simple, but not the choice reaction time task. The CNV, however, failed to differentiate between groups as well as between tasks.

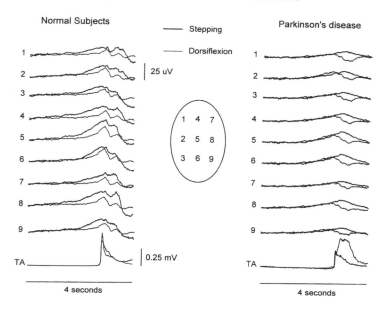

Figure 3. Bereitschaftspotentials preceding voluntary stepping (thick traces) and preceding dorsiflexion movements of the foot while seated (thin traces). Left panel shows recording in normal subjects; right panel illustrates group averaged results for Parkinson patients. The premovement modulation due to different types of movement is absent in Parkinson's disease. Rectified EMG from the tibialis ant. muscle. (From Vidailhet et al., 1993. Reproduced with permission).

Cunnington et al. (1997) used an approach to probe SMA function that circumvented explicit assumptions on the time course of SMA and primary motor cortex contributions to the BP. In their approach, BP components related to movement preparation and execution were examined separately by comparing BPs associated with imagined and actual movements. In one condition of a sequential button press task, subjects pressed buttons arranged in a 2 x 10 array. The actual path to follow through the array was indicated by lights underneath the buttons and the time to move to the next button was signalled by the lights. The BPs collected in this active movement condition were compared to the signals obtained in a condition where subjects imagined themselves moving their fingers along the path illuminated by the buttons, and a condition in which they just watched the sequentially illuminated buttons. To isolate the BP component relating only to movement preparation, the signals from the watching-cues condition were subtracted from those for the imagined-movement condition. Similarly, to separate the BP component for movement execution alone, signals for the imagined-movement condition were subtracted from the signals for the performed-movement condition. In brief, the results showed differences between Parkinson patients and control subjects only in the movement preparation component and not in the movement execution-related component of the BP. Assuming that the execution-related component comes mainly from primary motor cortex and that the movement preparation component involves SMA activation, the results were interpreted as generally consistent with previous studies reporting electrophysiological (Dick et al., 1989

Jahanshahi et al., 1995) or imaging (Playford et al., 1992; Jahanshahi et al., 1995) evidence for SMA dysfunction in Parkinson's disease.

One particular aspect of the data led Cunnington et al. (1997) to infer that impaired basal ganglia output to the SMA leads to a deficit in the termination of pre-movement preparatory phase activity (see also Brotchie et al., 1991 and Cunnington et al., 1995). The authors analysed the post-peak slope of the BP and established that after reaching peak amplitude, the BP is slower to return to baseline in later-stage Parkinson's patients compared to early-stage Parkinson's. Cunnington and co-workers pointed out that a prolonged peak of the BP is also apparent in BP traces of earlier studies, although not explicitly reported (e.g. Deecke et al., 1977; Dick et al., 1989). However, as noted in the previous section, the same feature characterised the BPs recorded in normal subjects after taking L-dopa. The decreased post-peak slope of the BP, may therefore be related to the fact that they examined patients under dopaminergic medication, instead of indicating an intrinsic SMA deficit in terminating premovement preparatory activity.

Internally generated vs. externally triggered movements

The single most important feature of the BP that has made it attractive for investigating movement preparation in Parkinson's disease, is its particular association with internally generated movements, i.e., the type of movement that is usually considered to be most impaired in Parkinson's disease. By implication, the BP is not ideally suited to investigate the differences between internally generated and externally triggered movements, as, taken strictly, there is no BP preceding externally triggered movements. When recording event-related EEG activity preceding the presentation of a stimulus that calls for a motor response, the movement preceding negativity in the EEG is inevitably mixed with activity associated with the anticipation of the stimulus. This aspect of stimulus anticipation makes it probably more appropriate to designate the recorded EEG potentials, in such situations, as CNV-like activity than as BPs. That the choice between these labels is not arbitrary will become clear later in this section. In this section we will use "movement-related potentials" (MRP) to designate both types of potentials.

Jahanshahi and co-workers pioneered a combined MRP and PET-imaging approach to investigate patients with Parkinson's disease during the performance of self-initiated and externally triggered movements (Jahanshahi et al., 1995). The self-initiated task involved extension of the right index finger at an approximate rate of once every 3 seconds. In the externally triggered condition subjects made the same finger lifting movements in response to a tone presented at an identical rate to that generated by the subject in the self-initiated condition. A tone was also presented 100 ms after the self-initiated movements, to control for the tone effect in the triggered condition. The results showed for both Parkinson patients and controls significant differences between the EEG potentials accompanying self-initiated and externally triggered movements, i.e., higher amplitude of late BPs and peak BPs for self-initiated movements. When the groups were compared, separately for each movement type, their EEG activity preceding externally triggered movements was identical, whereas for self-initiated movements the control group had higher amplitudes for early BPs and peak BPs.

The patterns of rCBF for the two types of movement showed differences that could be mapped onto the MRP findings. Matching the EEG results, the groups showed identical patterns of rCBF for externally triggered movements relative to rest. For self-initiated movements (relative to rest), normals showed greater activation of the SMA, anterior cingulate, left putamen, left insular cortex, right dorsolateral prefrontal cortex, and right parietal cortex. The significant underactivation of the SMA in Parkinson's disease patients supported the generation of the early BP in the SMA, as it was specifically in this component that patients and controls differed in the EEG study.

While the MRP results of the study by Jahanshahi et al. agreed well with previous studies of the BP in Parkinson's disease, the extent to which the PET part of the study revealed differences between self-initiated and externally triggered movements, was surprisingly small. Hence, the notion that these two types of movement are realised by distinct functional systems, i.e., a medial and a lateral premotor system, was treated with caution by the authors. Importantly, they emphasised that the common assumption that the SMA constitutes the major cortical projection area of the putamen is in need of revision, as the areas of the thalamus that receive basal ganglia input also project to the lateral premotor cortex and the primary motor cortex (Holsapple et al., 1991; Hoover and Strick, 1993; Matelli et al., 1989).

Notwithstanding the critical discussion of the distinction between medial vs. lateral premotor systems, the MRP data of Jahanshahi and co-workers suggested a deficit especially of self-initiated movements in Parkinson's disease. Strong claims regarding the specificity of such a deficit were made by Cunnington et al. (1995), who suggested that impaired internal control mechanisms, operating via the SMA, can be bypassed in Parkinson's disease when external cues are given. Cunnington and co-workers recorded MRPs using the same "tapping board", mounted with an array of 2 x 10 response buttons, as was used in the study that we discussed in the previous section (Cunnington et al., 1997). Medicated patients with Parkinson's disease and control subjects performed several different tasks in which they moved their index finger sequentially from button to button. The path followed along the two rows of buttons, as well as the timing of individual movements from one button to the next, was either self-determined or guided by spatial and temporal cues. These cues, in turn, could be predictable or unpredictable in timing and/or position. The results showed pre-movement MRPs preceding non-cued movements, both in Parkinson patients and in controls, which were interpreted as indicating involvement of the SMA in movements that must be internally determined. In control subjects, robust MRPs were also observed preceding externally cued predictable movements, but not with temporally unpredictable movements. In Parkinson's disease patients, however, pre-movement activity was greatly reduced for externally cued predictable movements. The authors, who equated the presence of an MRP with SMA activity, concluded that where external temporal cues were present, Parkinson patients relied on these cues to guide movement, thus bypassing defective internal control mechanisms operating via the SMA. Related observations have been made by Ikeda et al. (1997), who reported a dissociation between an almost absent CNV and relatively preserved BP in late stage Parkinson's disease.

The view that external cues allow Parkinson patients to bypass defective basal ganglia-thalamocortical circuits through the SMA is not the only possible interpretation of the data presented by Cunnington et al. (1995). Note that Parkinson's disease patients generated MRPs of considerable size preceding movements not guided by any cues, but no appreciable pre-movement activity preceding temporally predictable movements. As suggested earlier by Praamstra et al. (1996c), the complete absence of premovement EEG activity prior to predictable external cues indicates that the Parkinson's disease patients did not engage in preparatory activity as such. Would they have demonstrated premovement MRPs if they had been pressed harder or instructed in a different way? The answer seems to be confirmatory, as another study by Cunnington and co-workers reproduced the flat premovement activity in Parkinson's disease patients, while showing that an explicit instruction to use the available temporal cues resulted in sizeable MRPs, albeit smaller than in controls (see Figure 4; Cunnington et al., 1999). Moreover, the emergence of MRPs was accompanied by an improved performance. Together, these studies serve as a reminder to the importance of distinguishing between the BP preceding voluntary self-initiated movements, and CNV-like activity preceding externally cued movements; the latter is sensitive to differences in strategy to a degree not known of the BP.

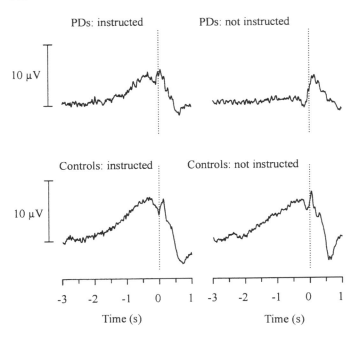

PDs: instructed   PDs: not instructed

10 μV

Controls: instructed   Controls: not instructed

10 μV

-3   -2   -1   0   1   -3   -2   -1   0   1

Time (s)   Time (s)

Figure 4. Movement-related potentials preceding predictably timed visual cues presented at Time 0 sec. Recordings from electrode Cz in Parkinson patients (upper row) and control subjects (lower row). An explicit instruction to anticipate the cue markedly changed the premovement activity in Parkinson patients. (From Cunnington et al., 1999. Reproduced with permission).

LRP studies in Parkinson's disease

The lateralised readiness potential (LRP) has greatly enriched the investigative potential of EEG, not only in the chronometric domain of reaction time studies in normals, but also in the areas of motor control and movement disorders. The first study exploring the feasability of obtaining LRPs from Parkinson's disease patients used a movement precueing task (Praamstra et al., 1996c). In this task, the subject pressed a button with the left or right index or middle finger. The reaction signal was preceded by either a non-informative precue or a precue giving the response side but not the response finger. In the 1 sec interval between precue stimulus and reaction or "go" signal, a CNV was recorded with higher amplitude for control subjects than for Parkinson's disease patients. The derivation of LRPs enabled the authors to determine whether the reduced CNV, besides reflecting deficient activation of the SMA/medial premotor structures, also indicated delayed activation of the primary motor cortex. The LRP developed in a similar fashion for control subjects and Parkinson's disease patients, starting shortly after the presentation of the precue when the precue provided information concerning the response side (see Figure 5). The preactivation of the motor cortex contralateral to the response hand, evidenced in the LRP, thus demonstrates that Parkinson's disease patients used the advance information to prepare a response. Moreover, as the onset latencies of the LRP were similar in patients and controls, it can also be inferred that the patients did not need more time than controls to process the available information and select the response side.

Wascher et al. (1997) also recorded CNV and LRP measures in movement precueing experiments, introducing two additional elements. In one experiment they varied the precue-reaction signal interval, providing information on how patients maintain preparation for a cued response over variable time intervals. In a separate experiment the cue validity was manipulated giving an opportunity to look at how subjects switch from a prepared response, indicated by the cue, to a different response demanded by the reaction signal. Response times were generally delayed in Parkinson's disease patients and showed, in addition, increased switch-costs. The CNV, as in many earlier studies, showed a reduced amplitude for Parkinson's disease patients in both tasks. As in the study by Praamstra et al. (1996), advance preparation on the basis of the cue information was reflected in the LRP. The increased switch-cost for patients seemed reflected in the LRP, but the latency of the LRP was not evaluated.

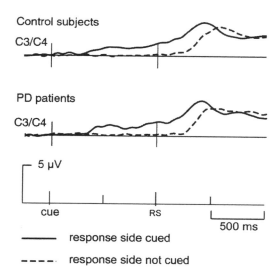

Figure 5. Lateralised readiness potentials (LRPs) recorded when the cue provided information about the response side (continuous line) and when the cue was non-informative (dashed line). Informative cues resulted in preactivation of the motor cortex contralateral to the response hand well in advance of the reaction signal. Note the similarity of the recordings in patients and control subjects. (Adapted, with permission, from Praamstra et al., 1996c).

The LRP data reviewed sofar might suggest that lateralised movement-related activity, as reflected in the LRP, is normal in Parkinson's disease. If so, this would seem to corroborate earlier BP and CNV studies claiming that Parkinson's disease especially affects the non-lateralised SMA contribution to these components and that motor cortex activation (presumably measured by the LRP) is normal in Parkinson's disease. The picture is not that simple, however. In the study by Praamstra et al. (1996c), the CNV amplitude was decreased in Parkinson's disease patients, but it was at the same time more strongly modulated by the information provided by the precue than in normals; i.e., the amplitude being higher when the precue allowed partial preparation of the response, than when the precue was uninformative. This extra activity was partly reflected in the LRP developing between precue and reaction signal, which was not of higher amplitude overall in Parkinson's disease, but which did demonstrate a more broadly distributed maximum that extended to more anterior electrode sites compared to control subjects. Two possible explanations were entertained. The altered distribution of the LRP might reflect compensatory recruitment of lateral premotor areas, a possibility recently emphasised in a PET study (Samuel et al., 1997a). Alternatively, it might be that the preparatory cortical activity in Parkinson's disease is inadequately 'focused', resulting in a broader distribution and net increase of neural activity preceding movement.

More recent LRP studies have yielded other examples of increased neural activity, recorded at electrode sites overlying the motor cortex, in patients with Parkinson's disease performing movements in reaction time tasks. The emerging picture is that the neural circuitry subserving motor responses to external stimuli is not spared in Parkinson's disease. To investigate the mechanisms underlying Parkinson's disease patients' reliance on external cues, Praamstra et al. (1998a) used a flanker task. Stimuli consisted of an array of arrows, of which the central one indicated the required response (left or right hand), while the flanking arrows were irrelevant distractors (see Figure 6a). In this task, a larger reaction time difference between compatible and incompatible conditions was obtained for Parkinson's disease patients than for healthy controls. This increased interference effect in Parkinson's disease was related to a stronger effect of incompatible flankers. More precisely, the incompatible flankers, (arrows pointing to the incorrect response side), induced stronger activation of the incorrect response, as evidenced by the LRP (see Figure 6b). Presumably, compatible flankers also had a stronger effect in Parkinson's disease, as the LRP in the compatible condition had a shorter onset latency for patients than for controls. The authors acknowledged that the results might reflect Parkinson patients' reliance on visual stimulus information for the initiation of movement. As an equally likely explanation, however, it was suggested that in Parkinson's disease impaired function of local inhibitory circuits in the motor cortex (cf. Ridding et al., 1995a) may make the motor cortex more susceptible to sensory input. According to the latter explanation, sensory neurons in the motor cortex, partaking in the process of sensory-motor translation, respond stronger, while the translation to motor output is less efficient. This explanation received some support from subsequent studies documenting an altered relation between the neural activity of the motor cortex, as reflected in the LRP, and motor output recorded by EMG or force measures, when movements are triggered by visual stimuli (Praamstra et al., 1999; Plat et al., 2000).

The line adopted by Praamstra and co-workers is based on evidence from single-cell neurophysiology that neural activity within the motor cortex is not by definition activity that has to be functionally characterised as purely motor. Presumably, the same applies to the LRP recorded by means of scalp electrodes above the motor cortex (Miller et al., 1992). While EEG does not allow a separation of the activity coming from corticospinal output neurons and activity from sensory and sensorimotor type neurons in the motor cortex (e.g., Shen and Alexander, 1997), it is not necessarily blind to the processes in the motor cortex leading up to corticospinal neuron activation. Praamstra and Plat (2001) recorded the LRP in a spatial stimulus-response (S-R) compatibility task where the task-relevant stimulus information, instructing for a left or a right hand response, was displayed randomly either to the left or to the right of a central fixation cross. The lateralised position of the stimulus required an attentional shift that elicited an attention-related ERP component just preceding the movement-related LRP, i.e. the N2pc. Praamstra and Plat observed that the N2pc extended far more anterior than could be explained on the basis of volume conduction from its presumed source in the occipital lobe. Given the N2pc distribution in their data, they inferred that the S-R compatibility task induced simultaneous attention-related activity in occipital and motor areas of the cortex (see also Oostenveld et al., 2001). Interestingly, the attention-related activity attributed to motor areas was of significantly higher amplitude in Parkinson's disease patients compared to age-matched controls. The authors speculated that this enhancement might be due to abnormal interaction of the attention-related activity with movement-related activity in the motor cortex. A spurious "ignition" of movement-related cells in the motor cortex by attention-related activity might be related to the decreased selectivity of pallidal and motor cortex neuronal discharges assumed in pathophysiological models of parkinsonian motor function (Alexander and Crutcher, 1990; see also Doudet et al., 1990; Filion et al., 1988).

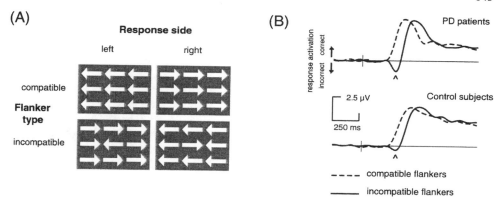

Figure 6. A. Stimuli of flanker task (Praamstra et al., 1998). In each of the four stimuli, the central arrow represented the task-relevant information, indicating whether a left or right hand response was required. The surrounding flankers influenced the response times depending on whether the indicated direction of the arrows was compatible or incompatible with the central arrow's direction. **B.** Lateralised readiness potentials for Parkinson patients and control subjects. Note the brief activation of the incorrect response hand in the incompatible condition (arrowheads). The amplitude of the incorrect activation was higher in patients than in controls, with corresponding differences in reaction times. (Reproduced with permission).

The impact of subcortical surgery in Parkinson's disease on the BP

The revival of neurosurgery for treatment of late stage Parkinson's disease has provided a unique opportunity to examine the role of subcortical structures in generating the BP. Limousin et al. (1999) examined the BP prior to self paced randomly selected joystick movements in 12 patients before and 3 months after unilateral pallidotomy. They found that pallidotomy had no effect on the BP accompanying movements of the hand ipsilateral to the lesion, but that it increased the amplitude of the late component of the BP during movements of the contralateral hand. There was no effect on the early component of the BP.

They were puzzled by the differential effect on early and late components of the BP. Metabolic imaging studies had shown that pallidotomy increases blood flow in the SMA during freely selected joystick movements (Ceballos-Baumann et al., 1994; Samuel et al., 1997b). Since the SMA contributes activity to both early and late phases of the BP, it had been expected that both phases of the BP would be affected. The fact that only the late phase was increased suggested a preferential effect on lateral premotor activity. Indeed, other imaging studies have also suggested increased activity in lateral premotor areas after pallidotomy in Parkinson's disease (Ceballos-Baumann et al., 1999; Eidelberg et al., 1996). This has been viewed as a compensatory process that may replace to some extent the underactivation of medial motor areas in Parkinson's disease (Dick et al., 1989).

There has been only one study of the effect of chronic deep brain stimulation on the BP in patients with Parkinson's disease. Brown et al. (1999) examined 6 patients with stimulators implanted bilaterally in the internal globus pallidus, and 6 with stimulators in the subthalamic nucleus. They recorded the BP prior to freely selected joystick movements with the stimulators turned off (bilaterally) and then with the stimulators turned on at their most effective clinical setting. In this small group of patients there was no significant effect on the BP. However, until a larger group is studied this data must be regarded as preliminary. The variability of the BP means that larger group sizes should be employed if sufficient statistical power is to be obtained to justify negative results.

In contrast to the rather small effects on the BP, STN stimulation has been found to increase the amplitude of the CNV over frontal and fronto-central areas in an S1-S2 task (Gerschlager et al., 1999). The reason for this could be that in such a task, patients may wait until receiving the S2 signal before they start to prepare to move. Effectively they may prefer to trigger their movements with an external cue, as in the study of Cunnington et al. (1995). After the stimulators are turned on, the patients' behaviour may approach that of healthy subjects who use S1 as a signal to prepare anticipatory activity in advance of the S2. This activity is then reflected in the larger CNV during stimulation.

In some centres, deep brain stimulating electrodes are implanted in two stages. The electrodes themselves are implanted first, and the leads are left exposed on the scalp for the next few days so that they can be tested for electrical continuity and clinical efficacy. If all is well, the leads are then internalised and fed under the skin to a stimulator box implanted in the subclavicular region. The period when the leads are exposed provides a unique opportunity to both stimulate and record from electrodes in deep brain structures. Recently, Brown et al. (2001) have carried out a series of experiments in which they have recorded STN and pallidal activity from such leads at rest and during voluntary contraction of the contralateral arm. Their recordings were particularly interesting because they had access to an unusual group of patients who had electrodes implanted simultaneously in both structures for the treatment of Parkinson's disease. They found that the activity in the two nuclei was synchronised at 20-30Hz when they recorded patients after overnight withdrawal of medication, whereas the frequency of coupling increased to 60-70Hz when patients were given L-DOPA. Given the present interest in the possible importance of coherent oscillatory activity in coding between brain structures, this provides an important insight into the role of deep brain structures in such effects.

DYSTONIA AND CHOREA

To explore the electrocortical correlates of dystonia, investigators face a variety of different ways to approach this heterogeneous movement disorder. Nevertheless, the picture emerging from studies using movement-related cortical potentials is relatively homogeneous. As noted by Deuschl et al. (1995), BP amplitude tends to increase rather than decrease with increasing EMG activity. Therefore, if anything, a higher rather than a smaller BP amplitude would be expected in dystonia. This prediction refers to the phenomena of inappropriate agonist-antagonist co-contraction and overflow of muscle activation in dystonia (e.g., Van der Kamp et al., 1989). Although this prediction is straightforward, the results from various studies have borne out the opposite.

Deuschl and co-workers (1995) investigated patients with a task-specific focal dystonia, i.e., writer's cramp. The task that was used to elicit the BP, however, was not writing but a simple finger movement that did not provoke symptoms during the experimental session. The BP was recorded with a relatively dense array of electrodes over sensorimotor areas, and showed a reduced amplitude in patients. The amplitude reduction was confined to contralateral and midline central electrodes in an interval of 300-200 ms before EMG onset, suggesting deficient contralateral motor cortex activation just preceding the initiation of movement. Notably, the BP differences were found for movements made by the affected arm as well as for movements by the unaffected side. Given the very small effects and their apparent independence of symptom manifestation during recordings, one could question the meaning and significance of the results. However, convergent results have been reported by others. Van der Kamp et al. (1995) also investigated patients with primary dystonia affecting the arm, using self-paced wrist movements. The results showed a similar attenuation of the late BP (see Figure 7). In dystonia secondary to basal ganglia lesions, Fève et al. (1994), found a reduced slope of the late as well as the early BP. A feature of the latter two studies not reported by Deuschl et al. was a reduction of the contralateral predominance of the late BP.

Based on evidence from TMS studies showing deficient intracortical inhibition in dystonia (Ridding et al., 1995b), Yazawa et al. (1999) used the BP to investigate the inhibitory mechanisms invoked when a sustained muscle contraction is terminated. Thus, they compared the BP preceding voluntary muscle relaxation with the BP preceding voluntary muscle contraction in patients with focal hand dystonia. In normal subjects, BP peak latencies differed between conditions, with shorter latencies for contractions than for relaxations. Muscle relaxation proceeded faster in the patient group, resulting in similar latencies for relaxation and contraction. In addition, the normal distribution of the BP with a contralateral maximum changed to a more symmetric distribution in the relaxation task. Together, the results were interpreted as support for deficient inhibitory processes in dystonia.

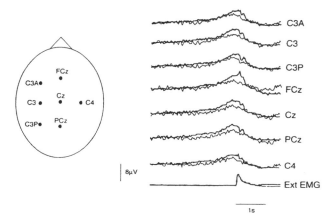

Figure 7. Bereitschaftspotential preceding voluntary wrist movements in patients with primary dystonia affecting the arm vs. control subjects. (From Van der Kamp et al., 1995. Reproduced with permission).

Kaji et al. (1995) used the CNV instead of the BP for investigating patients with cervical dystonia, reasoning that movement preparatory activity contingent on sensory stimuli might be more relevant to the pathophysiology of dystonia than the BP. Another aspect that distinguishes their approach is that they contrasted rotatory movements of the head, involving the affected muscle groups, with finger movements. While the head movements were accompanied by a reduction in amplitude of the late CNV, in patients compared to healthy controls, the same component was of normal magnitude with finger movements. A reverse dissociation was reported subsequently, by the same investigators, for patients with writer's cramp (Hamano et al., 1999). This dissociation between movements of different effectors fits the task specific or focal nature of the type of dystonia investigated and incriminates, according to the authors, the programming level of movement organisation.

Assuming that the late CNV and the BP are related (Rohrbaugh et al., 1986), the results from the study by Kaji et al. are in agreement with the BP studies that have been performed (Deuschl et al, 1995; Van der Kamp et al., 1995). Matching the premovement potential data to the data obtained in rCBF studies is less straightforward. In idopathic dystonia, activation studies have demonstrated a pattern of overactivation in prefrontal and premotor areas, combined with underactivation of sensorimotor executive areas (Ceballos-Baumann et al., 1995a, 1997). The underactivation possibly corresponds to the amplitude reduction seen in premovement EEG potentials, reflecting deficient activity in motor cortex inhibitory circuits. Given that PET integrates activity over longer periods of time, the excess frontal activity might represent recruitment of inhibitory activity during movement execution, invisible in premovement potentials. An observation of overactivity in sensorimotor cortex in patients with acquired dystonia (Ceballos-Baumann et al., 1995b) does not fit in with this pattern, however, as this group had premovement potentials comparable to patients with primary dystonia (cf. Berardelli et al., 1998, for a similar discussion).

Attempts at explaining dystonia in terms of current concepts of the basal ganglia-thalamocortical circuitry have drawn upon, among other sources, investigations on L-dopa induced dyskinesia and dystonia in parkinsonian monkeys (Berardelli et al., 1998; Mitchell et al., 1990). The induction of both dystonic and choreatic phenomena by L-dopa has suggested a related pathophysiological background, also expressed by their co-occurrence in choreoathetosis. Unfortunately, there is little to be added to this hypothesis from investigations of movement-related potentials in choreatic movement disorders. Shibasaki et al. (1982) investigated the movement-related potential correlates of involuntary choreatic movements in three patients with chorea-acanthocytosis and three patients with Huntington's disease. In two of the chorea-acanthocytosis patients a slow negative wave was observed with predominance over the contralateral hemisphere. In contrast, no movement preceding activity was found in the Huntington's disease patients. Preceding voluntary movements, on the other hand, Johnson et al. (2001) found a reduced amplitude of the BP in Huntington's disease, on the basis of which they inferred deficient SMA function.

The rare disorder of paroxysmal kinesigenic choreoathetosis has only infrequently been investigated using movement-related potentials. In a review of 26 patients, Houser et al. (1999) briefly referred to BP investigations in two cases who demonstrated a reduced amplitude of the early BP with a relatively steep late negativity. In a better documented case report, Franssen et al. (1983) studied the CNV, which showed a strongly enhanced early CNV that normalised under phenytoin treatment. The late CNV, more closely associated with movement preparation, was normal. The authors proposed that the abnormal early CNV might reflect a functional disturbance in the subcortical and thalamic regulation of cortical tone.

CEREBELLAR DYSFUNCTION

Investigations of cerebellar function by means of BP recordings in patient groups have not been pursued as frequently as the same approach in Parkinson's disease. Yet, the framework of a lateral and a medial premotor system, which has had such a strong impact on BP studies in Parkinson's disease, entails a distinct hypothesis as to how the BP might be affected in cerebellar disease. If, as originally proposed by Goldberg (1985), the cerebellum and the basal ganglia participate in relatively segregated lateral and medial premotor systems, each of which are capable of activating the motor cortex, then cerebellar dysfunction should be expected not to compromise the early BP, attributed to SMA activation. On the other hand, the later BP components, reflecting primary motor and possibly lateral premotor cortex activity, might demonstrate abnormalities as a result of cerebellar dysfunction. Indeed, alterations of rCBF patterns found in patients with cerebellar degeneration were interpreted along these lines by Wessel et al. (1995). Their findings included decreased activity of the lateral premotor cortex (PMv) and increased SMA activity, attributed to a dependence on the medial premotor system.

The results obtained in BP studies of cerebellar patients fit the above picture to some extent. That is, in patients with degenerative disorders of the cerebellum, as well as in patients with cerebellar hemisphere lesions, the BP has been found normal in its entirety (Shibasaki et al., 1978a) or to have at least a normal or enhanced early segment (Wessel et al., 1994; Gerloff et al., 1996; Tarkka et al., 1993). As to the late BP, these same studies observed a substantial reduction in amplitude and a loss of topographical differentiation manifested in a greater spread over the ipsilateral hemisphere. Although the number of investigated patients in these studies was small, the consistency across studies argues for their reliability. In addition, different types of movement were examined, ranging from simple finger movements (Tarkka et al., 1993) and rapid alternating movements in a single joint (Gerloff et al., 1996), to sequential and goal directed movements (Wessel et al., 1994).

This makes it unlikely that the altered topography was related to poor execution of the movements or to the movements being more effortful, rather than to changes in cortical motor processing.

The BP abnormalities in patients with vascular lesions of the cerebellum were resolved on re-examination after 8-10 months (Gerloff et al., 1996). It is not clear whether the same occurs when the lesion involves the dentate nucleus. Shibasaki and co-workers (Shibasaki et al., 1986; Ikeda et al., 1994; Kitamura et al. 1999) reported a complete absence of the BP in cases where lesions involved the dentate nucleus or when this structure was affected by a degenerative process as in Ramsay Hunt syndrome. In contrast, their data suggest that when lesions spare the dentate nucleus, the component structure of the BP can be retained, albeit with a reduced amplitude of both early and late BP.

In a case report by Ikeda et al. (1994), concerning a patient with a cerebellar efferent lesion, absence of the BP was contrasted with a normal CNV at the frontocentral midline. The authors inferred that CNV generation in frontal lobe structures, including the motor cortex, is not dependent on input to these areas coming from the cerebellum. More extensive information was provided by Verleger et al. (1999), who studied the CNV in patients with cerebellar degeneration during the performance of movements that varied in coordination difficulty. The CNV amplitude was reduced overall in cerebellar patients, but was still modulated dependent on how taxing the motor task was (see Figure 8). The amplitude reduction was independent of performance, as it was found both with normally executed simple movements as with impaired movements requiring bimanual coordination. Interestingly, the CNV recordings in cerebellar patients demonstrated the same loss of topographical specificity as has been reported for the BP (Verleger et al., 1999).

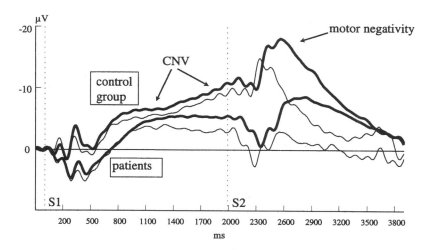

Figure 8. Recordings of the CNV at electrode Cz in control subjects and cerebellar patients. Thick and thin traces refer to recordings in different tasks. Differences between tasks are outweighed by marked group differences of the CNV as it develops between the S1 and S2 stimuli. (From Verleger et al., 1999. Reproduced with permission).

Together, the various studies in patients with cerebellar dysfunction or cerebellar lesions unambiguously show that the cerebellum has a profound influence on cortical motor physiology as expressed in the BP and the CNV. As motor areas 4 and 6 are the main termination sites of cerebellar efferents, the changes of these EEG potentials presumed to reflect the activity of these areas are relevant to the motor dysfunction resulting from pathology within the cerebellum, even though the abnormalities seem independent of the manifestation of coordination impairments (Verleger et al., 1999). Exactly what it means is yet difficult to delineate. The cerebellar output from the dentate nuclei through the ventrolateral thalamic nucleus to the motor cortex is commonly taken to exert a tonic facilitatory effect on the motor cortex. This facilitatory effect is checked by an inhibitory influence of the Purkinje cells onto the dentate nuclei. Within this framework, an attenuation of late BP and CNV components might be understood as resulting from a decreased facilitatory output of the cerebellum (Wessel et al., 1994, 1995). Note that this explanation incompletely captures the altered physiology, given that distant effects of cerebellar pathology on the motor cortex can also take the form of enhanced electrocortical activity, as in cortical myoclonus, which we will consider next.

MYOCLONIC DISORDERS

The technique of backaveraging used to obtain the BP has been widely used to investigate movement disorders characterised by involuntary myoclonic movements (e.g. Shibasaki et al., 1978b, 1985; Obeso et al., 1985). Both action, stimulus sensitive, and spontaneous forms of myoclonic movements, as occurring in a variety of different conditions, have thus been investigated and found to be often accompanied by abnormal sensorimotor cortical discharges. However, only infrequently does the myoclonus preceding cortical activity consist of a slow negative wave with a resemblance to the BP (e.g. cases 3 and 4 of Shibasaki and Kuroiwa, 1975; case 10 of Obeso et al., 1985). The majority of patients with discrete EEG events time-locked to the myoclonic jerks show instead a myoclonus-related cortical spike. Several detailed studies on these spikes have convincingly demonstrated that they closely match the form of giant somatosensory evoked potentials (SEPs) recorded in the same patients. These giant SEPs usually show grossly enhanced P1 and N2 (P25 and N33) components. Regardless of whether they are averaged relative to a stimulus or backaveraged relative to a myoclonic jerk, the first positive component has been found to precede (upper extremity) myoclonus by about 20 ms, consistent with conduction along direct corticomotoneuron pathways.

An origin of the myoclonus related cortical discharges in sensorimotor cortex is supported by polarity reversals of the enhanced P1 and N2 between electrode sites anterior and posterior of the central sulcus (Shibasaki et al., 1978b). In spite of the apparently localised dysfunction, some of the conditions associated with cortical myoclonus involve the cerebellar system rather than the sensorimotor cortex. For instance, pathology within the cerebellum and sparing of sensorimotor cortex was recently demonstrated for cases with coeliac disease with cortical myoclonus (Bhatia et al., 1995; Tijssen et al., 2000). Similarly, for dyssynergia cerebellaris myoclonica (Ramsay Hunt syndrome), the pathology is considered to be located in the dentate nuclei. The latter condition is accompanied by the enhanced cortical discharges preceding myoclonus, described above (Obeso et al., 1985). In contrast, preceding voluntary movements the EEG shows an absence of the BP, as described by Shibasaki and co-workers (Shibasaki et al., 1978b, 1986). To our knowledge, no reports exist that document these pathological changes in the same patient.

Given that in cortical myoclonus BP-like potentials are extremely rare, while there are no EEG events time-locked to the involuntary movements in myoclonus of subcortical or reticular origin, the presence of a BP preceding myoclonus can be of diagnostic significance. As suggested by Toro and Torres (1986) and by Terada et al. (1995), it may indicate psychogenic myoclonus.

Tics and restless legs, for convenience included in this section, share with reticular forms of myoclonus an absence of specific EEG events preceding the involuntary movements. Obeso et al. (1981) investigated patients with Gilles de la Tourette and found a BP preceding voluntary movements, but no premovement EEG potentials preceding tics. A similar pattern of results was reported for restless legs by Trenkwalder et al. (1993), consistent with subcortical loci of activation in an fMRI study of this patient group (Bucher et al. 1997).

CONCLUDING REMARKS

Much of the original appeal of the BP, as a means of studying movement disorders, has been the simple fact that it provided a way to investigate the time course of movement-related brain activity. As knowledge concerning the BP in normal subjects evolved, this allowed the investigation of more specific questions, by means of BP recordings, in neurological patients. Once the SMA became regarded as a likely source of the BP, a considerable part of the BP studies in Parkinson's disease focused on the question of whether BP abnormalities in this disorder signalled deficient SMA function. Nevertheless, a significant part of BP studies in movement disorders have remained exploratory in nature, sometimes yielding unanticipated and stimulating results. Where research using the BP has been motivated by strong assumptions regarding its neural sources, its results should be interpreted with some caution. Although contributions from mesial frontal cortex (SMA, cingulate motor areas) and primary motor cortex to the BP can hardly be doubted, it is unlikely that their respective contributions are as segregated in time as suggested by subdivisions of the BP in early and late components. Abnormalities of the BP cannot be unambiguously attributed to either mesial premotor cortex or primary motor cortex. With its constantly improving temporal resolution, evidence from event-related fMRI can be of value in this respect.

No less important than converging evidence from imaging techniques, is the comparison of BP results with information from other approaches to the analysis of EEG and comparison with other neurophysiological techniques. Defebvre et al. (1994) combined a BP analysis of EEG data with an analysis of event-related EEG desynchronisation (ERD), yielding comparable changes of their time course in Parkinson patients compared to control subjects. Such a result is consistent with the hypothesis of Brown and co-workers that the basal ganglia have a crucial role in the regulation of cortical oscillatory activity, controlling the release from idling rhythms and the formation of synchronous activity underlying the selection and performance of voluntary movements (Brown and Marsden, 1998). As well as opening new theoretical perspectives, EEG or MEG (de)synchronisation obviously complements information derived from BP recordings, sometimes enabling analysis of movement-related brain activity in experimental situations where it is not well feasible to record a BP (Brown and Marsden, 1999; Wang et al., 2000).

A significant step forward in movement-related potentials research has been the introduction of the LRP. Its value in electrophysiological analyses of movement disorders was considered above. Originally proposed as a marker for the timing of response choice, reflecting differential activation of the left and right motor cortex and hence recorded with only few electrodes, it is now recognised as a somewhat more complex signal. Depending on task requirements and stimuli, areas other than the primary motor cortex and neural activity other than movement execution-related activity may generate lateralised event-related potentials, making it worthwhile to study the LRP with a more extensive array of electrodes. Where the BP is inherently associated with self-paced voluntary movements, the LRP extends the investigative repertoire to experimental situations where the external context guides action. It is therefore to be expected that the LRP will gain an equally important place in the neurophysiological investigation of movement disorders.

Transcranial magnetic stimulation (TMS) investigations have revealed functional changes of the motor cortex in a range of different neurological movement disorders (e.g., Ridding et al., 1995a,b). Such findings, in some instances, contrast with earlier conclusions drawn from transcranial electrical stimulation of the motor cortex, and emphasise that the motor cortex is more than a collection of corticomotoneuronal output cells that merely relay the signals elaborated at a higher level. It is as yet unclear how changes of motor cortical excitability, as measured with TMS in movement disorders, are reflected in movement-related potentials. Research in healthy subjects, however, has already established that temporary changes in excitability, induced by TMS, have measurable effects on the BP

(Rossi et al., 2000). Such combined approaches may provide new information on how the field potentials recorded at the scalp relate to neural events in the motor cortex.

Recording of the BP, the LRP, ERD, TMS and functional imaging each have advantages and disadvantages in the study of the processes of movement preparation and response selection in movement disorders. Complementary use of these techniques in the study of movement disorders will allow investigators to explore the strengths of each technique to gain the spatial and temporal resolution necessary to understand the mechanisms of motor impairment in various movement disorders.

REFERENCES

Alexander, G.E. and Crutcher, M.D. (1990) Functional architecture of basal ganglia circuits: neural substrates of parallel processing. Trends in Neuroscience 13, 266-271.

Amabile, G., Fattapposta, F., Pozzessere, G., Albani, G., Sanarelli, L., Rizzo, P.A. and Morocutti, C. (1986) Parkinson disease: Electrophysiological (CNV) analysis related to pharmacological treatment. Electroenceph. clin. Neurophysiol. 64, 521-524.

Barrett, G., Shibasaki, H. and Neshige, R. (1986) Cortical potential shifts preceding voluntary movement are normal in parkinsonism. Electroenceph. clin. Neurophysiol. 63, 340-348.

Berardelli, A., Rothwell, J.C., Hallett, M., Thompson, P.D., Manfredi, M. and Marsden, C.D. (1998) The pathophysiology of primary dystonia. Brain 121, 1195-1212.

Bhatia, K.P., Brown, P., Gregory, R., Lennox, G.G., Manji, H., Thompson, P.D., Ellison, D.W. and Marsden C.D. (1995) Progressive myoclonic ataxia associated with coeliac disease. The myoclonus is of cortical origin, but the pathology is in the cerebellum. Brain 118, 1087-1093.

Bötzel, K., Mayer, M., Oertel, W.H. and Paulus, W. (1995) Frontal and parietal premovement slow brain potentials in Parkinson's disease and aging. Mov. Disord. 10, 85-91.

Brotchie, P., Iansek, R. and Horne, M.K. (1991) Motor function of the monkey globus pallidus. 2. Cognitive aspects of movement and phasic neuronal activity. Brain 114, 1685-1702.

Brown, P. and Marsden, C.D. (1998) What do the basal ganglia do? Lancet 351, 1801-1804.

Brown, P., Oliviero, A., Mazzone, P., Insola, A., Tonali, P. and Di Lazzaro, V. (2001) Dopamine dependency of oscillations between subthalamic nucleus and pallidum in Parkinson's disease. J. Neurosci. 21, 1033-1038.

Brown, R.G., Limousin-Dowsey, P., Brown, P., Jahanshahi, M., Pollak, P., Benabid, A.L., Rodriguez-Oroz, M.C., Obeso, J. and Rothwell, J.C. (1999) Impact of deep brain stimulation on upper limb akinesia in Parkinson's disease. Ann. Neurol. 45, 473-488.

Bucher, S.F., Seelos, K.C., Oertel, W.H., Reiser, M. and Trenkwalder C. (1997) Cerebral generators involved in the pathogenesis of the restless legs syndrome. Ann. Neurol. 41, 639-645.

Ceballos-Baumann, A.O., Obeso, J.A., Vitek, J.L., Delong, M.R., Bakay, R., Linazasoro, G. and Brooks, D.J. (1994) Restoration of thalamocortical activity after posteroventral pallidotomy in Parkinson's disease. Lancet 344, 814.

Ceballos-Baumann, A.O., Passingham, R.E., Warner, T., Playford, E.D., Marsden, C.D. and Brooks, D.J. (1995a) Overactive prefrontal and underactive motor cortical areas in idiopathic dystonia. Ann. Neurol. 37, 363-372.

Ceballos-Baumann, A.O., Passingham, R.E., Marsden, C.D. and Brooks, D.J. (1995b) Motor reorganisation in acquired hemidystonia. Ann. Neurol. 37, 746-757.

Ceballos-Baumann, A.O., Sheean, G., Passingham, R.E., Marsden, C.D. and Brooks, D.J. (1997) Botulinum toxin does not reverse the cortical dysfunction associated with writer's cramp. A PET study. Brain 120, 571-582.

Ceballos-Baumann, A.O., Boecker. H., Bartenstein, P., von Falkenhayn, I., Riescher, H., Conrad, B., Moringlane, J.R. and Alesch, F. (1999) A positron emission tomographic study of subthalamic nucleus stimulation in Parkinson disease: enhanced movement-related activity of motor-association cortex and decreased motor cortex resting activity. Arch. Neurol. 56, 997-1003.

Cunnington, R., Iansek, R., Bradshaw, J.L. and Phillips, J.G. (1995) Movement-related potentials in Parkinson's disease: Presence and predictability of temporal and spatial cues. Brain 118, 935-950.

Cunnington, R., Iansek, R., Johnson, K.A. and Bradshaw, J.L. (1997) Movement-related potentials in Parkinson's disease: Motor imagery and movement preparation. Brain, 120 1339-1353.

Cunnington, R., Iansek, R. and Bradshaw, J.L. (1999) Movement-related potentials in Parkinson's disease: External cues and attentional strategies. Mov. Disord. 14, 63-68.

Deecke, L., Englitz, H.G., Kornhuber, H.H. and Schmitt, G. (1977) Cerebral potentials preceding voluntary movement in patients with bilateral or unilateral Parkinson akinesia. In: Desmedt, J.E., (Eds.) Attention, voluntary contraction and event-related cerebral potentials. Progress in Clinical Neurophysiology, Vol. 1, pp. 151-163. Basel: Karger.

Deecke, L. and Kornhuber, H.H. (1978) An electrical sign of participation of the mesial 'supplementary' motor cortex in human voluntary finger movement. Brain Res. 159, 473-476.

Deecke, L. (1985) Cerebral potentials related to voluntary actions: parkinsonian and normal subjects. In: Delwaide, P.J. and Agnoli, A., (Eds.) Clinical neurophysiology in parkinsonism, pp. 91-105. Amsterdam: Elsevier.

Defebvre, L., Bourriez, J.L., Dujardin, K., Derambure, P., Destee, A. and Guieu, J.D. (1994) Spatiotemporal study of Bereitschaftspotential and event-related desynchronization during voluntary movement in Parkinson's disease. Brain Topography 6, 237-244.

Deiber, M.P., Passingham, R.E., Colebatch, J.G., Friston, K.J., Nixon, P.D. and Frackowiak, R.S.J. (1991) Cortical areas and the selection of movement: a study with positron emission tomography. Exp. Brain Res. 84, 393-402.

Deuschl, G., Toro, C., Matsumoto, J. and Hallett, M. (1995) Movement-related cortical potentials in writer's cramp. Ann. Neurol. 38, 862-868.

Dick, J.P.R., Cowan, J.M.A., Day, B.L., Berardelli, A., Kachi, T., Rothwell, J.C. and Marsden CD (1984) The corticomotoneurone connection is normal in Parkinson's disease. Nature 310, 407-409.

Dick, J.P.R., Cantello, R., Buruma, O., Gioux, M., Benecke, R., Day, B.L., Rothwell, J.C., Thompson, P.D. and Marsden, C.D. (1987) The Bereitschaftspotential, L-Dopa and Parkinson's disease. Electroenceph. clin. Neurophysiol. 66, 263-274.

Dick, J.P.R., Rothwell, J.C., Day, B.L., Cantello, R., Buruma, O., Gioux, M., Benecke, R., Berardelli, A., Thompson, P.D. and Marsden, C.D. (1989) The Bereitschaftspotential is abnormal in Parkinson's disease. Brain 112, 233-244.

Doudet, D.J., Gross, C., Arluison, M. and Bioulac, B. (1990) Modifications of precentral cortex discharge and EMG activity in monkeys with MPTP-induced lesions of DA nigral neurons. Exp. Brain Res. 80, 177-188.

Eidelberg, D., Moeller, J.R., Ishikawa, T., Dhawan, V., Spetsieris, P., Silbersweig, D., Stern, E., Woods, R,P., Fazzini, E., Dogali, M. and Beric, A. (1996) Regional metabolic correlates of surgical outcome following unilateral pallidotomy for Parkinson's disease. Ann. Neurol. 39, 450-459.

Fattapposta, F., Pierelli, F., Traversa, G., My, F., Mostarda, M., D'Alessio, C., Soldati, G., Osborn, J. and Amabile, G. (2000) Preprogramming and control activity of bimanual self-paced motor task in Parkinson's disease. Clin. Neurophysiol. 111, 873-883.

Fève, A.P., Bathien, N. and Rondot, P. (1992) Chronic administration of L-dopa affects the movement-related cortical potential in patients with Parkinson's disease. Clin. Neuropharm. 15, 100-108.

Fève, A.P., Bathien, N. and Rondot, P. (1994) Abnormal movment-related potentials in patients with lesions of basal ganglia and anterior thalamus. J. Neurol. Neurosurg. Psychiatry 57, 100-104.

Filion, M., Tremblay, L. and Bedard, P.J. (1988) Abnormal influences of passive limb movement on the activity of globus pallidus neurons in parkinsonian monkeys. Brain Res. 444, 165-176.

Filipovic, S.R., Covickovic-Sternic, N., Radovic, V,M., Dragasevic, N., Stojanovic-Svetel, M. and Kostic, V.S. (1997) Correlation between Bereitschaftspotential and reaction time measurements in patients with Parkinson's disease. Measuring the impaired supplementary motor area function? J. Neurol. Sci. 147, 177-183.

Franssen, H., Fortgens, C., Wattendorf, A.R. and Van Woerkom, T.C.A.M. (1983) Paroxysmal kinesigenic choreoathetosis and abnormal contingent negative variation. Arch. Neurol. 40, 381-385.

Gerloff, C., Altenmüller, E. and Dichgans, J. (1996) Disintegration and reorganization of cortical motor processing in two patients with cerebellar stroke. Electroenceph. clin. Neurophysiol. 98, 59-68.

Gerschlager, W., Alesch, F., Cunnington, R., Deecke, L., Dirnberger, G., Endl, W., Lindinger, G. and Lang, W. (1999) Bilateral subthalamic nucleus stimulation improves frontal cortex function in Parkinson's disease. An electrophysiological study of the contingent negative variation. Brain 122, 2365-73.

Goldberg, G. (1985) Supplementary motor area structure and function: review and hypotheses. Behav. Brain Sci. 8, 567-615.

Hamano, T., Kaji, R., Katayama, M., Kubori, T., Ikeda, A., Shibasaki, H. and Kimura, J. (1999) Abnormal contingent negative variation in writer's cramp. Clin. Neurophysiol. 110, 508-515.

Holsapple, J.W., Preston, J.B. and Strick P.L. (1991) The origin of thalamic inputs to the 'hand' representation in the primary motor cortex. J. Neurosci. 11, 2644-54.

Hoover, J.E. and Strick, P.L. (1993) Multiple output channels in the basal ganglia. Science 259, 819-21.

Houser, M.K., Soland, V.L., Bhatia, K.P., Quinn, N.P. and Marsden, C.D. (1999) Paroxysmal kinesigenic choreoathetosis: a report of 26 patients. J. Neurol. 246, 120-6.

Ikeda, A., Lüders, H.O., Burgess, R.C. and Shibasaki, H. (1992) Movement-related potentials recorded from supplementary motor area and primary motor area. Brain 115, 1017-43.

Ikeda, A., Shibasaki, H., Nagamine, T., Terada, K., Kaji, R., Fukuyama, H. and Kimura, J. (1994) Dissociation between contingent negative variation and Bereitschaftspotential in a patient with cerebellar efferent lesion. Electroenceph. clin. Neurophysiol. 90, 359-364.

Ikeda, A., Shibasaki, H., Nagamine, T., Xu, X., Terada, K., Mima, T., Kaji, R., Kawai, L., Tatsuoka, Y. and Kimura, J. (1995) Peri-rolandic and fronto-parietal components of scalp-recorded giant SEPs in cortical myoclonus. Electroenceph. clin. Neurophysiol. 96, 300-309.

Ikeda, A., Shibasaki, H., Kaji, R., Terada, K., Nagamine, T., Honda, M., Hamano, T. and Kimura, J. (1996) Abnormal sensorimotor integration in writer's cramp: study of contingent negative variation. Mov. Disord. 11, 683-690.

Ikeda, A., Shibasaki, H., Kaji, R., Terada, K., Nagamine, T., Honda, M. and Kimura, J. (1997) Dissociation between contingent negative variation (CNV) and Bereitschaftspotential (BP) in patients with parkinsonism. Electroenceph. clin. Neurophysiol. 102, 142-151.

Jahanshahi, M., Brown, R.G. and Marsden, C.D. (1992) Simple and choice reaction time and the use of advance information for motor preparation in Parkinson's disease. Brain 115, 539-64.

Jahanshahi, M., Jenkins, I.H., Brown, R.G., Marsden, C.D., Passingham, R.E. and Brooks, D.J. (1995) Self initiated versus externally triggered movements. I. An investigation using measurement of regional cerebral blood flow with PET and movement-related potentials in normal and Parkinson's disease subjects. Brain 118, 913-933.

Johnson, K.A., Cunnington, R., Iansek, R., Bradshaw, J.L., Georgiou, N. and Chiu, E. (2001) Movement related potentials in Huntington's disease: movement preparation and execution. Exp. Brain Res. 138, 492-499.

Kaji, R., Ikeda A., Ikeda, T., Kubori, T., Mezaki, T., Kohara, N., Kanda, M., Nagamine, T., Honda, M., Rothwell, J.C. Shibasaki, H. and Kimura J. (1995) Physiological study of cervical dystonia. Task-specific abnormality in contingent negative variation. Brain 118, 511-522.

Kitamura, J., Shabasaki, H., Terashi, A. and Tashima, K. (1999) Cortical potentials preceding voluntary finger movement in patients with focal cerebellar lesion. Clin. Neurophysiol. 110, 126-132.

Limousin, P., Brown, R.G., Jahanshahi, M., Asselman P., Quinn, N.P., Thomas, D., Obeso, J.A. and Rothwell, J.C. (1999) The effects of posteroventral pallidotomy on the preparation and execution of voluntary hand and arm movements in Parkinson's disease. Brain 122, 315-327.

Matelli, M., Luppino, G., Fogassi, L. and Rizzolatti, G. (1989) Thalamic input to inferior area 6 and area 4 in the macaque monkey. J. Comp. Neurol. 280, 468-88.

Miller, J.O., Riehle, A. and Requin, J. (1992) Effects of preliminary perceptual output on neuronal activity of the primary motor cortex. J. Exp. Psychol. Hum. Percept. Perform. 18, 1121-1138.

Mima, T., Nagamine, T., Ikeda A., Yazawa, S., Kimura, J. and Shibasaki, H. (1998) Pathogenesis of cortical myoclonus studied by magnetoencephalopathy. Ann. Neurol. 43, 598-607.

Mima, T., Nagamine, T., Nishitani, N., Mikuni, N., Ikeda, A., Fukuyama, H., Takigawa, T., Kimura, J. and Shibasaki, H. (1998) Cortical myoclonus: sensorimotor hyperexcitability. Neurology 50, 933-942.

Mitchell, I.J., Luquin, R., Boyce, S., Clarke, C.E., Robertson, R.G., Sambrook, M.A. and Crossman, A.R. (1990) Neural mechanisms of dystonia: evidence from a 2-deoxyglucose uptake study in a primate model of dopaminergic agonist-induced dystonia. Mov. Disord. 5, 49-54.

Obeso, J.A., Rothwell, J.C. and Marsden, C.D. (1981) Simple tics in Gilles de la Tourette's syndrome are not prefaced by a normal premovement EEG potential. J. Neurol. Neurosurg. Psychiatry 44, 735-738.

Obeso, J.A., Rothwell, J.C. and Marsden, C.D. (1985) The spectrum of cortical myoclonus. Brain 108, 193-224.

Oostenveld, R., Praamstra, P., Stegeman, D.F. and Van Oosterom, A. (2001) Overlap of attention and movement-related activity in lateralized event-related potentials. Clin. Neurophysiol. 112, 477-481.

Passingham, R.E. (1993) The frontal lobes and voluntary action. Oxford: Oxford University Press.

Plat, F.M., Praamstra, P. and Horstink, M.W.I.M. (2000) Redundant signals effects on reaction time, response force, and movement-related potentials in Parkinson's disease. Exp. Brain Res. 130, 533-539.

Playford, E.D., Jenkins, I.H., Passingham, R.E., Nutt, J., Frackowiak, R.S.J. and Brooks, D.J. (1992) Impaired mesial frontal and putamen activation in Parkinson's disease: A positron emission tomography study. Ann. Neurol. 32, 151-161.

Praamstra, P., Cools, A.R., Stegeman, D.F. and Horstink, M.W.I.M. (1996a) Movement-related potential measures of different modes of movement selection in Parkinson's disease. J. Neurol. Sci. 140, 67-74.

Praamstra, P., Stegeman, D.F., Horstink, M.W.I.M. and Cools, A.R. (1996b) Dipole source analysis suggests selective modulation of the supplementary motor area contribution to the readiness potential. Electroenceph. clin. Neurophysiol. 98, 468-477.

Praamstra, P., Meyer, A.S., Cools, A.R., Horstink, M.W.I.M. and Stegeman, D.F. (1996c) Movement preparation in Parkinson's disease: time course and distribution of movement-related potentials in a movement precueing task. Brain 119, 1689-704.

Praamstra, P., Stegeman, D.F., Cools, A.R. and Horstink, M.W.I.M. (1998) Reliance on external cues for movement initiation in Parkinson's disease. Evidence from movement-related potentials. Brain 121, 167-177.

Praamstra, P., Plat, E.M., Meyer, A.S. and Horstink, M.W.I.M. (1999) Motor cortex activation in Parkinson's disease: Dissociation of electrocortical and peripheral measures of response generation. Mov. Disord. 14, 790-799.

Praamstra, P. and Plat F.M. (2001) Failed suppression of direct visuomotor activation in Parkinson's disease. J. Cogn. Neurosci. 13, 1-13.

Rascol, O., Sabatini, U., Brefel, C., Fabre, N., Raj, S., Senard, J.M., Celsis, P., Viallard, G., Montastruc, J.L. and Chollet, F. (1998) Cortical motor overactivation in parkinsonian patients with L-dopa induced peak-dose dyskinesia. Brain 121, 527-533.

Ridding, M.C., Inzelberg, R. and Rothwell, J.C. (1995a) Changes in excitability of motor cortical circuitry in patients with Parkinson's disease. Ann. Neurol. 37, 181-188.

Ridding, M.C., Sheean, G., Rothwell, J.C., Inzelberg, R. and Kujirai, T. (1995b) Changes in the balance between motor cortical excitation and inhibition in focal task specific dystonia. J. Neurol. Neurosurg. Psychiatry 59, 493-498.

Rohrbaugh, J.W., McCallum, W.C., Gaillard, A.W.K., Simons, R.F., Birbaumer, N. and Papakostopoulos, D. (1986) ERPs associated with preparatory and movement-related processes. A review. In: McCallum, W.C., Zappoli, R. and Denoth, F., (Eds.) Cerebral Psychophysiology: Studies in event-related potentials (EEG Suppl. 38), pp.189-229. Amsterdam: Elsevier.

Rossi, S., Pasqualetti, P., Rossini, P.M., Feige, B., Ulivelli, M., Glocker, F.X., Battistini, N., Lucking, C.H., Kristeva-Feige, R. (2000) Effects of repetitive transcranial magnetic stimulation on movement-related cortical activity in humans. Cer. Cortex 10, 802-808.

Samuel, M., Ceballos-Baumann, A.O., Blin, J., Uema, T., Boecker, H., Passingham, R.E. and Brooks, D.J. (1997a) Evidence for lateral premotor and parietal overactivity in Parkinson's disease during sequential and bimanual movements. A PET study. Brain 120, 963-976.

Samuel, M., Ceballos-Baumann, A.O., Turjanski, N., Boecker, H., Gorospe, A., Linazasoro, G., Holmes, A.P., DeLong, M.R., Vitek, J.L., Thomas, D.G., Quinn, N.P., Obeso, J.A. and Brooks, D.J. (1997b) Pallidotomy in Parkinson's disease increases SMA and prefrontal activation during performance of volitional movements: an H_2^{15}O PET study. Brain 120, 1301-13.

Schell, G.R. and Strick, P.L. (1984) The origin of thalamus inputs to the arcuate premotor and supplementary motor areas. J. Neurosci. 4, 539-560.

Shen, L. and Alexander, G.E. (1997) Neural correlates of a spatial sensory-to-motor transformation in primary motor cortex. J. Neurophysiol. 77, 1171-1194.

Shibasaki, H. and Kuroiwa, Y. (1975) Electroencephalographic correlates of myoclonus. Electroenceph. clin. Neurophysiol. 39, 455-463.

Shibasaki, H., Shima, F. and Kuroiwa, Y. (1978a) Clinical studies of the movement-related cortical potential (MP) and the relationship between the dentatorubrothalamic pathway and readiness potential (RP). J. Neurol. 219, 15-25.

Shibasaki, H., Yamashita, Y. and Kuroiwa, Y. (1978b) Electroencephalographic studies of myoclonus. Brain 101, 447-460.

Shibasaki, H., Sakai, T., Nishimura, H., Sato, Y., Goto, I. and Kuroiwa, Y. (1982) Involuntary movements in chorea-acanthocytosis: a comparison with Huntington's chorea. Ann. Neurol. 12, 311-314.

Shibasaki, H., Yamashita, Y., Neshige, R., Tobimatsu, S. and Fukui, R. (1985) Pathogenesis of giant somatosensory evoked potentials in progressive myoclonic epilepsy. Brain 108, 225-240.

Shibasaki, H., Barrett, G., Neshige, R., Hirata, I. and Tomoda, H. (1986) Volitional movement is not preceded by cortical slow negativity in cerebellar dentate lesion in man. Brain Res. 368, 361-365.

Simpson, J.A. and Khuraibet, A.J. (1987) Readiness Potential of cortical area 6 preceding self paced movement in Parkinson's disease. J. Neurol. Neurosurg. Psychiatry, 50, 1184-91.

Tarkka, I.M., Reilly, A. and Hallett, M. (1990) Topography of movement-related cortical potentials is abnormal in Parkinson's disease. Brain Res. 522, 172-175.

Tarkka, I.M., Massaquoi, S. and Hallett, M. (1993) Movement-related cortical potentials in patients with cerebellar degeneration. Acta Neurol. Scand. 88, 129-135.

Tassinari, C.A., Rubboli, G. and Shibasaki, H. (1999) Neurophysiology of positive and negative myoclonus. Electroenceph. clin. Neurophysiol. 107, 181-195.

Terada, K., Ikeda, A., Van Ness, P.C., Nagamine, T., Kaji, R., Kimura J. and Shibasaki, H. (1995) Presence of Bereitschaftspotential preceding psychogenic myoclonus: clinical application of jerk-locked back averaging. J. Neurol. Neurosurg. Psychiatry 58, 745-7.

Tijssen, M.A.J., Thom, M.D., Ellison, D.W., Wilkins, P., Barnes, D., Thompson, P.D. and Brown, P. (2000) Cortical myoclonus and cerebellar pathology. Neurology 54, 1350-1356.

Toro, C. and Torres, F. (1986) Electrophysiological correlates of a paroxysmal movement disorder. Ann. Neurol. 20, 731-4.

Touge, T., Werhahn, K.J., Rothwell, J.C. and Marsden CD. (1995) Movement-related cortical potentials preceding repetitive and random-choice hand movements in Parkinson's disease. Ann. Neurol. 37, 791-799.

Trenkwalder, C., Bucher, S.F., Oertel, W.H., Proeckl, D., Plendl, H. and Paulus, W. (1993) Bereitschaftspotential in idiopathic and symptomatic restless legs syndrome. Electroenceph. clin. Neurophysiol. 89, 95-103.

Van der Kamp, W., Berardelli, A., Rothwell, J.C., Thompson, P.D., Day, B.L. and Marsden, C.D. (1989) Rapid elbow movements in patients with torsion dystonia. J. Neurol. Neurosurg. Psychiatry 52, 1043-1049.

Van der Kamp, W., Rothwell, J.C., Thompson, P.D., Day, B.L. and Marsden, C.D. (1995) The movement-related cortical potential is abnormal in patients with idiopathic torsion dystonia. Mov. Disord. 10, 630-633.

Verleger, R., Wascher, E., Wauschkuhn, B., Jaskowski, P., Allouni, B., Trillenberg, P. and Wessel, K. (1999) Consequences of altered cerebellar input for the cortical regulation of motor coordination, as reflected in EEG potentials. Exp. Brain Res. 127, 409-422.

Vidailhet, M., Stocchi, F., Rothwell, J.C., Thompson, P.D., Day, B.L., Brooks, D.J. and Marsden, C.D. (1993) The Bereitschaftspotential preceding simple foot movement and initiation of gait in Parkinson's disease. Neurology 43, 1784-8.

Vidailhet, M., Atchison, P.R., Stocchi, F., Thompson, P.D., Rothwell, J.C. and Marsden, C.D. (1995) The Bereitschaftspotential preceding stepping in patients with isolated gait ignition failure. Mov. Disord. 10, 18-21.

Wang, H.C., Lees, A.J. and Brown, P. (1999) Impairment of EEG desynchronisation before and during movement and its relation to bradykinesia in Parkinson's disease. J. Neurol. Neurosurg. Psychiatry 66, 442-446.

Wascher, E., Verleger, R., Vieregge, P., Jaskowski, P., Koch, S. and Kömpf, D. (1997) Responses to cued signals in Parkinson's disease. Distinguishing between disorders of cognition and of activation. Brain 120, 1355-1375.

Wessel, K., Verleger, R., Nazarenus, D., Vieregge, P. and Kömpf, D. (1994) Movement-related cortical potentials preceding sequential and goal-directed finger and arm movements in patients with cerebellar atrophy. Electroenceph. clin. Neurophysiol. 92, 331-341.

Wessel, K., Zeffiro, T., Lou, J.S., Toro, C. and Hallett, M. (1995) Regional cerebral blood flow during a self-paced sequential finger opposition task in patients with cerebellar degeneration. Brain 118, 379-393.

Yazawa, S., Ikeda, A., Kaji, R., Terada, K., Nagamine, T., Toma, K., Kubori, T., Kimura, J. and Shibasaki, H. (1999) Abnormal cortical processing of voluntary muscle relaxation in patients with focal hand dystonia studied by movement-related potentials. Brain 122, 1357-66.

THE BEREITSCHAFTSPOTENTIAL IN SCHIZOPHRENIA AND DEPRESSION

Klaus Peter Westphal

Practice for Neurology and Psychiatry,
Neuer Graben 21,
D-89073 Ulm, Germany
e-mail: neuro@dr-k-westphal.de

INTRODUCTION

The earliest investigation of the Bereitschaftspotential (BP) in schizophrenia was published three decades ago by M. Timsit (Timsit, 1970, Timsit-Berthier et al., 1973a). At about the same time, abnormal movement-related cortical potentials in psychiatric patients had also been identified by Dongier (1973). Ten years later, Kornhuber (1983) described findings of lower amplitudes and longer latencies in the BP of patients with schizophrenia. These findings were later followed by studies conducted by Chiarenza et al. (1985) and Papakostopoulos et al. (1985a). To date, a number of journal papers have been published reporting studies investigating the BP in schizophrenia. Other related cortical potentials such as the contingent negative variation (CNV) and the post-imperative negative variation (PINV) have also been recorded in schizophrenia (eg Tismit-Berthier et al, 1973b; Klein et al, 1996; Rockstroh et al, 1994; Wagner et al, 1996; Verleger et al, 1999;). A handful of studies have also examined the BP or CNV in depression (eg Their et al, 1986; Khanna et al, 1989; Knott et al, 1991; Haag et al, 1994; Heimberg et al, 1999). The BP, the CNV and the lateralized readiness potential have also been recorded in other psychiatric disorders such as anxiety states (eg Boudarene & Tismit-Berthier, 1997) and alcoholism (Marinkovic et al, 2000) in a few studies. In this chapter, I will focus on the studies of the BP in schizophrenia and depression, but do not attempt to produce an exhaustive or all-inclusive review. I will first describe the studies which have investigated the BP in patients with schizophrenia and depression, followed by a summary and critical evaluation of their results before finally considering their implication for the pathophysiology of these disorders. However, before reviewing the experimental evidence, let us consider the rationale behind studying the BP in psychiatric disorders such as schizophrenia and depression.

MOTOR POTENTIALS IN SCHIZOPHRENIA: WHAT IS THE RATIONALE?

Emil Kraepelin (1905) believed that the fundamental process in schizophrenia (dementia praecox) was a weakening of volition. He also postulated that the illness is caused by damage to the frontal lobes of the brain.   Subtle motor abnormalities such as clumsiness, disorganization and slowness of movement have been described in patients with schizophrenia (King, 1965; Manschreck, 1986).   Patients particularly those with chronic schizophrenia exhibit negative symptoms such as poverty of speech and movement and flattening of affect. It is probably fair to say that these observations/findings/symptoms have formed the rationale behind most studies of the BP in schizophrenia to date.

The BP was the first brain potential to be described that could be connected to volition or performance of a voluntary action (Kornhuber und Deecke, 1964, 1965). Kornhuber proposed a model for the organization of voluntary movement and studied movement-related EEG changes before voluntary movements in order to find a correlate for the hypothesized volitional deficits in schizophrenia (1983, 1984a, 1984b, 1984c). Together with Deecke and W. Lang, Kornhuber proposed a special role for the supplementary motor area (SMA) in the control of timing of voluntary movements (Deecke et al., 1976, Lang, W. et al., 1984, Kornhuber et al., 1989).

At the same time as this work on the BP, there was increasing knowledge of frontal dysfunction in schizophrenia based on use of functional imaging (reviews by Ingvar, 1987; Weinberger, 1988; Bunney & Bunney, 2000).   A better understanding of the SMA, especially concerning its role in control of sequential motor tasks (Deecke et al, 1985) and increasing knowledge of motor disturbances in schizophrenia (King 1965; Manschrek, 1986; Boks, 2000) was achieved over the last three decades.

The two earliest studies of BP in schizophrenia were partly driven by an attempt to identify cortical potentials 'characteristic' of these patients. Timsit-Berthier (1973), was one of the first investigators to conduct studies of motor potentials in patients with psychiatric disorders. She investigated the CNV as well as the BP in different psychiatric populations since "certain potential configurations may occur principally in psychotic subjects". Kornhuber and co-workers (1983) considered it interesting to investigate the BP in schizophrenia because of the lack of volition, deficits in concentration and motor disturbances typically found in these patients. Their goal was to find an earlier biological 'marker' for the diagnosis of schizophrenia with negative symptoms.

Chiarenza et al. (1985) based their study of the BP on psychomotor deficits in schizophrenia as described by King (1965). Singh et al (1992) noted the presence of movement-related deficits such as abnormal saccadic eye movements and failure of synchronous tapping in schizophrenia. This, together with evidence for frontal dysfunction in schizophrenia, led Singh et al (1992) to consider the study of the BP as an index of motor preparation, engaging the motor cortex and the SMA, to be of interest in these patients. The studies of Kubota et al. (1999) and Karaman et al. (1997) were also based on motor dysfunctions associated with schizophrenia, and found lower BP amplitudes in patients suffering from more pronounced negative symptoms (Andreasen, 1984). Adler et al.'s (1989) study of BP in tardive dyskinesia was based on the hypothesis that if the BP reflected dopaminergic activity in the basal ganglia, one should find an increased amplitude of BP in patients with tardive dyskinesia.

Along with increasing knowledge of SMA function and volitional processes, the hypotheses on which the studies of BP are based have become more and more distinct over the last five years.   Consistent with Kraepelin's formulation of lack of volition in schizophrenia,

Frith (1987, 1992) proposed that in patients with chronic schizophrenia, intentions of will are not formed which gives rise to the negative behavioral signs such as poverty of action and speech, flattening of affect and lack of volition. Fuller et al. (1999) set out to use movement-related potentials recorded prior to self-initiated or externally-triggered movements to test the hypothesis advanced by Frith (1992) that while willed actions are impaired in schizophrenia particularly those with predominance of negative symptoms, stimulus-driven actions are intact. Willed actions are self-selected or self-initiated voluntary movements. Differences in brain activation have been shown with PET between self-initiated and externally triggered movements (Jahanshahi et al. 1995). Fuller et al. (1999) predicted significantly lower BP in patients with schizophrenia with a high ranking for negative symptoms and it was hypothesized that particularly the volitional process involved to a greater extent in self-initiated than externally-triggered actions, which contributes to the BP, should be disturbed. The SMA which is considered to be involved in the generation of the BP, is also noted to play a role in the control of self-generated, sequential and more complex movements. More recent studies such as those of Dreher et al. (1999) and Löw et al. (2000) have used movement-related potentials to examine the differential impairment of such self-generated, sequential or complex movements in patients with schizophrenia.

In summary, despite its gradual refinement, the rationale for investigating motor-potentials in schizophrenia has in some ways remained the same over the last three decades: motor disturbances especially those concerned with volitional movements known to occur in schizophrenia. Due to these deficits, changes in the BP were expected. Since these motor deficits occur mainly as part of negative symptoms, several studies hypothesized that the concomitant changes in the BP should be more pronounced in chronic patients or in those with negative symptoms.

EARLIER STUDIES OF THE BP IN SCHIZOPHRENIA

The earlier studies discussed in this section, conducted between 1973 and 1989, were important in establishing that the BP is impaired in schizophrenia compared to normal control subjects.

Timsit-Berthier et al. (1973) studied the BP and CNV in 94 psychotic patients (average age: 27 years) compared with 86 neurotics and 72 controls. For recording the BP, subjects were requested to perform a button press every 4-5 seconds. In addition to the EEG, the EMG and EOG and electrodermal responses were recorded. Sweep time analyses at four-second intervals, with a premovement epoch of 1.4 seconds was used. The electrode position Cz (vertex) versus mastoid was selected for further qualitative analysis. All courses of the BP were divided into two types. The type N is a persisting negative BP after the end of motor activity. The type P was most often negative and flat in character; when it was biphasic, the main characteristic was the inversion of polarity after the muscular contraction. Type P was seen significantly more often in psychotics (P < 0.05). Sometimes, the slow negative component was flat. Overall, Timsit-Berthier noted that a very different type of motor potential appeared mainly among psychotics.

This early study of the BP by Tismit-Bethier et al. (1973) was followed a decade later by a series of studies by Kornhuber and his group. In the first study (Westphal et al., 1984), 16 psychotic patients on neuroleptics were compared with 15 matched controls. Diagnoses were confirmed by two independent raters. Neurological deficits such as parkinsonism or tardive dyskinesia were not described. A self-paced fast voluntary flexion movement of the fingers was performed. EMG of the flexor digitorum longus muscle was used as the trigger for back-averaging the BP. Other measurements included EOG and galvanic skin response. The EEG was recorded from Cz, Pz, and the left and the right central and parietal positions with linked ears as reference. Epoch duration was 4.0 seconds. Baseline was set during the first 500 ms. Amplitudes and latencies were measured. The results showed that the BP lasted twice as long in treated patients with schizophrenia than in normal controls. At Cz, the latency was 1500 ms in patients with schizophrenia versus 800 ms in controls. Other work by this group found that in some untreated patients, the BP showed a positive course (Westphal et al., 1986). The lower surface negative shift together with earlier onset of this negative shift has been related to the fact that patients with schizophrenia exhibit a longer motor reaction time (Kornhuber, 1983).

In the second study (Grözinger et al 1986), the Kornhuber group tried to replicate the earlier findings. The investigation was performed in 31 medicated patients and matched normal controls. The mean age was 34.2 ± 3.9 years. The mean duration of illness was 8.5 years (range 1-24 years). The mean chlorpromazine-equivalent dose was 1475 mg. The sample comprised 17 paranoid, 10 hebephrenic, one catatonic and three residual patients (AMDP, DSM-III). The BP started had a longer mean latency in patients compared to controls at Cz (1143 versus 871 ms, $P < 0.10$). However, no significant group differences were obtained for the amplitudes. Details of these results are given in Grözinger et al. (1986) and Westphal et al (1985).

The third study (Westphal et al., 1996) comprises the results of unmedicated patients (n = 13). Compared to controls, significantly lower amplitudes were obtained in patients who had never undergone treatment or were non-medicated at the time of assessment at 800, 400, 200, 100 and 0 ms prior to movement onset, when averaging across all electrodes. Topographical comparison between the electrodes in each group showed that the maximal amplitude of the BP over Cz was less pronounced in the untreated patients with schizophrenia compared to the normal controls. Also, the frequency of a the BP having a positive course was significantly higher in the patient group.

In 1985, two further groups published studies of the BP in schizophrenia. The first was by Chiarenza et al. (1985) who investigated 13 patients with chronic schizophrenia (seven paranoid, six hebephrenic). The patients were free of medication for 20 days. No evidence of parkinsonism or tardive dyskinesia was described. The patients were classified according to the DSM-III classification and the Brief Psychiatric Rating Scale (BPRS). Mean age was 28 years and the average duration of the illness was about five years. The motor task consisted of pressing a button held in the left hand with the thumb. EMG was recorded from the forearm flexor muscle. The EEG was recorded from nine scalp electrodes (Fz F7, F8, Cz C3, C4 and PZ, where C3 and C4 were positioned 2 cm anteriorly to the 10-20 location over the motor cortex). The BP and the motor cortex potential (MCP) were measured. As evident from Figure 1, in patients with schizophrenia, the amplitude of the BP was significantly reduced compared to controls. For example, the BP in normal controls was about -7.56 mV at Cz, compared with -4.90 mV at Cz for the patients. The MCP measured about 100 ms before movement onset was -16.10 mV in controls, but only -9.82 mV in patients. These differences were found to be statistically significant. A higher inter- and intra-individual variability in patients is described. At the parietal electrode position the maximal amplitude of negativity (MCP) was about -5.67 mV (sd $\pm$ 6.70) in patients. The standard deviation shows that some patients would have had a BP with a positive course. Correlations were evaluated between the symptom ratings and the amplitudes of the different brain potentials. No notable correlations between clinical symptoms or scores of the BPRS with the BP or the MCP were found.

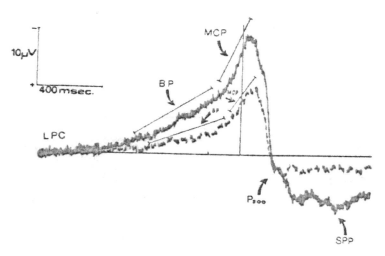

Figure 1: The Bereitschaftspotential (BP) and motor cortex potential (MCP) for patients with schizophrenia (dotted line) and normal controls (continuous line). The vertical line indicates the trigger point. Figure reproduced from Chiarenza et al (1985) with permission.

Papakostopoulos and colleagues were the second group of investigators to publish work on the BP in schizophrenia in 1985. Papakostopoulos et al. (1985a) used a self-paced, goal-directed task. The study sample comprised nine patients. Two were new cases, while the other seven had not undergone treatment for two years. Following recruitment for the study, patients received neuroleptics and then participated in the study protocol. Results were compared to those obtained from controls. Mean patient age was 45.4 ± 15 years. Extrapyramidal symptoms, including rigidity, tremor, gait changes, facial expression, hypokinesia and speech disorders were documented. EEG investigations were carried out when parkinsonism appeared. After this investigation, treatment with the anticholinergic agent orphenadrine was instituted. EEG assessment was repeated under this treatment four weeks later. Movement consisted of a thumb movement of the left hand and then the right hand, or pressing with both hands simultaneously. Electrodes were placed at FPz, Pz and Cz pre- and post-centrally, as well as over the left and right hemispheres. EMG was taken from the left and right forearm flexor muscles. The sampling epoch lasted from 1.2 s preceding to 0.824 s following initiation of the movement. Inter-trial intervals ranged from 8-30 s. The results of clinical psychiatric evaluation and the mean chlorpromazine units were not given. The authors did, however, provide a rating of extrapyramidal symptoms. The results show a significantly smaller amplitude of the BP in patients compared to normal subjects. MCP developed in both groups, but was smaller in the patients. The second investigation included administration of the anticholinergic agent and showed that while the amplitude of the BP was larger in patients with than without the drug, it still did not attain the value observed in the normal control group. Factor-analysis was performed. The increase in the BP (negativity) in treated patients following supplemental administration of the anticholinergics suggests improved function of the neuronal system. Decreased negativity in patients receiving neuroleptics might be due to cholinergic hyperactivity. Thus, increased cholinergic activity could lead to increased cortical tonic negativity which cannot be increased any further despite adequate performance of motor tasks. This would explain the lower amplitudes of the BPs and MCP in patients with schizophrenia treated with neuroleptics. In the second paper published by Papakostopoulos et al. (1985b) 48 artifact-free trials were available for each subject and were used for analysis. The BP in the patient group was significantly (P < 0.01) smaller than in normal controls.

MORE RECENT STUDIES OF THE BP IN SCHIZOPHRENIA

Perhaps partly based on imaging results suggesting frontal dysfunction in schizophrenia, the studies of the BP in schizophrenia conducted since the late 1980s have been more hypothesis driven. For example, greater impairment of the BP over frontal sites and greater deficits of the BP in patients with higher negative than positive symptoms have been predicted.

From the results presented above it is evident that the effect of neuroleptic medication used to treat schizophrenia and the resulting tardive dyskinesia on the BP is unclear. This question was more specifically addressed in a study by Adler et al. (1989) who noted that if the BP reflected dopaminergic activity, then an increased BP amplitude in patients with schizophrenia and tardive dyskinesia would be predicted. They investigated 11 medicated patients together with 11 other patients with schizophrenia exhibiting symptoms of tardive dyskinesia and a group of 11 matched controls. Mean age was 37 ± 10 years for medicated patients without tardive dyskinesia, 39 ± 13 years for the group with tardive dyskinesia, and 31 ± 6 years for normal controls. Time since onset of the disease was over five years in both groups. The mean chlorpromazine equivalent dosage was 210 ± 112 mg/day. Patients' neurological examination was normal, including rating with the Abnormal Involuntary Movement Scale (AIMS). Patients with symptoms of parkinsonism were excluded. Patients were receiving anticholinergics. Patients were instructed to gently press a telegraph key with the index finger and the second finger of the right or left hand. Electrodes were positioned 2 cm anterior to the C3 and C4 in the 10-20 system linked ear lobes electrodes served as reference. Sweep duration was 4.0 s. The sampling epoch began 2.0 s prior to movement onset. EMG was measured in the superficial flexor digitorum of the third finger. The baseline was set at the 300 ms period, beginning at 1500-1000 ms before the movement. BP amplitudes were averaged during right hand movement over the left hemisphere at C3'. Amplitudes were 5.4 ± 3.0 mV in normal controls, 5.2 ± 2.1 in schizophrenics without tardive dyskinesia, but 18.4 ± 18.4 in schizophrenics with tardive dyskinesia. The latter group exhibited amplitudes of 18.2 ± 13.8 mV over the rifht hemisphere during right hand movements. Student's t-test showed significant differences between the patients with or without tardive dyskinesia over both hemispheres for both right or left hand movements. The increased amplitude of the BP significantly correlated with the severity of tardive dyskinesia as measured on the AIMS. There was high amplitude variability in the schizophrenic group with tardive dyskinesia. The authors did not report the latencies of the BP. The authors interpret their findings to suggest that the amplitude of the BP reflects the dopaminergic activity of the basal ganglia. The authors speculate that the higher BP amplitude in patients with tardive dyskinesia may be due to a supersensitivity of the dopamine receptors or to a relative dopaminergic excess in tardive dyskinesia. The finding that there were no differences between normals and medicated patients with schizophrenia without tardive dyskinesia is also discussed, with the authors postulating that this might be due to the patients' neuroleptic (dopamine antagonist) treatment.

Singh et al. (1992) investigated nine medicated schizophrenics matched with nine normal controls. Subjects' mean age was 49.3 ± 13.6 years. Eight of the patients exhibited active psychotic symptoms including auditory hallucinations and delusions. Patients were classified according to DSM-III-R. All patients were on neuroleptics, with a mean chlorpromazine equivalent dose of 431 ± 255 mg/day. Mean duration of illness was 24.7 ± 10.5 years. Patients were rated for global severity of symptoms. Patients with tardive dyskinesia or drug-induced parkinsonism were excluded. Neurological examination was conducted at the time of testing. None of the patients had extrapyramidal symptoms. All patients were right handed. Patients were instructed to briskly press a button with the thumb. The button press was self-initiated. Inter-movement interval ranged between 3-10 s. The EEG epoch lasted from 1400 ms before to 600 ms after movement onset. EMG was obtained from the thenar muscle and EOG was also recorded. Midline electrodes were located at FPz, Fz, Cza, Cz, with FP1, F3, C3a, C3 and P3 over the left hemisphere and corresponding positions over the right hemisphere. Linked earlobes served as reference. 150 trials for each hand in each group were included. The BP was identified as the initial portion of premovement negativity. About 450 ms prior to movement onset, the second component (NS') occurred. The third component (motor potential, MP) refers to the additional negativity generated immediately preceding the movement. The amplitudes of these different parameters were calculated in restricted windows, i.e. BP at -100 to -500 ms, NS' at -500 to -100 ms, and MP at -100 to 0 ms. The amplitude of the BP and the NS' component was either reduced or abolished in a majority of patients with schizophrenia at frontal and central scalp sites (Figure 2). The MP component was also reduced. The scalp topographical distribution of late components in schizophrenia was similar to those of controls (see Figure 2). No relationship could be established between duration of symptoms and amplitudes, or between drug dosage and amplitudes. Abnormal motor-related potentials in schizophrenia might relate to deficits in different processes in schizophrenia. These include readiness, preparation, initiation, planning, volition, and attention to a voluntary act. These latter psychological constructs have been associated with amplitude variation. The findings are discussed as signs of frontal lobe dysfunction. They are compared to lower BP amplitudes in Parkinson's disease and in patients with prefrontal cortex lesions. This similarity of findings in schizophrenia and patients with prefrontal lesion provide further support for frontal dysfunction in schizophrenia.

Karaman et al. (1997) investigated 30 patients with schizophrenia, eight of whom were not on medication. Subgroups were formed, including nine patients suffering from predominantly positive and seven from negative schizophrenia (based on the criteria of Andreasen, 1984), while 14 showed a mixed type. In the treated group, the chlorpromazine equivalent dosage was 376 mg/day. Mean age was 34.2 years and the mean duration of illness was 10.0 years. Neurological examination was performed. Diagnoses were made according to DSM-III-R. Patients were instructed to perform a rapid extension movement of the right wrist. The EMG was obtained from the extensor digitorum communis muscle. EEG was recorded from C3, C4 and Cz.. Epoch duration was 2000 ms, ranging from 1600 ms before to 400 ms after movement onset. 50 trials were recorded. BP amplitude was measured at peak negativity (N1), the early component of BP (NS1) 650 ms before EMG onset, as well as the late lateralized component of the BP (NS2), derived by subtracting NS1 from peak negativity. Mean peak negativity (N1) in controls was -10.7 mV, and -7.2 mV in patients with schizophrenia. The mean early component, NS1, was -2.9 ± 2.0 mV in the control group and -1.3 ± -1.6 mV in schizophrenics. The latter included instances of positive or flat traces in the early component. Further analysis showed that lower amplitudes of the early component may be associated with positive symptoms in schizophrenia, while reduction of the late component is likely associated with negative symptoms. The analysis did not reveal any medication-related effects: there were no significant effects of medication on the BP amplitude and there were no difference between the medicated and drug-free patient groups.

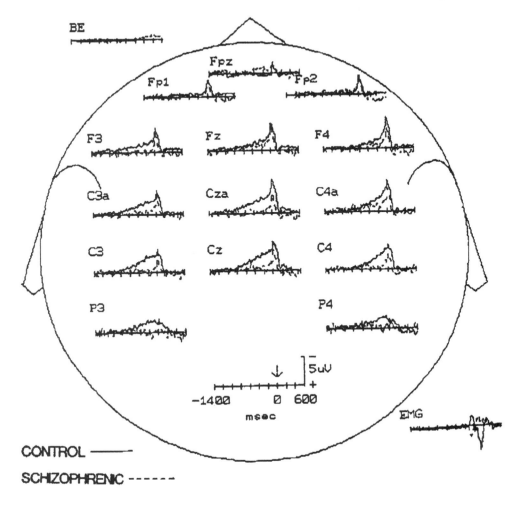

Figure 2: Group grand-averaged waveforms from control (solid lines) and patients with schizophrenia (dotted lines) for right hand (top) and left hand (bottom) button press. All subjects were right-handed and contralateral enhancement of the movement-related potentials is evident for both groups. The decrement or absence of the BP and the NS' components in patients with schizophrenia particularly over the frontal leads (F3 and F4) and a relative preservation of the MP are clear. The arrow indicates the trigger point from which back-averaging was done. Reproduced from Singh et al (1992) with permission.

Twenty-sever medicated schizophrenic patients diagnosed according to DSM-IV criterial were compared to 31 healthy volunteers by Kubota et al. (1999). All patients received antipsychotic drugs and antiparkinsonian agents. The schizophrenic group consisted of three catatonic, ten disorganized, 11 paranoid and three undifferentiated types. Patients were evaluated using the positive and negative syndromes scale (Andreasan, 1984). Mean age in the schizophrenia group was 30.5 ± 6.4 years and mean duration of the illness was 9.7 ± 5.8 years. No evidence for neurological or spinal root disease was seen in any patient or volunteer. Subjects performed a push button exercise by flexing the right thumb. Inter-trial interval was about 5 s. EEG was recorded from C3, CZ and C4. Earlobes served as the reference. The epoch lasted 3000 ms. The results were described mainly qualitatively. The patterns of negativity in the 27 patients with schizophrenia were divided into different waveform types The motor-related cortical potentials (MRCPs) were classified as normal or abnormal according to the Timsit-Berthier classification (1973). Six subjects had normal waveforms, while 21 showed abnormal MRCPs. In the control group of 31 subjects, 27 had normal waveforms. The incidence of abnormal waveforms was significantly higher in schizophrenics than in the control group.

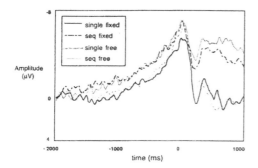 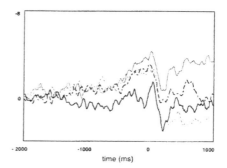

Figure 3: Grand averages of the mean BP at Cz for the single fixed, single-free, sequence-fixed, sequence-free conditions. Reproduced with permission from Dreher et al. (1999).

In the study of Dreher et al (1999), eleven medicated schizophrenics were matched to controls according to age and gender. The mean age of the patients was 39.1 ± 10.04 years, and the mean age of the controls was 38.9 ± 7.98 years. All subjects were right-handed. Patients with tardive dyskinesia were excluded. Some patients showed mild drug-induced Parkinson symptoms. Mean duration of hospitalization was 97.4 ± 54.70 days. Most patients suffered from paranoid schizophrenia (ICD 10 F 20.0). Movements were performed with the index, middle, third and small fingers of the right hand and were self-paced intervals of every 4-5 s. Four different motor tasks were performed, combining different levels of task complexity (single vs sequence) and mode of movement selection (fixed vs free). The four tasks were: fixed single-key press, fixed sequence of four key presses, freely selected single key press, freely selected sequence of four key presses. Fifty artifact-free responses in each task were used for analysis. Electrode positions were Fz, Cz, Pz, C1, F3, C3, P3 and C2, F4, C4, P4 with linked earlobes as the reference. EMG was obtained from the flexor side of the right forearm. Epoch duration was from -2000 ms to +1200 ms after movement onset. The baseline was from -2000 ms to -1900 ms. Higher BP and NS' were obtained in both groups with sequential than single movements, and for freely selected than fixed movements Patients with schizophrenia exhibited lower and more irregular BP amplitudes, which held true for all four motor tasks (see Figure 3). The onset of the BP occurred significantly later in patients with schizophrenia than in controls. For example, when performing a simple self-paced single key-press, peak amplitudes at Cz were -5.05 mV in controls and -1.43 ± 1.76 mV in schizophrenia. The lower BP amplitudes at Cz in schizophrenia is interpreted as a motor preparation deficit. The lower amplitude of NS' in schizophrenia is considered to be a deficit in the decision making process. The results were considered to reflect a focal deficit of activation of the SMA in schizophrenia, which was interpreted in terms of a fronto-striatal dysfunction.

Fuller et al (1999) compared five patients with the diagnosis of schizophrenia with high ratings for negative symptoms with six patients with the diagnosis of schizophrenia with high ratings for positive symptoms as well as with six normal controls. Groups were matched on age and handedness. They did not differ on the Beck Depression Inventory or the Mini Mental State Examination (MMSE) and none of the subject scored below the cut off of 25 on the MMSE. The patients were rated for current positive symptoms and negative signs differed significantly between the mean scores for positive symptoms and negative signs. The two patient groups did not differ in terms of duration of illness and or with regard to the mean dose of anticholinergic or neuroleptic medication. Electrodes were over F3, Fz, F4, Fc3, Fcz, Fc4, C3, Cz, C4, P3, Pz, P4. Linked earlobes served as reference. EOG and EMG from the extensor indices proprius were obtained. Self-initiated movements were executed once every 3 s (brisk lifting of the right index finger) and patients also performed externally triggered movements (finger lifting movement in response to a tone). As shown in Figure 4, patients with schizophrenia and high ratings for negative symptoms exhibited reduced amplitudes of motor-related potentials for the late and the peak component and reduced slopes of the early and late motor-related potentials prior to self-initiated movements. These differences were not found prior to externally triggered movements. Patients with higher ratings for positive symptoms did not differ significantly from the normal controls in terms of amplitude or slope of motor-related potentials prior to self-initiated or externally triggered movement. Differences, especially in patients with high ratings for negative signs, are interpreted as showing impairment of willed actions, mediated by impaired functioning of the fronto-striatal loops. The results suggested that motor-preparatory processes are more profoundly impaired in patients with schizophrenia in whom negative symptoms predominate.

Catatonia is a psychomotor syndrome which results in an inability to execute and terminate movements giving rise to akinesia and posturing, both of which respond to benzodiazepines. Therefore, Northoff et al. (2000) recorded movement-related potentials in ten patients with akinetic catatonic schizophrenia compared with 10 patients with schizophrenia without catatonia as well as in 20 healthy controls before and after administration of lorazepam. The groups were matched for age, sex, medication, and psychiatric disease. The only difference between the two patient groups was presence or absence of catatonia. Patients with dyskinesias or hyperkinesias and/or neuroleptic-induced hypokinesias were excluded from the study. All patients were right-handed. EEG evaluation was performed eight days after hospital admission and withdrawal of lorazepam administered for catatonia after admission. Patients were treated with antidepressive drugs or neuroleptics. The chlorpromazine equivalent dosage was about 180 mg/day in catatonics and about 167 mg/day in the psychiatric controls. Seven patients were treated with anticholinergics. Duration of illness was about 7.5 years in both groups. The motor task consisted of a rapid lifting (extension) of the right index finger, which rested on a fixed surface. EMG was recorded from the right extensor indicis proprius muscle. Inter-trial interval was about 4-5 s. Electrodes were positioned at Fz, Cz, Pz. F3, T3, P3 over the left hemisphere and the corresponding positions over the right hemisphere. Linked earlobes served as the reference. The EEG epoch lasted 3000 ms, from 2500 ms before to 500 ms after EMG onset. About 50 trials were recorded in each group. The latency of onset of the late BP over Fz, Cz and Pz was significantly longer in catatonics than in healthy controls. Similarly, the latency of the movement potential at Fz, Cz and C3 was significantly longer in catatonics than in psychiatric and healthy controls. No differences in amplitudes between catatonic patients and psychiatric controls were found. However, both psychiatric groups showed a tendency towards lower amplitudes than the healthy controls. After receiving lorazepam, catatonics show a significantly later onset of the

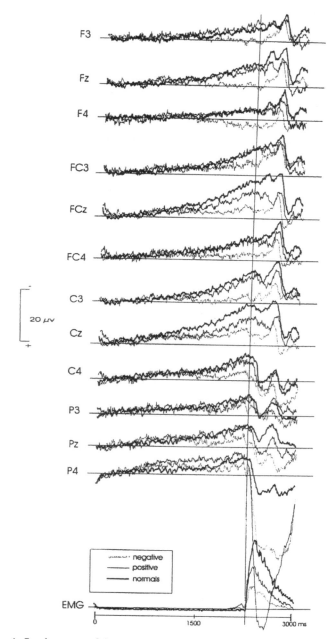

Figure 4: Grand averages of the movement-related potentials prior to self-initiated movements for the normal controls, patients with schizophrenia and predominance or positive or negative symptoms. Reproduced from Fuller et al. (1999) with permission.

late BP and a significantly earlier onset of the movement potential. There were significant correlations between motor symptoms, anxiety, and latency of the late BP over Cz in catatonia. The results are discussed in terms of the BP as a trait marker of cortical motor function in post-acute catatonia patients.

The most recent study is by Löw et al. (2000), in which the BP was investigated in 16 schizophrenic patients matched to controls for age and sex. EEG was derived from 61 scalp locations. Subjects were instructed to execute simple sequences of button pressing or to produce self-generated sequences. Sequences were self-initiated at intervals of about 10 s. Early BP, late BP, negative slope (NS') and motor-potential were measured. Preceding self-generated sequences the controls showed larger amplitudes of the motor-potentials than preceding simple sequences. This difference was not significant in the group with schizophrenia. The findings were interpreted as supporting a view of dysfunctional action planning in schizophrenia.

STUDIES OF THE BP IN DEPRESSION

Slow brain potentials have been relatively extensively studied in patients with depression. However, only a limited number of studies used the BP paradigm. Some of these studies will be considered below. The studies of BP in depression have been largely based on the observed psychomotor retardation characteristic of some patients with depressive illness, leading to the prediction that motor preparatory processes as measured by the BP would be impaired in these patients.

In the study by Thier et al. (1986), patients with primary depression were compared with eleven healthy controls. Patients were instructed to press a micro-switch to initiate a voluntary movement, followed by an acoustic 'go' or 'no go' signal. Depression was scored using the Hamilton depression scale. EEG was recorded from Fz, Cz, Pz linked to the mastoids. EMG was derived from the flexor indicis muscle of the right hand. Epoch duration was over 4 s, starting 1.52 s prior to movement for BP measurements. Although depressed patients had considerably longer reaction times, their BPs began at the same time as those of the controls, developed in the same way, reached equal amplitudes, and showed a similar decline after the movement had been completed.

Khanna et al. (1989) compared twenty-six subjects with DSM-III criteria for major depression without psychotic features to 26 healthy controls. Mean age for patients was 31.8 years, and for controls 30.2 years. EEG was recorded from Fz and Cz, with linked ear-lobes as the reference. EMG was derived from the right thenar eminence and forearm. Subjects were asked to observe the central 2 cm of the screen, which contained bright yellow vertical lines. They were instructed to press two micro-switches in response to a dot moving slowly across the screen. 49 trials were recorded. The amplitude of the BP was measured as the maximum negativity reached before pre-motion positivity began. There was a statistically lower amplitude in depression. (see Figure 5) The results were discussed in terms of a frontal deficit in depression.

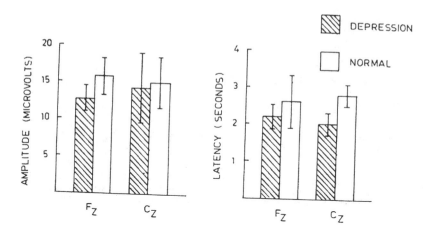

Figure 5: The amplitude and latency of the BP at Fz and Cz for patients with depression and normal controls. Reproduced from Khanna et al. (1989) with permission.

Fourteen patients with unipolar endogenous depression were compared to 18 matched healthy volunteers by Haag et al. (1994). Subjects had to clench and open their right fist at self-paced intervals. EEG was recorded from 19 standard positions of the 10-20 system. EMG was derived from the flexor carpi ulnaris muscle of the right arm. The authors measured the onset of negative potential rise and BP amplitude prior to movement onset with BP integral (calculation of the area under the curve). Onset of BP in patients was at -1287 ms (mean), and at -1345 ms in controls. Mean BP amplitude at Cz was -9.5 μV in patients and -12.1 μV in controls (mean). Lateralization of BP amplitude in normal subjects occurred only during the last 80 ms before movement onset with a shift of the maximal amplitude from Cz versus C3. Patients showed a significant asymmetry of the maximal BP integral to the left. The lower amplitudes of the mean BP in patients were not statistically significant. However, significant differences were found between the groups when comparing an area ratio. The results were discussed as a deactivation of the right ("emotional") hemisphere in depression. Three patients showed a normalization of the abnormal BP distribution when in remission immediately prior to hospital discharge.

Over the past three decades, 14 journal articles have investigated the BP in schizophrenia. Their results are summarized in Table 1. Significantly lower amplitudes of the BP in patients with schizophrenia who did not suffer from tardive dyskinesia or catatonia were found by Timsit-Berthier (1973), Chiarenza et al. (1985), Papakostopoulos et al. (1995), Singh et al. (1992), Karaman et al. (1997), Kubota et al. (1999), Dreher et al. (1999), Fuller et al (1999), and Löw et al (2000). No significant differences from controls in BP amplitudes for the patients were found by Northoff et al. (2000), Adler et al. (1989) and Grözinger et al. (1986). Nevertheless, in all studies, there was a tendency toward lower amplitudes and higher variability than in controls. The results of latencies are difficult to compare since not all authors used these parameters.

OVERVIEW

Table 1: Summary of the results of the studies which have recorded the BP in patients with Schizophrenia

Author	Patients (n) Subgroups (n)	Medicatio	Bereitschaftspotential NS1 Early comp.	NS2 Late comp.	N1 Peak pot.	Clinical Correlation	variability BP in schizo-phrenics
Timsit-Berthier 1973	94	+	Sign. more flat or "negative" waveform single pat. With positive polarity			Compared to normals and neurotics	higher
Westphal et al. 1984	16	+	Longer latency of early comp., lower peak potential				higher
Grözinger et al. 1986	31	+	longer latency sign. diff. Amplitudes				higher
Westphal et al. 1996	13	–	Sign. lower single pat. With positive polarity	Sign. Lower	Sign. lower		higher
Chiarenza et al. 1985	13	–	Sign. lower	sign. Lower	Sign. lower	no correlation between BPAS and BP	higher
Papakosto-poulos et al. 1985	9	–			Sign. lower		higher
Adler et al. 1989	22 subgroup 11 tardive dyskinesia	+	no sign. diff. in tardive dyskinesia. No sign. diff. in schizophrenic early comp.; late comp. peak pot	higher in tardive dyskinesia		Positive correlation between AIMS scale and BP	higher
Singh et al. 1992	9	+	Sign. lower Shorter latency	sign. Lower	Sign. lower	no correlation drug dosage and duration of symptoms	higher
Karaman et al. 1997	30		Sign. lower	sign. Lower	Sign. lower	- positive correlation at C4 for NS1 and positive sympt.	higher
	Subgroup 8	+	Sign. lower	–	Sign. lower	- positive correlation C4, NS2, N1 and negative sympt.	
	22	–	Sign. lower	sign. Lower	Sign. lower		
	9 positive	±	Sign. lower	–	Sign. lower		
	7 negative	±	–	sign. Lower	Sign. lower		
	14 mixed type	±	Sign. lower	–	Sign. lower		
Kubota et al. 1999	27	+	flat and abnormal waveforms single pat. with positive polarity			tendency of lower PANSS score in pat. With abnormal BP	higher
Dreher et al. 1999	11	+	Sign. lower. Fixed versus free movements (single and sequences) with sign. group, complexity and mode of movement effect	sign. Lower	Sign. lower		higher
Fuller et al. 1999	11	+				- positive correlation positive symptoms at Fz and F4 for NS2 and N1	
	Subgroup -5 negative	+	ns	sign. Lower	Sign. lower		
	-6 positive	+	ns	Ns	ns	- patients with higher ratings of positive symptoms have more normal amplitudes	
			no sign. diff. in externally triggered movements				
Northoff et al. 2000	20	+				correlation with catatonic symptoms and axiety in NS2 and MP	higher in latencies
	Subgroup 10 catatonic	+	- no sign. amplitude findings sign delayed onset of NS2 and MP				
	10 schizo-phrenic	+					
Löw et al. 2000	16	+	Schizophrenics no sign diff. In NS1, NS2 and N1 beween self generated sequences and repetitive responses; controls differed significantly. Lower amplitudes in schizophrenics in self generated sequences.				

As evident from Table 1, while the results are relatively consistent in that with exceptions (Grozinger et al., 1986; Adler et al. 1989; Northoff et al., 2000) the majority show lower amplitudes of one or more of the BP components in patients with schizophrenia, there are also a number of inconsistencies in the results of these studies. BPs can show normal amplitudes in untreated and as well as in treated patients with schizophrenia, depending on the specific constellation of the symptoms that the patients have and the specific types of task used for recording the BP. Therefore, it is important to consider some of the factors that may have contributed to these discrepanices.

Recording procedures

In terms of technical aspects of the recordings, almost all studies were beyond reproach. All movements were self-paced with inter-trial intervals usually exceeding 5 s. The movements were brisk, with triggering done by EMG onset. Visual inspection and artifact control including EOG were used to exclude sweeps contaminated by artifacts. The number of trials in most studies was about 50-60 sweeps. Electrodes were positioned according to the 10-20 EEG-system, with six to nine electrode positions being the norm in most studies. Others such as Karaman et al. (1997), Adler et al. (1989), Kubota et al. (1999) as well as Timsit-Berthier (1973) only used central positions.

BP measurement

The measurements of premovement negativity were not consistent. The study of Timsit-Berthier (1973) described the results more qualitatively, dividing patients into a type N and a type P. A qualitative analysis was also used by Kubota et al. (1999) and partly by Westphal et al. (1996). Most authors, however, used amplitude measurements and tried to distinguish between the different components of the negativity arising before initiation of voluntary movements. All used peak negativity and most tried to distinguish the two main components of the BP i.e. the early and the late components. Some were able to define latencies.

Experimental tasks

The specific motor tasks differed across studies. Simple, voluntary self-paced single movements were performed in the studies by Timsit-Berthier (1973, button press), Kornhuber (1983, flexion movement of the fingers), Adler et al. (1989, press a telegraph key with index finger and second finger), Singh et al. (1992, briskly press a button with the thumb), Karaman et al. (1997, brisk extension movement of the right wrist), Kubota et al. (1999, push button by flexing the right thumb), and Northoff et al. (2000, brisk lifting [extension] of the right index finger).

Dreher et al (1999), Fuller et al (1999) and Löw et al. (2000) compared movement-preceding negativity prior to different types of movement which differed in terms of the degree of volitional control necessary. For example, in the study of Dreher et al. (1999), differences from normals became more pronounced when producing free sequential voluntary movements. Fuller et al. (1999) found impaired movement potentials prior to self-initiated movements in patients with schizophrenia and high negative signs but no differences between patients and controls when movements were externally-triggered. Löw et al. (2000) described differences between self-generated sequences and repetitive responses, with lower amplitudes in patients with schizophrenia during self-generated sequences.

Patient characteristics

In most studies, standardized diagnostic criteria such as DSM-III, DSM-III-R or ICD were used. Neurological examinations were performed in most cases and some studies rated neurological deficits. This was most prominent in the work of Adler et al. (1989), who studied patients with schizophrenia with tardive dyskinesia, as well as in Northoff's study (2000) which investigated patients with catatonia. The patients with tardive dyskinesia (Adler et al., 1989) showed very high negativity of about -18.4 μV over the contralateral motor cortex prior to right finger movements. The special case of patients with catatonic schizophrenia revealed a delay in the onset of the late readiness and the movement potential (Northoff et al., 2000). Differences in movement-related potentials have also been found between patients with predominance of positive versus negative symptoms. Karaman et al (1997) found the NS1 component to be reduced in patients with positive symptoms and the late BP component in patients with negative symptoms. Prior to self-initiated movements, Fuller et al (1999) reported that the amplitude of the late and peak BP and the slope of the early and late BP were significantly lower than controls for patients with schizophrenia and predominance of negative symptoms but not for patients with positive symptoms. The results of these more recent studies clearly demonstrate that the specific constellation of symptoms observed in the patients included in a sample are important determinants of the type of deficits observed in the BP.

Medication effects

In certain studies (Chiarenza et al. [1985], 8 patients; Karaman et al. [1997], Westphal et al. [1996]) patients were either not currently undergoing treatment or had never been treated. In all other studies, patients were receiving medication. In most studies where patients were on treatment, equivalent doses of neuroleptics were given   However, it was not always clear whether or not patients may have experienced some mild Parkinson symptoms due to neuroleptic treatment. The effect of neuroleptic medication on the BP is unclear. The results of Adler et al (1989) for patients with schizophrenia with or without tardive dyskinesia suggests that the dopamine-antagonist medication that most patients are treated with affects the amplitude of the BP. Others (eg Karaman et al, 1997) did not find significant differences in the amplitude of the BP between medicated and drug-free patients. Fuller et al (1999) found a negative correlation between the dose of neuroleptic medication and the slope of the early BP, suggesting that patients on higher doses of medication had lower slopes of the early BP. Given these inconsistent results across studies, the impact of neuroleptic medication on the BP clearly requires further investigation.

FUTURE RESEARCH

Kraepelin considered the weakening of volition as a fundamental pathological process in schizophrenia. The concept of volition and its possible neural basis in the frontal cortex has proved an interesting framework for the study of the BP in schizophrenia. Kornhuber (1984 a,b,c, 1987, 1989) described the role of the frontal lobes and their connections to limbic and basal ganglia areas in the control of voluntary decision-making. More recently, Jahanshahi and Frith (1998) reviewed evidence suggesting that the fronto-striatal circuits may form the physiological basis of volitional action. The limbic system has been implicated in the pathophysiology of schizophrenia (Czermansky et al, 1991; Tamminga et al, 2000; Wright et al, 2000). Based on these ideas, development of new experimental paradigms to evaluate limbic-frontal interplay (Mogenson, 1987) and the effects of this interplay on performance of volitional action may prove fruitful.

In assessing possible deficits of willed action, besides the BP and other slow brain potentials, investigation of changes in EEG frequency before and during voluntary movements using spectral analysis may prove of value. Disturbances in activation in the theta range over the midfrontal cortex before voluntary movements (Westphal et al. 1992, 1990A) and alpha activity have been shown. Changes in the topographical pattern of the EEG in the theta band has been described in schizophrenia (Westphal et al. 1990b) and has been confirmed by means of magnetoencephalography (MEG) (Fehr et al, in press). This work could be further developed by examining possible EEG frequency differences between self-generated and triggered actions or simple versus complex movements in schizophrenia.

A goal for future research dealing with the BP in schizophrenia should be to differentiate motor dysfunctions, particularly in untreated patients, and to quantify these deficits and then examine how they may relate to the impairment of the BP. In this respect, it is also important to note that although the rationale for studying the BP and CNV in schizophrenia and depression has been largely the existence of motor symptoms such as negative signs or psychomotor retardation characterizing these patients, as noted by Fuller et al (1999) motivational deficits (such as apathy) could also contribute to the impairment of movement-related potentials in these disorders. This clearly needs to be investigated in future studies. As noted above, the role of medication and further refinement of the effect of symptom subtypes on the BP are also worthy of further exploration.

This article is dedicated to Prof. H. H. Kornhuber, Prof. L. Deecke, and Dr. Berta Grözinger

REFERENCES

Adler, L.E., Pecevick, M. and Nagamoto, H. (1989) Bereitschaftspotential in Tardive Dyskinesia. *Movement Disorders*, **4/2**, 105-112

Andreasen, N.C. (1984). *Scale for assessment of negative symptoms (SANS)*. Iowa City. University of Iowa.

Boks, M. P.M. (2000) The specificity of neurological signs in schizophrenia: a review. *Schizophrenia Research*, **43**, 109-116

Boudarene, M., Timsit-Berthier, M. (1997). Interest of events-related potentials in assessment of posttraumatic stress disorder. *Ann N Y Acad Sci*, **821**, 494-8

Bunney, W.E. and Bunney, B.G. (2000). Evidence for a compromised dorso-lateral pre-frontal cortical parallel circuit in schizophrenia. In: Sedvall, G. and Terenius, L. (eds) *Schizophrenia: Pathophysiological Mechanisms*. Elsevier, Amsterdam, pp. 138 – 146

Chiarenza, G.A., Papakostopoulos, D., Dini, M. and Cazzullo, C.L. (1985) Neurophysiological correlates of psychomotor activity in chronic schizophrenics. *Electroencephal. Clin. Neurophysiol.*, **61**, 218-228

Czernansky, J.G., Murphy, G.M., and Faustman, W.O. (1991) Limbic/Mesolimbic Connections and the Pathogenesis of Schizophrenia. *Biol Psyhiatry*, **30**, 383-400

Deecke, L., Grözinger, B. and Kornhuber, H.H. (1976). Voluntary finger movement in man: Cerebral potentials and sere . *Biol. Cybern.*, **23**, 99 – 119

Deecke, L., Kornhuber, H.H., Lang, W., Lang, M., and Schreiber, H. (1985) Timing function of the frontal cortex in sequential motor and learning tasks. *Human Neurobiol*, **4**, 143-154

Dongier, M. (1973) Event related slow potential changes in psychiatry. In: Bogoch, S. (Ed.) *Biological Diagnosis of Brain Disorders, pp. 47-59*. New York: Spectrum

Dreher, J.-C., Trapp, W., Banquet, J.-P., Keil, M., Günther, W. and Burnod, Y. (1999)Planning dysfunction in schizophrenia: impairment of potentials preceding fixed/free and single/sequence of self-initiated finger movements. *Exp Brain Res*, **124**, 200-214

Fehr, T., Kissler, J., Moratti, S., Wienbruch, C. and Rockstroh, B. (in press), Source distribution of neuromagnetic slow waves and MEG-delta activity in schizophrenic patients. *Biological Psychiatry* (in press).

Frith, C.D. (1987). The positive and negative symptoms of schizophrenia reflect impairments in the perception and initiation of action. *Psychol Med*, **17**, 631-48

Frith, C.D. (1992). *The cognitive neuropsychology of schizophrenia*. Hove. Lawrence Earlbaum Associates.

Fuller, R., Nathaniel-James, D. and Jahanshahi, M. (1999). Movement-related potentials prior to self-initiated movements are impaired in patients with schizophrenia and negative signs. *Exp. Brain Res.*, **126**, 545 – 555

Grözinger, B., Westphal, K.P., Diekmann, V., Frech, M.M., Nitsch, J., Andersen, C., Scherb, W., Neher, K.D. and Kornhuber, H.H. (1986) Schizophrene Patienten und Gesunde: EEG-Unterschiede bei Willkürbewegungen. In: W. Keup (ed) *Biologische Psychiatrie*. Springer, Berlin, pp. 181-186

Haag, C., Kathmann, N., Hock, C., Günther, W., Vorderholzer, U. and Laakmann, G. (1994)Lateralization of the Bereitschaftspotential to the Left Hemisphere in Patients with Major Depression. *Biol Psychiatry*, **36**, 453-457

Heimberg, D.R., Naber, G., Hemmeter, U., Zechner, S., Witzke, W., Gerhard, U., Dittmann, V., Holsboer-Trachsler, E., Hobi, V. (1999). Contingent negative variation and attention in schizophrenic and depressed patients. *Neuropsychobiology*, **39**, 131-40

Ingvar, D.H. (1987). Evidence for frontal/pre-frontal cortical dysfunction in chronic schizophrenia: The phenomenon of "hypofrontality" reconsidered. In: Helmchen, H. and Henn, S.A. (eds) *Biologic perspectives of schizophrenia. Life sciences research report 40*, pp. 201 – 213. Chichester: John Wiley and sons

Jahanshahi, M., Jenkins, I.H., Brown, R.G., Marsden, C.D., Passingham, R.E. and Brooks, D.J. (1995) Self-initiated versus externally triggered movements. I. An investigation using measurement of regional cerebral blood flow with PET and movement-related potentials in normal and Parkinsons's disease subjects. *Brain*, 118, 913-933

Jahanshahi, M., Frith, D. (1998) Willed action and its impairments *Cognitive Neuropsychology*, 15, 483-533

Karaman, T., Özkaynak, S., Yaltkaya, K. and Büyükberker, C. (1997) Bereitschaftspotential in schizophrenia. *British Journal of Psychiatry*, 171, 31-34

Khanna, S., Mukundan, C.R., and Channabasavanna, S.M. (1989) Bereitschaftspotential in Melancholic Depression. *Biol Psychiatry*, 26, 526-529

King, H.E. (1965). Reaction time and speed of voluntary movement by normal and psychotic subjects. *J. psychol.*, 59, pp. 219 – 227

Klein, C., Rockstroh, B., Cohen, R., Berg, P. (1996) Contingent negative variation (CNV) and determinants of the post-imperative negative variation (PINV) in schizophrenic patients and healthy controls. *Schizophr Res*, 21, 97-110

Knott, V.J., Lapierre, Y.D., de Lugt, D., Griffiths, L., Bakish, D., Browne, M., Horn, E. (1991) Preparatory brain potentials in major depressive disorder. *Prog Neuropsychopharmacol Biol Psychiatry*, 15, 257-62

Kornhuber, H.H. (1983) Chemistry, Physiology and Neuropsychology of Schizophrenia: Towards an earlier diagnosis of schizophrenia I. *Arch. Psychiatr. Nervenkrankheit*, 233, 415-422

Kornhuber, H.H. (1984 a) Mechanisms of Voluntary Movement. In: *Cognition and Motor Processes*, Prinz, W. and Sanders, A.F. (eds.), Springer Verlag Berlin Heidelberg, pp. 163-173

Kornhuber, H.H. (1984 b) Attention, Readiness for Action, and the Stages of Voluntary Decision – Some Electrophysiological Correlates in Man. In: *Experimental Brain Research*, Suppl. 9, Springer Verlag Berlin Heidelberg, pp. 420-429

Kornhuber, H.H. (1984 c) Mechanisms of Voluntary Movement. In: *Cognition and Motor Processes*, Prinz, W. and Sanders, A.F. (eds.), Springer Verlag Berlin Heidelberg, pp. 163-173

Kornhuber, H.H. (1987) Handlungsentschluß, Aufmerksamkeit und Lernmotivation im Spiegel menschlicher Hirnpotentiale. Mit Bemerkungen zu Wille und Freiheit. In: *Der Wille in den Humanwissenschaften*. Heckhausen et al. (Hrsg.), Springer Verlag, Berlin Heidelberg, pp. 376-401

Kornhuber, H.H. und Deecke, L. (1964) Hirnpotentialveränderungen beim Menschen vor und nach Willkürbewegungen. *Pflügers Arch. ges. Physiol.*, 281, 52

Kornhuber, H.H. und Deecke, L. (1965) Hirnpotentialänderungen bei Willkürbewegungen und passiven Bewegungen des Menschen: Bereitschaftspotential und reafferente Potentiale. *Pflügers Arch. ges. Physiol.*, 284, 1-17

Kornhuber, H.H., Deecke, L., Lang, W., Lang, M. and Kornhuber A. (1989) Will, Volitional Action, Attention and Cerebral Potentials in Man: Bereitschaftspotential, Performance-Related Potentials, Directed Attention Potential, EEG Spectrum Changes. In: *Volitional Action*, Hershberger, W.A. (Ed.), Elsevier Science Publishers B.V. (North-Holland), pp. 107-168

Kraepelin, E. (1905). *Einführung in die psychiatrische Klinik*. Barth. Leipzig.

Kubota, F., Hiroshi, M., Shibata, N. and Yarita, H. (1999) A Study of Motor Dysfunction Associated with Schizophrenia Based on Analyses of Movement-Related Cerebral Potentials and Motor Conduction Time. *Biol Psychiatry*, 45, 412-416

Lang, W., Lang, M., Heisse, B., Deecke, L. and Kornhuber, H.H. (1984). Brain potentials related to voluntary hand tracking, motivation and attention. *Human Neurobiol.* 3, 235 – 240

Löw, A., Eckert, S., Cohen, R., Berg, P. and Rockstroh, B. (2000). *Action planning in schizophrenia – a readiness potential study*. Poster presented at the 40th Annual Meeting of the Society for Psychophysiological Research (SPR), San Diego, CA.

Manschreck, T.C. (1986) Motor abnormalities in schizophrenia. In: Nasrallah, H.A., Weinberger, D.R. (eds.) *Handbook of Schizophrenia*, Vol. 1, Elsevier Amsterdam, New York, Oxford, pp. 65-96

Marinkovic, K., Halgren, E., Klopp, J., Maltzman, I. (2000) Alcohol effects on movement-related potentials: a measure of impulsivity? *J Stud Alcohol*, 61, 24-31

Mogenson, G.J. (1967) Limbic-Motor Integration. *Progress in Psychobiology and Physiological Psychology*, 12, 117-170

Northoff, G., Pfennig, A., Krug, M., Danos, P., Leschinger, A., Schwarz, A. and Bogerts, B. (2000) Delayed onset of late movement-related cortical potentials and abnormal response to lorazepam in catatonia. *Schizophrenia Research*, 44, 193-211

Papakostopoulos, D., Banerji, N.K., Pocock, P.V., Newton, P. and Kelly, N.J. (1985a) Drug Induced Parkinsonism and Brain Electrical Activity. In: Papakostopoulos, D., Butler, S. and Martin, J. (Eds.) *Clinical and Experimental Neuropsychophysiology*, pp. 256-285. London: Croom Helm

Papakostopoulos, D., Banerji, N.K., Pocock, P.V., Newton, P. and Kelly, N.J. (1985b) Brain Macropotentials During Goal Directed Behavior in Patients with Schizophrenia. *Biological Psychology* 20, p. 207 (abstract)

Rockstroh, B., Muller, M., Wagner, M., Cohen, R., Elkert, T. (1994). Event-related and motor responses to probes in a forewarned reactive time task in schizophrenic patients. *Schizophrenia Research*, 1, 23-24

Sedvall, G. and Terenius, L. (eds.), (2000) Schizophrenia: Pathophysiological Mechanisms. *Brain Research Reviews*, Vol. 31/2.3, Elsevier Amsterdam

Singh, J., Knight, R.T, Rosenlicht, N., Kotun, J.M., Beckley, D.J. and Woods, D.L. (1992) Abnormal premovement brain potentials in schizophrenia. *Schizophrenia Research* 8, 31-41

Tamminga, C.A., Vogel, M., Gao, X.-M., Lahti, A.C., Holcomb, H. (2000) The limbic cortex in schizophrenia: focus on the anterior cingulate. In: Sedvall, G. and Terenius, L. (eds.) *Schizophrenia: pathophysiological mechanisms*. Elsevier, Amsterdam, pp. 364-370

Thier, P., Axmann, D. and Giedke, H. (1986) Slow brain Potentials and psychomotor retardation in depression. *Electroencephalography and clinical Neurophysiology*, **63**, 570-581

Timsit, M., (1970) Etude du Phenomène de Kornhuber chez les Schizophrenes. *Rev. Neurol.*, **122**, 449-451

Timsit-Berthier, M., Delaunoy, J. and Rousseau, J.C. (1973a) Slow Potential Changes in Psychiatry. II. Motorpotential. *Electroencephalography and Clinical Neurophysiology*, **35**, 363-367

Timsit-Berthier, M., Delaunoy, J., Konincky, N., Rousseau, J.C. (1973b) Slow Potential Changes in Psychiatry. I. Contingent negative variation. *Electroencephalogr Clin Neurophysiol*, **35**, 355-61

Verleger, R., Wascher, E., Arolt, V., Daase, C., Strohm, A., Kompf, D. (1999) Slow EEG potentials (contingent negative variation and post-imperative negative variation) in schizophrenia: their association to the present state and to Parkinsonian medication effects. *Clin Neurophysiol*, **110**, 1175-92

Wagner, M., Rendtorff, N., Kathmann, N., Engel, R.R. (1996) CNV, PINV and probe-evoked potentials in schizophrenics. *Electroencephalogr Clin Neurophysiol*, **98**, 130-43

Weinberger, D.R. (1988). Schizophrenia and the frontal lobe. (1988) *Trends in neurosiences*. **11**, 367-370

Westphal, K.P., Neher, K.G., Grözinger, B., Diekmann, V. and Kornhuber, H.H. (1984) Movement correlated differences in the EEG of schizophrenics and normals. *Electroencephalogr Clin Neurophysiol*, **58**, 98P (Abstract)

Westphal, K.P., Grözinger, B., Diekmann, V., Frech, M.M., Nitsch, J., Andersen, C., Scherb, W., Neher, K.D. und Kornhuber, H.H. (1985), Unterschiedliche EEG-Aktivität bei Schizophrenen und Gesunden vor und während willkürlicher Fingerbewegungen. In: Genzhirt, H., Berlit u. Haack, G., (Eds.) *Verhandlung der Deutschen Gesellschaft für Neurologie, Band 3: Kardiovaskuläre Erkrankungen und Nervensystem. Neurotoxikologie. Probleme des Hirntodes*, pp. 878-880. Springer Verlag Heidelberg.

Westphal, K.P., Neher, K.G., Grözinger, B., Diekmann, V. and Kornhuber, H.H. (1986), Differences between schizophrenic patients and normal controls in Bereitschaftspotential, alphaactivity and other EEG signs. In: Eds. W.C. McCallum, R. Zappoli and F. Denoth *Cerebral Psychophysiology: Studies in Event-Related Potentials (EEG Suppl. 38)*, pp. 464-465. Elsevier Science Publishers B.V.

Westphal, K.P., Grözinger, B., Diekmann, V., Kornhuber, A.H. (1990a), EEG-Zeichen gestörter Willkürmotorik bei Schizophrenen. In: Huber, G. (Ed.) *Idiopathische Psychosen, Psychopathologie, Neurobiologie, Therapie*. pp. 291-304, Schattauer, Stuttgart.

Westphal, K.P., Grözinger, B., Diekmann, V., Scherb, W., Reeß, J., Leibing, U., Kornhuber, H.H. (1990b), Slower Theta-activity over the midfrontal cortex in schicophrenic patients. *Acta Psychiatr Scand*, **81**, 132-138

Westphal, K.P., Grözinger, B., Becker, W., Diekmann, V., Scherb, W., Reeß, J., Leibing, U., Kornhuber, H.H. (1992) Spectral analysis of EEG during self-paced movements: differences between untreated schizophrenics and normal controls. *Biol Psychiatry*, **31**, 1020-1037

Westphal, K.P., Grözinger, B., Kotchoubey, Diekmann, V., Schreiber, H. und Kornhuber, H.H. (1996) Hypofrontalität bei unbehandelten Schizophrenen vor Willkürbewegungen. In: Möller und Müller-Spahn (Eds.) *Aktuelle Perspektiven der biologischen Psychiatrie*, pp. 317-319.

Wright, I.C., Rabe-Hesketh, S., Woodruff, P.W.R., David, A.S., Murray R.M., and Bullmore, E.T. (2000) Meta-Analysis of Regional Brain Volumes in Schizophrenia. *Am J Psychiatry*, **157**, 16-25

MOVEMENT-RELATED CORTICAL POTENTIALS IN PATIENTS WITH FOCAL BRAIN LESIONS

Christian Gerloff

Cortical Physiology Research Group
Department of Neurology
Eberhard-Karls University
72076 Tuebingen, Germany

INTRODUCTION

Some of the unanswered questions regarding movement-related cortical potentials (MRCPs) in general, and the Bereitschaftspotential in particular, concern their likely generators and whether these are active in parallel or sequentially. At the behavioral level, it still needs to be clarified, whether preparation, intention to act, anticipation or a combination of these contribute to the generation of different MRCP components. Alterations of MRCPs due to focal brain lesions may help to answer some of these questions.

In this chapter, the Bereitschaftspotential is regarded as one particular form of MRCPs, i.e., the slow rising surface negativity related to self-paced movements carried out at a rate of approximately one every 5 or 10 seconds or slower. This is the classical "Bereitschaftspotential" as first described by Kornhuber & Deecke (Kornhuber and Deecke, 1964, Kornhuber and Deecke,1965) , with its components BP, NS', and MP as characterized later (Deecke et al., 1980, Shibasaki et al., 1980) (Fig.1). The available data on focal brain lesions demonstrate effects on these three components, but mostly refrain from further differentiation of effects on subcomponents like IS (Barrett et al., 1986), fpMP or ppMP (Tarkka and Hallett, 1991, Hallett, 1994, Toro et al., 1993), PMP or P-50 (Deecke et al., 1980, Shibasaki et al., 1980). For this reason, the latter components will not be considered further in this chapter.

Ideally, the effects of lesions should be evaluated in follow-up studies, because the initial dysfunction tends to be compensated, for example, as a result of function being taken over by homologous or other brain regions (Weiller et al., 1992, Hamdy and Rothwell, 1998, Rossini et al., 1998, Gerloff et al., 1996a). Since in the acute phase, motor function is often insufficient, MRCPs cannot always be obtained. These limitations, however, apply to all neuroimaging procedures in which the signal is obtained in relation to active performance (e.g., functional MRI, PET), and need to be kept in mind throughout this chapter.

The Bereitschaftspotential
Edited by Jahanshahi and Hallett, Kluwer Academic/Plenum Publishers, New York, 2003

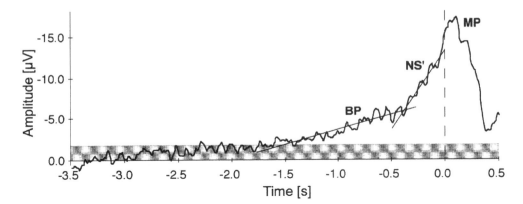

Figure 1. Three components of the Bereitschaftspotential: BP (early negative slope of the Bereitschaftspotential), NS' (late negative slope), MP (motor potential).

FOCAL BRAIN LESIONS IN HUMANS

The human supplementary motor area (SMA, mesial part of Brodmann area {BA} 6), the lateral premotor cortex (PMC, lateral aspect of BA6), the primary motor cortex (M1, BA4), and the primary somatosensory cortex (S1, BA3,1,2) have been considered key players in the generation of MRCPs (Toro et al., 1993, Deecke, 1987, Nagamine et al., 1994, Shibasaki et al., 1980, Hallett and Toro, 1996, Gerloff et al., 1998a, Ikeda et al., 1992). The proper functioning of these cortical areas depends on the integrity of basal ganglia and cerebellum as we know from patients with Parkinson's disease or cerebellar degeneration. Thus, in addition to focal cortical lesions, lesions of subcortical structures are likely to affect MRCP components as well.

Focal lesions of the supplementary motor area

The SMA has been proposed the crucial generator of the early BP component (see Fig.1) (Deecke, 1987, Ikeda et al., 1992, Mackinnon et al., 1996, Barrett et al., 1986, Deecke et al., 1985, Lang et al., 1991, Praamstra et al., 1996, Erdler et al., 2000). This view has been challenged on the basis of brain electric source analyses suggesting that the scalp-recorded BP could also be a result of concurrent bilateral activation of the sensorimotor cortices (Boetzel et al., 1993). If the SMA contributes significantly to the BP, then this part of the MRCP should be desintegrated after SMA lesions – at least to some extent.

Data on the acute phase after SMA lesion are not available, and would be difficult to obtain, because the patients tend to suffer from contralesional reduction of spontaneous movement. Deecke and colleagues described a series of 12 patients with chronic unilateral lesions of the SMA (Deecke et al., 1987). In all patients, a unilateral lesion of the mesial frontal cortex involving the SMA was confirmed by CT scan. The lesions were either caused by ischemia in the territory of the anterior cerebral artery (ACA) or by operative treatment of a tumor. Eight lesions were located in the right, 4 in the left hemisphere. All patients were right-handed. The time between neurological event (ischemia, operation) and experiment was 4.5 ± 4.5 years (mean ± SD). The Bereitschaftspotential was recorded in relation to unimanual index finger movements of the right and left hand, and the results for contralesional and ipsilesional movements were compared statistically. The main results were (i) the absence of the BP maximum at Cz with movements of either side, and (ii) a greater relative reduction of the BP amplitude at Cz and FCz with contralesional than with ipsilesional movements. These data suggest that (i) the SMA contributes to the generation of the BP and – as Deecke and colleagues concluded – that (ii) each SMA controls movements of either side of the body with preference for the contralateral half of the body.

The BP in the patients with chronic unilateral SMA lesions had not vanished. In fact, the amplitude reduction was rather small, and most prominent in comparison to the relative amplitudes over the contralateral sensorimotor cortex (C3, C4). This may have several reasons. Foremost, there is plenty of clinical evidence suggesting that the prognosis of unilateral SMA lesions is excellent. As a rule, the initial reduction of spontaneous activity resolves to near complete recovery within weeks (Laplane et al., 1977, Zentner et al., 1996). Since this is not true for bilateral lesions of the SMA, in which patients fall into severe akinetic mutism, the most plausible explanation is that the intact SMA can take over function once its counterpart has been damaged. The persistence of a reduced BP is, therefore, not surprising in the chronic stage after a unilateral SMA lesion. Another reason is that the SMA might not be the sole contributor to the early BP component. This is suggested by subdural grid recordings showing the complete Bereitschaftspotential including the early negativity prior to self-paced movements also over the M1 and in vicinity to the PMC (Ikeda et al., 1992). In the same study, the BP amplitudes were similarly high over SMA and M1/PMC, but higher in the hemisphere contralateral to the moving limb. At the behavioral level, several invasive and non-invasive stimulation data support the concept that both planning and initiation of self-paced movements and the organization of complex motor acts is compromised if SMA function is disturbed (Lim et al., 1994, Gerloff et al., 1997a).

Additional information is available from magnetoencephalography (MEG). The magnetic readiness field (RF) is thought to correspond to the BP component. The RF is less consistent and of lower amplitude than the BP. This has been attributed to the fact that MEG is insensitive to radial sources, so it can only pick up SMA activity arising from the mesial aspects of the SMA in the interhemispheric fissure. Because the SMA tends to be activated bilaterally, the resulting dipole sources are likely to have opposite orientations and cancel each other out. Only small amounts of bilateral SMA activity would thus be measurable with MEG. Valuable information in favor of this concept has been provided by Lang and colleagues (1991) who unequivocally demonstrated a current dipole source in the left SMA starting about 1200 ms prior to the initiation of voluntary movements of the right thumb in a patient with infarction of the right SMA.

The results in patients with unilateral SMA lesions plausibly demonstrate a contribution of the SMA to the early Bereitschaftspotential, but do not allow for the conclusion that the BP component (see Fig. 1) arises exclusively from there.

Focal lesions of the basal ganglia and internal capsule

Basal ganglia (BG) input is crucial for normal function of the SMA as evident from the clinical picture in Parkinson's disease or after hypoxic lesions of the BG. Exclusive lesions of globus pallidus internus (GPi) or externus (GPe), or putamen (Put) are rare. The available MRCP data refer to ischemic strokes and, as a rule, to combined lesions of basal ganglia and the pyramidal tract in the internal capsule (IC) (Fig. 2). In line with the clinical picture of these stroke patients, the IC lesion dominates the deficit with spastic hemiparesis.

Shibasaki provided the first information on unilateral cerebral lesions and their effects on MRCPs (Shibasaki, 1975). Among his patients, 7 were classified as having deep-seated lesions of varying etiology (e.g., ischemic, neoplastic). The location of these lesions were thalamus in 4, and the deep frontal lobe or the IC in 3 patients. In 3 of these patients, the MRCP was normal, in 2 patients the maximum negativity (usually corresponding to MP, see Fig.1) was reduced over the affected hemisphere, and in the remaining 2 patients, bilateral abnormalities of the MRCP were found. The 2 patients with bilateral MRCP changes had thalamic tumors. In contrast, none of the patients with cortical lesions (n=10) had bilateral MRCP alterations, but 90% of these patients exhibited pronounced amplitude reduction of the negativity normally maximal over the affected hemisphere when the paretic hand was moved. Abnormalities of the MRCP were more frequent in patients with more severe weakness and pyramidal signs, and the MRCP tended to be more affected in patients with tumors than in patients with ischemic lesions. These were qualitative observations in a relatively small patient sample, which did not reach statistical significance, and the potential influence of the size of the lesions was not evaluated in detail. Nevertheless, this was a crucial first study demonstrating that MRCPs are affected by focal brain lesions at the cortical and at the subcortical level.

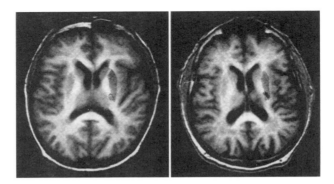

Figure 2. Typical ischemic lesions in the (left) basal ganglia. In both patients, the ischemic lesion (marked with a black line) affects both, the posterior limb of the internal capsule and the posterior aspect of the globus pallidus (enhanced structure with darker grey). Radiologic conventions, right corresponds to left and vice versa.

More than 10 years later, four well documented cases of "capsular ataxic hemiparesis" were published (Saitoh et al., 1987). CT scanning was used to confirm a singular lesion in the posterior limb of the left IC in each patient (slightly bigger than the ones in Fig. 2). All patients were right-handed. The time between ischemia and experiment was greater than 4 weeks (exact time not given), and at the time of the experiment all patients had recovered full muscle strength but were left with mild "ataxia" of the right upper extremity. MRCPs were recorded in relation to index finger flexions from 5 scalp electrodes. As controls, the healthy hand of the patients was used as well as 2 healthy control subjects and 2 patients with multi-system atrophy (olivo-ponto-cerebellar atrophy, OPCA). None of the patients with the lesion in the posterior limb of the IC had a significant amplitude reduction of the MRCP. It is possible that this result reflects efficient post-lesional re-establishment of cortico-cortical and cortico-spinal ciruits with excellent recovery of function at the behavioural level. It would have been interesting to see MRCP data from the acute phase. Since these are not available, the absence of abnormal MRCPs in the 4 patients reported by Saitoh and colleagues (1987) cannot be taken as evidence that IC lesions have no effect on the MRCP. The more likely conclusion is that the degree of abnormality of the MRCP correlates better with the extent of the motor deficit than with the site of the lesion. This interpretation would also be in line with the results of Shibasaki (1975).

Further support for this view comes from a recent study on 13 stroke patients. The average time between the onset of neurological symptoms and the experiment was 9 weeks. Nine patients had subcortical lesions without major cortical involvement (Platz et al., 2000), and 7 of these had mild to moderate hemiparesis. MRCPs were recorded from 27 scalp electrodes. Ten healthy subjects served as controls. While the patient cohort in this study was well described both clinically and in terms of motor performance, the data on the MRCPs are more difficult to appreciate. No original MRCP traces are shown, and the BP, NS', and MP components were not analyzed. Instead, a single equivalent dipole was fitted for an early part of the MRCP (750 to 500 ms before movement onset) and for the later preparatory phase of the MRCP (250 to 0 ms before movement onset). According to the authors, this procedure was used for data reduction since it allowed them to differentiate EEG maps in the group of hemiparetic patients (subcortical, n=7; 3 left IC, 4 right IC; cortical, n=1; data pooled, n=8), patients with somatosensory deficits (n=3), and patients with ideomotor apraxia (n=2). In the hemiparetic patients, the location of the equivalent dipole during the early MRCP phase was more anterior and more lateral than in controls. These changes were absent in patients with pure somatosensory deficits, and had a different pattern in patients with apraxia. Since a detailed evaluation of MRCP components was not part of this study, further conclusions cannot be drawn in this regard. However, what seems clear from the results of this study is that the degree of abnormality of the MRCP correlates better with the extent of the motor deficit than with the site of the lesion. This is illustrated by patient H.E. of Platz et al., who had a lesion in the posterior limb of the IC but only sensory deficits (light touch, thermal sense). He showed no abnormality in the MRCP modelling analysis, in contrast to all other patients with lesions of the IC, who suffered from hemiparesis and had impaired MRCP dipole results.

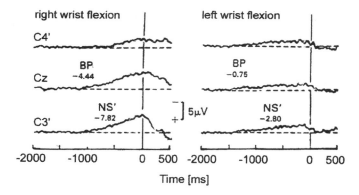

Figure 3. Reduced amplitudes (BP, NS', MP) and absence of late lateralization of MRCPs in a patient with chronic unilateral lesion of the right basal ganglia. Modified from Fève et al. (1994) with permission.

In another study, 10 stroke patients with various lesions were included (Green et al., 1999). Only one of these patients had a lesion of the (right) IC 2 years prior to the experiment. In this patient who had a good recovery after left hemiparesis, high-resolution EEG (128 channels) was recorded, dipole models were computed and the results coregistered with anatomical MRI. The modelling data in this patient suggested a mesial displacement of the MP generator for movements of the affected hand. The main generator of this component is normally localized in the sensorimotor hand region contralateral to the moving hand (Toro et al., 1993, Gerloff et al., 1996b, Gerloff et al., 1998a, Gerloff et al., 1998b). With the modelling algorithm used in this paper, reduction of the lateralization could be due to both true displacement of the generator inside the affected hemisphere or increased relative contribution of an ipsilateral generator in the sensorimotor region of the intact hemisphere, or both. The latter interpretation is supported by two more MRCP studies on patients with chronic neurological deficits (Fève et al., 1994, Kitamura et al., 1996). Feve and colleagues (1994) studied bilateral (n=9) and unilateral (n=6) lesions of the BG and anterior thalamus (vascular, hypoxic, encephalitic, traumatic). EEG was recorded from 3 channels (C3, Cz, C4), and the onset and slope of the BP and NS' components were analyzed. The main result was reduced gradients of BP and NS' compared with 10 healthy control subjects. This effect was seen with movements of the affected hand (bilaterally in bilateral lesions) and resulted in a less lateralized NS' (Fig.3). Kitamura and colleagues (1996) investigated two patients with vascular lesions of the IC and BG. While the NS' component was lateralized normally preceeding elbow movements of the intact arm, the same component was distributed bilaterally over both hemispheres with movements of the recovered paretic upper extremity.

An increased and beneficial engagement of the hemisphere ipsilateral to the paretic hand after good recovery has been proposed on the basis of various imaging studies (Weiller et al., 1992, Chollet et al., 1991, Cramer et al., 1997), but has been questioned by others because direct ipsilateral cortico-spinal responses are usually present in patients with poor outcome after unilateral stroke (Netz et al., 1997, Turton et al., 1996). At this point, what is lacking is the demonstration of a clearcut association of ipsilateral overactivity in patients with excellent recovery and without contaminating mirror movements. The available evidence from MRCP studies on focal (mostly ischemic) lesions of the IC and BG so far suggest (i) that all components can be affected (the evidence for alterations of the later components, NS' and MP, is somewhat stronger), (ii) that alterations of the MRCP comprise a more bilateral distribution of NS', amplitude reductions and more bilateral distribution of MP, and perhaps an anterior shift of the early amplitude maximum of the BP, and (iii) that MRCP impairment is more representative of the patient's behavioral performance than of the lesions site, extent, or etiology.

Focal lesions of the cerebellum

The cerebellar input through the ventrolateral (VL) and intermedioventral (Vim) nucleus of the thalamus to PMC and M1 has been well established. Already in the mid seventies, Shibasaki and colleagues became interested in the contribution of cerebellar input to the generation of MRCPs. They recorded MRCPs from 20 patients with cerebellar ataxia (Shibasaki et al., 1978) and found that 55% of them had abnormal MRCPs with reduced amplitudes. Remarkably, 6 of the 8 patients with cerebellar cortical degeneration and 2 of the 3 patients with cerebellar hemispheric lesions had normal MRCPs. However, all 5 patients with dyssynergia cerebellaris myoclonica (presumed Ramsay Hunt syndrome) had impaired MRCPs, as had 2 of the 3 patients with vascular midbrain lesions (Benedikt's syndrome) in this study. The authors concluded that the more extensive neurodegeneration with involvement of the dentato-rubro-thalamic pathway in Ramsay Hunt patients or patients with Benedikt's syndrome was the likely cause for the disintegration of the MRCP. Further support for this concept comes from 4 patients who underwernt Vim thalamotomy for treatment of intractable tremor of the hand. The MRCPs in these patients were present before operation and abolished after the intervention on the side of the thalamotomy.

A later report provided complementary information stating that the severe amplitude reduction in Ramsay Hunt patients concerned all MRCP components, in particular both the early BP, and the late NS' and MP (Shibasaki et al., 1986). The observations made in patients with various system degenerations involving the cerebellar nuclei and dentatorubrothalamic pathway were important because they proved that the cerebello-cerebral interaction is critical for the generation of the normal MRCP and, thus, that cerebellar dysfunction induces measurable changes in cortical movement preparation and execution. Interestingly, slowly progressive degeneration of cerebellar cortex only (sparing the deep nuclei) seemed to have a less profound effect on MRCPs. This was confirmed by

Tarkka and colleagues who found only mild alterations of MRCPs in 6 patients with cerebellar degenerative disease (Tarkka et al., 1993). In 4 of these patients, MRCP amplitudes and latencies were normal, in 2 the NS' component was reduced. In all 6 patients, however, the topography of the late MRCP components was less focussed than in normal controls. One problem with analyzing patients with system degenerations is the difficulty of being sure about the structures affected by the neurodegenerative process. As we know now, "cerebellar ataxias" have numerous variants which can be distinguished on the basis of molecular genetic testing (SCA1-14, etc.) and affect different systems outside the cerebellum. Another problem is the chronicity of degenerative processes that may allow for some reorganization at the cerebellar, but also at the cerebral cortex level and thereby compensate the evolving motor deficit. What can be measured in the chronic stage is a combination of primary pathology and secondary adaptive reorganization, and it is hard to predict how this might influence the individual MRCP components.

Prompted by these considerations, we studied two patients with cerebellar stroke in the subacute phase (7-13 days) after the event and on follow-up in the chronic stage (8-10 months later) after complete recovery of function (Gerloff et al., 1996a). The lesions are illustrated in Fig.4. MRCPs were recorded from 26 scalp eletrodes, and onset, peak latencies, and amplitudes of the BP, NS' and MP components (see Fig.1) were determined and compared to a control group of 10 healthy volunteers. In the subacute and chronic stages, patients were able to execute the required finger movement. Kinematic analyses, however, showed that the maximum acceleration in the ataxic hand was significantly lower than in the control group, even after near complete clinical recovery in the chronic stage. The main result was a reduction of the slope and amplitude of the late MRCP components NS' and MP when the affected hand was moved. The peak amplitude was -10.4 ± 1.6 µV (mean±SEM) for the affected hand, and -15.6 ± 1.4 µV for the unaffected hand. The characteristic increase of the slope of the Bereitschaftspotential between BP and NS' was also absent when the affected hand was moved (difference of slopes "NS' minus BP", unaffected, -13.3 µV/s; affected, +8.7 µV/s). Most importantly, these effects were temporary and the NS' and MP components re-occurred on follow-up 8-10 months later. The early BP component had different characteristics in the two patients. It was preserved and even augmented in patient 1, and abolished in patient 2. This observation is in line with Shibasaki's description of MRCP changes in dentate lesions (presumed Ramsay Hunt syndrome) compared with pure cerebellar cortical lesions. Our patient 1 had no evident involvement of the deep cerebellar nuclei, while in patient 2 the center of the lesion was just there. It is therefore plausible to conclude that cerebellar lesions that also include the deep cerebellar nuclei desintegrate the whole MRCP (BP, NS', MP), while pure cerebellar cortical lesions target mostly the late components NS' and MP. In either situation, the topography of NS' and MP tends to be less well focused. The MRCP waveforms in the two patients with cerebellar stroke are displayed in Fig.4. These data extend the results of Shibasaki and colleagues (1978, 1986) in that they emphasize the restorative capacities of the cerebello-thalamo-cortical loop and the necessity to take the temporal characteristics of the destructive process into account.

The restorative capacity of the cerebello-thalamo-cortical loop is also supported by observations of Shibasaki et al. (1978) in one patient with Parkinson's disease. This patient had intact MRCPs, went through Vim thalamotomy for intractable tremor, and the MRCPs disappeared postoperatively. Follow-up recordings demonstrated a re-occurrence of MRCPs in this patient 1.5 years after thalamotomy in spite of the continuing effect on the tremor (Fig. 5).

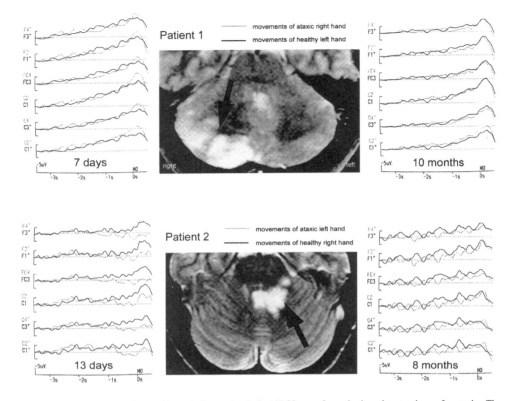

Figure 4. MRCPs in two patients with cerebellar stroke. *Left,* MRCP waveforms in the subacute phase after stroke. The thin line in each graph corresponds to the cortical potentials related to movements of the ataxic hand. *Middle,* Ischemic cerebellar lesions on T2 MR images. Patient 1, infarction in the territory of the right posterior inferior cerebellar artery (PICA). Patient 2, infarction in the territories of both the left anterior inferior cerebellar artery (AICA) and the left PICA. In both cases, the brain stem territories of the PICA seemed unaffected. *Right,* MRCP waveforms in the chronic phase after stroke and with good motor recovery. The thin line in each graph corresponds to the cortical potentials related to movements of the formerly ataxic hand.

That the effects of cerebellar dysfunction on MRCPs are rather specific and not related to attentional processes, is suggested by data of Ikeda and colleagues (Ikeda et al., 1994). MRCPs and contingent negative variation (CNV) were recorded from in a patient with ischemic infarction in the mesial tegmentum of the midbrain involving the decussation of the superior cerebellar peduncle. MRCPs in association with hand movements were completely absent while the CNV was normally present at the frontocentral midline.

MRCP studies in patients with pathologies of cerebellar cortex, cerebellar nuclei and cerebellar efferent pathway to PMC and M1, relayed through the thalamus (VL, Vim), have yielded important information on motor physiology and cerebello-cerebral interaction: (i) the integrity of the MRCP depends upon the normal functioning of each part of the cerebello-cerebral loop, (ii) the late components such as NS' are particularly sensitive to (but not exclusively affected by) cerebellar input failure, (iii) in addition to amplitude reductions and slower gradients also scalp topography of late MRCP components tends to be less focussed without proper cerebellar input, (iv) acute cerebellar or Vim lesions have a more pronounced effect on the MRCP than chronic lesions of the same type; to some extent these effects are reversible, indicating perhaps that there is a considerable restorative potential at the level of cerebellum and thalamus, (v) the deep cerebellar nuclei are key structures in the production of a normal MRCP; MRCP regeneration is less efficient after lesions affecting the dentate nucleus which may be a pathophysiological correlate of the poorer functional outcome in patients with lesions affecting the deep cerebellar nuclei as opposed to pure cortical lesions, a clinical observation first made by G. Holmes (Holmes, 1917).

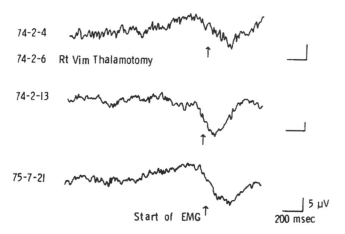

Figure 5. MRCP disappearance after stereotaxic Vim thalamotomy (February 6, 1974) and reappearance 1.5 years later (July 21, 1975) in a patient with Parkinson's disease (with permission from Shibasaki et al. 1978).

Other focal lesions

Frontal cortex. Besides the precentral gyrus, the frontal lobe contains a variety of premotor structures that are involved in preparation of voluntary movement and have been proposed to generate components of MRCPs. Modelling studies on MRCPs related to fast repetitive finger movements have provided evidence for a sequential, but overlapping activation of the bilateral PMC (BA6) and M1 (BA4, anterior wall of the central sulcus) (Gerloff et al., 1997b, Gerloff et al., 1998a, Gerloff et al., 1998b). In a group of 11 patients with lesions of BA8 and 46, extending into BA6, 9, 10, 11, 12, 44, and 45 to a variable degree (Fig.6), the amplitudes of the entire MRCP, but preferentially of the BP and NS' components were reduced (Singh and Knight, 1990). In a subgroup analysis, the abnormality of BP and NS' was maximal in patients with lesions of BA6 and 8. The MP component could be clearly identified in these patients. This supports the modelling results which predicted an important contribution of BA6 to the generation of MRCP premotor components. Since these patients did not have obvious lesions of the SMA, the reduction of the early BP component clearly favors the concept of a partially parallel processing of premotor information in mesial and lateral parts of BA6 (SMA and PMC) and possibly BA4 (M1) as suggested by Ikeda and colleagues on the basis of subdural grid recordings (Ikeda et al., 1992).

Honda and colleagues studied 2 patients with combined MRCP and PET analyses (Honda et al., 1997). One patient suffered from left hemispheric premotor (and parietal) cortex lesions, the other from lesions in the right Rolandic area (and several subcortical locations). A consistent finding in these patients was the absence of the contralateral predominance of late MRCP components when the affected hand was moved. The relatively larger contribution of the ipsilateral hemisphere was also reflected in increased regional cerebral blood flow measures.

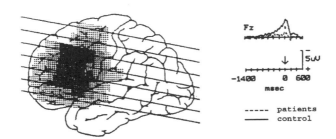

Figure 6. *Left,* pattern of frontal lobe lesions in 11 patients (mean lesion volume = 39.4 cc). In 5 patients the lesion was located in the right, in 6 patients in the left hemisphere. *Right,* amplitude reduction of the MRCP (in particular of the BP and NS' component) in this patient group. Grand average data, right thumb movement. Modified from Singh and Knight (1990) with permission.

Parietal cortex and temporo-parietal junction. Singh and Knight have provided a systematic evaluation of the effects of parietal and temporal lesions on MRCPs (Singh and Knight, 1993, Knight et al., 1989). All lesions were caused by ischemic infarctions of branches of the middle cerebral artery (MCA). The study included 7 patients with lesions of the temporo-parietal junction (BA22, caudal 39, 40, 42), 5 patients with lesions of the superior parietal cortex (BA5, 7, rostral 39, 40), and 5 patients with large combined lesions of the posterior association cortex involving both the temporo-parietal junction and the superior parietal cortex (BA7, 22, 39, 40, 41, 42). MRCPs were recorded from 14 electrodes, and the results compared with 14 healthy controls (Singh and Knight, 1993). The lesion patterns are summarized in Fig.7.

Singh and Knight addressed the question of whether different aspects of temporal and parietal cortex have a modulatory impact on MRCPs. From the topographic mapping data with MRCP maxima extending into the parietal cortex, this appeared to be quite likely. Also, recent imaging data have demonstrated contributions of parietal regions to movement selection, preparation and execution (Deiber et al., 1991, Deiber et al., 1996, Weeks et al., 1999). The MRCP analyses of Singh and Knight were unpreceded in the accuracy with which the lesion patterns were analyzed. Also, there were no significant differences in the clinical capabilities of the patients to execute the required button press movements. Lesions of the temporo-parietal junction had no effect on the MRCPs. Lesions of similar volume but located in the superior parietal cortex altered the MRCPs significantly. These patients generated a slowly rising negativity as late as about 500 ms prior to movement onset, and hardly produced any discernable early BP component. The NS' and MP components could be identified in this group, but were of lower amplitude and lacked the typical lateralization. In the third group with substantially larger lesions of the posterior association cortex, the BP and NS' components were reduced or absent. The MP component could be identified in all 5 patients of this group, but again with a lack of the characteristic lateralized topography in unimanual movements. The data of Singh and Knight (1993) demonstrate that in addition to the standard generators, namely, the SMA, PMC, M1, S1 and the input of basal ganglia and cerebellum, MRCPs are also subject to the modulating influence of 'higher' posterior parietal association areas. It appears as though the MP component is the most robust one and thus the one that is least modulated by parietal input. In other words, the parietal lobe contribution to the MRCP is preparatory or related to movement selection, and has less impact on movement execution. In line with the clinical observation that unilateral parietal damage may cause bilateral limb apraxia, the MRCP alterations after unilateral parietal lesions occurred with movements of both hands and over both hemispheres.

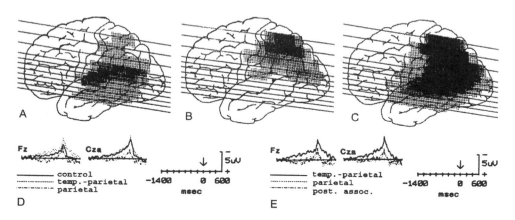

Figure 7. *Top,* patterns of damage to (A) temporo-parietal junction, mean lesion volume = 34.9 cc, (B) superior parietal cortex, mean lesion volume = 31.9 cc, and (C) posterior association cortex involving both the temporo-parietal junction and the superior parietal cortex, mean lesion volume = 95 cc. *Bottom,* MRCPs in these patient groups. Grand average data, right thumb movement. (D) Compared are healthy subjects ('control'), patients with lesions of the temporo-parietal junction ('temp.-parietal'; lesions as demonstrated in A), and patients with lesions of the superior parietal cortex ('parietal'; lesions as demonstrated in B). No significant MRCP change was observed in the 'temp.-parietal' group, but a marked amplitude reduction in the 'parietal group'. (E) Direct comparison of lesions of the temporo-parietal junction, lesions of the superior parietal cortex, and lesions of the posterior association cortex ('post. assoc.'; lesions as demonstrated in C; other conventions as in D). Note the marked reduction of MRCP amplitude in both patient groups in which the lesions involved posterior and superior parietal cortex ('parietal' and 'post. assoc.'). Modified from Singh and Knight (1993) with permission.

It should be noted that parietal and prefrontal lesions caused relatively similar abnormalities of the MRCP. One possible explanation is that movement preparation requires information processing in an extended parietal-prefrontal network (Sakai et al., 1998, Deiber et al., 1997, Roland, 1984) and that the final compound preparatory activity as reflected by BP and NS' can be compromised by lesions at either node of this cortico-cortical circuit.

EXPERIMENTAL LESIONS IN ANIMALS

The data in monkeys are particularly valuable in regard to the cerebellar contribution to cortical motor potentials and the flexibility of primary motor and primary sensory cortex contribution to MRCPs. Sasaki and colleagues have been a protagonist in this field of research (Sasaki et al., 1979, Sasaki and Gemba, 1984, Sasaki et al., 1989, Sasaki and Gemba, 1991, Tsujimoto et al., 1998, Tsujimoto et al., 1993).

Cerebellar lesions. Unilateral cerebellar hemispherectomy ipsilateral to the moving limb eliminates the characteristic surface-negative, depth-positive (s-N, d-P) field potentials in the contralateral motor cortex which precede visually initiated and self-paced movements in intact monkeys (*Macaca fuscata*) (Sasaki et al., 1979, Sasaki and Gemba, 1984). A similar but reversible effect can be produced by cooling and rewarming of the dentate nucleus of the cerebellum (Tsujimoto et al., 1993). In these monkeys, dentate cooling reduces the size of the intracortical motor potential. The effect occurred in all monkeys prior to visually initiated reaction time movements, and in 2 of 3 monkeys also prior to self-paced movements (Fig.8). The amplitude reduction was most pronounced in the last 500 ms before movement onset, i.e., in a time window similar to the NS' of human scalp MRCPs. These changes of field potential configuration were also relevant at the behavioural level. They resulted in a delay of the reaction times in the visually triggered movement task. One conclusion drawn by the authors is that the cerebellar input through thalamic nuclei into M1 produces excitatory postsynaptic potentials (EPSPs) in the apical dendrites of the pyramidal neurons in M1, and that the EPSPs in the superficial layers produce the s-N, d-P field potential. It is likely that the MRCP effects of acute cerebellar lesions in humans (Gerloff et al., 1996a) follow similar rules. This would also explain why the late MRCP components (NS', MP) tend to be more affected in cerebellar disease than the early BP.

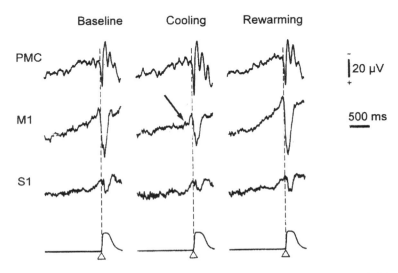

Figure 8. Effects of dentate cooling on cortical field potentials related to self-paced movements in *Macaca fuscata*. Movement onset is marked with the triangle in the bottom traces (mechanogram). Note the reversible reduction of the MRCP in the M1, in particular of the late negative slope (arrow), during cooling. Cooling scarcely affected the MRCP recorded from premotor cortex (PMC) or primary somatosensory cortex (S1). Modified from Tsujimoto and colleagues (1993) with permission.

Lesions of the primary motor cortex. Another fascinating experiment was undertaken by Sasaki and Gemba and published in 1984. Long before 'plasticity and reorganization' became one of the most attractive topics in neuroscience, these authors proposed that the tight intracortical connectivity between M1 and S1 might allow for some functional compensation in case of lesions (Sasaki and Gemba, 1984). In monkeys *(Macaca fuscata)*, motor cortex function was transiently blocked by local cooling during execution of visually initiated movements. M1 cooling caused reduction of the pre-movement s-N, d-P field potential in the forelimb motor area, followed by prolonged reaction times of the weakened contralateral wrist muscles. The most striking finding was a concomitant increase of the s-N, d-P field potential in S1. To test if this 'compensatory' increase has functional relevance, Sasaki and Gemba applied cooling simultaneously to both M1 and S1. This resulted in complete paralysis of the respective limb. These results not only improve our understanding of strategies of local short-term reorganization in the primate brain, but also emphasize that MRCPs are complex compound potentials, and that the relative contribution of different generators is subject to rapid changes in pathological conditions.

CONCLUDING REMARKS

Careful evaluations of MRCP pathology due to various forms of focal brain lesions have yielded important insights into the physiology of voluntary movement in humans and non-human primates. Perhaps some of the most intriguing data come from the thorough analysis of cerebellar input to human and monkey cortical motor areas. It has become unequivocally clear that each element from cerebellar cortex, to deep cerebellar nuclei and dentato-rubro-thalamic pathway through the midbrain, up to the thalamic nuclei needs to be intact to generate a normal MRCP and proper movement. The NS' component immediately before movement onset seems to be particularly sensitive to disturbances of the cerebellar input to lateral premotor and primary motor cortex. MRCP analysis has provided us with a unique, although small and indirect, window into assessing cerebellar function non-invasively in humans. And yet, the various studies reviewed here also highlight the complexity of the MRCP as a compound waveform whose exact generators are still to be determined. The NS' is quite sensitive to cerebellar dysfunction, the BP to lesions of the mesial premotor system including the SMA and basal ganglia, and the MP is the most robust element of the MRCP, hardly affected by small focal lesions at all. Nevertheless, it is not 'safe' to interpret any of these components as specific markers for the integrity of distinct brain regions. This limitation is well illustrated by the strikingly similar effects of prefrontal and parietal cortical lesions on the MRCP. The value of the MRCP in patients with focal brain lesions is that it provides us with further information on the pathophysiology of the motor deficits. Because of its millisecond resolution, it is possible to differentiate between deficits at the level of early or late movement planning and at the final levels of movement execution and reafferent feedback. Further lesion studies with high-resolution EEG/MEG, physiology-based source models, comparative neuroimaging data like functional MRI or PET, and coregistration with individual anatomy are necessary to elaborate the diagnostic value of MRCPs further.

REFERENCES

Barrett, G., Shibasaki, H. and Neshige, R. (1986) Cortical potentials preceding voluntary movement: evidence for three periods of preparation in man. *Electroencephalogr. Clin. Neurophysiol.,* **63**, 327-39.

Boetzel, K., Plendl, H., Paulus, W. and Scherg, M. (1993) Bereitschaftspotential: is there a contribution of the supplementary motor area? *Electroencephalogr. Clin. Neurophysiol.,* **89**, 187-96.

Chollet, F., DiPiero, V., Wise, R. J., Brooks, D. J., Dolan, R. J. and Frackowiak, R. S. (1991) The functional anatomy of motor recovery after stroke in humans: a study with positron emission tomography. *Ann. Neurol.,* **29**, 63-71.

Cramer, S. C., Nelles, G., Benson, R. R., Kaplan, J. D., Parker, R. A., Kwong, K. K., Kennedy, D. N., Finklestein, S. P. and Rosen, B. R. (1997) A functional MRI study of subjects recovered from hemiparetic stroke. *Stroke,* **28**, 2518-27.

Deecke, L. (1987) Bereitschaftspotential as an indicator of movement preparation in supplementary motor area and motor cortex. *Ciba Found. Symp.,* **132**, 231-50.

Deecke, L., Eisinger, H. and Kornhuber, H. H. (1980) Comparison of Bereitschaftspotential, pre-motion positivity and motor potential preceding voluntary flexion and extension movements in man. *Prog. Brain Res.,* **54**, 171-6.

Deecke, L., Kornhuber, H. H., Lang, W., Lang, M. and Schreiber, H. (1985) Timing function of the frontal cortex in sequential motor and learning tasks. *Hum. Neurobiol.,* **4**, 143-54.

Deecke, L., Lang, W., Heller, H. J., Hufnagl, M. and Kornhuber, H. H. (1987) Bereitschaftspotential in patients with unilateral lesions of the supplementary motor area. *J. Neurol. Neurosurg. Psychiatry,* **50**, 1430-4.

Deiber, M. P., Ibanez, V., Sadato, N. and Hallett, M. (1996) Cerebral structures participating in motor preparation in humans - a positron emission tomography study. *J. Neurophysiol.,* **75**, 233-47.

Deiber, M. P., Passingham, R. E., Colebatch, J. G., Friston, K. J., Nixon, P. D. and Frackowiak, R. S. J. (1991) Cortical areas and the selection of movement: a study with positron emission tomography. *Exp. Brain Res.,* **84**, 393-402.

Deiber, M. P., Wise, S. P., Honda, M., Catalan, M. J., Grafman, J. and Hallett, M. (1997) Frontal and parietal networks for conditional motor learning: a positron emission tomography study. *J. Neurophysiol.,* **78**, 977-91.

Erdler, M., Beisteiner, R., Mayer, D., Kaindl, T., Edward, V., Windischberger, C., Lindinger, G. and Deecke, L. (2000) Supplementary motor area activation preceding voluntary movement is detectable with a whole-scalp magnetoencephalography system. *Neuroimage,* **11**, 697-707.

Fève, A., Bathien, N. and Rondot, P. (1994) Abnormal movement related potentials in patients with lesions of basal ganglia and anterior thalamus. *J. Neurol. Neurosurg. Psychiatry,* **57**, 100-4.

Gerloff, C., Altenmüller, E. and Dichgans, J. (1996a) Disintegration and reorganization of cortical motor processing in 2 patients with cerebellar stroke. *Electroencephalogr. Clin. Neurophysiol., 98,* 59-68.

Gerloff, C., Corwell, B., Chen, R., Hallett, M. and Cohen, L. G. (1997a) Stimulation over the human supplementary motor area interferes with the organization of future elements in complex motor sequences. *Brain, 120,* 1587-1602.

Gerloff, C., Grodd, W., Altenmüller, E., Kolb, R., Nägele, T., Klose, U., Voigt, K. and Dichgans, J. (1996b) Coregistration of EEG and fMRI in a simple motor task. *Human Brain Mapping, 4,* 199-209.

Gerloff, C., Toro, C., Uenishi, N., Cohen, L. G., Leocani, L. and Hallett, M. (1997b) Steady-state movement-related cortical potentials: A new approach to assessing cortical activity associated with fast repetitive finger movements. *Electroencephalogr. Clin. Neurophysiol., 102,* 106-113.

Gerloff, C., Uenishi, N. and Hallett, M. (1998a) Cortical activation during fast repetitive finger movements in humans: Dipole sources of steady-state movement-related cortical potentials. *J. Clin. Neurophysiol., 15,* 502-13.

Gerloff, C., Uenishi, N., Nagamine, T., Kunieda, T., Hallett, M. and Shibasaki, H. (1998b) Cortical activation during fast repetitive finger movements in humans: Steady-state movement-related magnetic fields and their cortical generators. *Electroenceph. Clin. Neurophysiol., 109,* 444-53.

Green, J. B., Bialy, Y., Sora, E. and Ricamato, A. (1999) High-resolution EEG in poststroke hemiparesis can identify ipsilateral generators during motor tasks. *Stroke, 30,* 2659-65.

Hallett, M. (1994) Movement-related cortical potentials. *Electromyogr. Clin. Neurophysiol., 34,* 5-13.

Hallett, M. and Toro, C. (1996) In *Supplementary sensorimotor area,* Vol. 70 (Ed, Lüders, H. O.) Lippincott-Raven, Philadelphia, pp. 147-152.

Hamdy, S. and Rothwell, J. C. (1998) Gut feelings about recovery after stroke: the organization and reorganization of human swallowing motor cortex. *Trends Neurosci., 21,* 278-82.

Holmes, G. (1917) The symptoms of acute cerebellar injuries due to gunshot injuries. *Brain, 40,* 461-535.

Honda, M., Nagamine, T., Fukuyama, H., Yonekura, Y., Kimura, J. and Shibasaki, H. (1997) Movement-related cortical potentials and regional cerebral blood flow change in patients with stroke after motor recovery. *J. Neurol. Sci., 146,* 117-26.

Ikeda, A., Lüders, H. O., Burgess, R. C. and Shibasaki, H. (1992) Movement-related potentials recorded from supplementary motor area and primary motor area. Role of supplementary motor area in voluntary movements. *Brain, 95,* 323-334.

Ikeda, A., Shibasaki, H., Nagamine, T., Terada, K., Kaji, R., Fukuyama, H. and Kimura, J. (1994) Dissociation between contingent negative variation and Bereitschaftspotential in a patient with cerebellar efferent lesion. *Electroencephalogr. Clin. Neurophysiol., 90,* 359-64.

Kitamura, J., Shibasaki, H. and Takeuchi, T. (1996) Cortical potentials preceding voluntary elbow movement in recovered hemiparesis. *Electroencephalogr. Clin. Neurophysiol., 98,* 149-56.

Knight, R. T., Singh, J. and Woods, D. L. (1989) Pre-movement parietal lobe input to human sensorimotor cortex. *Brain Res., 498,* 190-4.

Kornhuber, H. H. and Deecke, L. (1964) Hirnpotentialänderungen beim Menschen vor und nach Willkürbewegungen, dargestellt mit Magnetband-Speicherung und Rückwärtsanalyse. *Pflügers Arch., 281,* 52.

Kornhuber, H. H. and Deecke, L. (1965) Hirnpotentialänderungen bei Willkürbewegungen und passiven Bewegungen des Menschen: Bereitschaftspotential und reafferente Potentiale. *Pflügers Arch., 284,* 1-17.

Lang, W., Cheyne, D., Kristeva, R., Beisteiner, R., Lindinger, G. and Deecke, L. (1991) Three-dimensional localization of SMA activity preceding voluntary movement. A study of electric and magnetic fields in a patient with infarction of the right supplementary motor area. *Exp. Brain Res., 87,* 688-95.

Laplane, D., Talairach, J., Meininger, V., Bancaud, J. and Orgogozo, J. M. (1977) Clinical consequences of corticectomies involving the supplementary motor area in man. *J. Neurol. Sci., 34,* 301-14.

Lim, S. H., Dinner, D. S., Pillay, P. K., Lüders, H., Morris, H. H., Klem, G., Wyllie, E. and Awad, I. A. (1994) Functional anatomy of the human supplementary sensorimotor area: results of extraoperative electrical stimulation. *Electroencephalogr. Clin. Neurophysiol., 91,* 179-93.

Mackinnon, C. D., Kapur, S., Hussey, D., Verrier, M. C., Houle, S. and Tatton, W. G. (1996) Contributions of the mesial frontal-cortex to the premovement potentials associated with intermittent hand movements in humans. *Human Brain Mapping, 4,* 1-22.

Nagamine, T., Toro, C., Balish, M., Deuschl, G., Wang, B., Sato, S., Shibasaki, H. and Hallett, M. (1994) Cortical magnetic and electric fields associated with voluntary finger movements. *Brain Topogr., 6,* 175-83.

Netz, J., Lammers, T. and Homberg, V. (1997) Reorganization of motor output in the non-affected hemisphere after stroke. *Brain, 120,* 1579-86.

Platz, T., Kim, I. H., Pintschovius, H., Winter, T., Kieselbach, A., Villringer, K., Kurth, R. and Mauritz, K. H. (2000) Multimodal EEG analysis in man suggests impairment-specific changes in movement-related electric brain activity after stroke. *Brain, 123 Pt 12,* 2475-90.

Praamstra, P., Stegeman, D. F., Horstink, M. W. and Cools, A. R. (1996) Dipole source analysis suggests selective modulation of the supplementary motor area contribution to the readiness potential. *Electroencephalogr. Clin. Neurophysiol., 98,* 468-77.

Roland, P. E. (1984) Organization of motor control by the normal human brain. *Hum. Neurobiol., 2,* 205-16.

Rossini, P. M., Caltagirone, C., Castriota-Scanderbeg, A., Cicinelli, P., Del Gratta, C., Demartin, M., Pizzella, V., Traversa, R. and Romani, G. L. (1998) Hand motor cortical area reorganization in stroke: a study with fMRI, MEG and TCS maps. *Neuroreport, 9,* 2141-6.

Saitoh, T., Kamiya, H., Mizuno, Y., Shimizu, N., Niijima, K., Ohbayashi, T. and Yoshida, M. (1987) Neurophysiological analysis of ataxia in capsular ataxic hemiparesis. *J. Neurol. Sci., 79,* 221-8.

Sakai, K., Hikosaka, O., Miyauchi, S., Takino, R., Sasaki, Y. and Pütz, B. (1998) Transition of Brain Activation from Frontal to Parietal Areas in Visuomotor Sequence Learning. *J. Neurosci., 18,* 1827-40.

Sasaki, K. and Gemba, H. (1984) Compensatory motor function of the somatosensory cortex for the motor cortex temporarily impaired by cooling in the monkey. *Exp. Brain Res., 55,* 60-68.

Sasaki, K. and Gemba, H. (1991) In *Event-related brain research (EEG Suppl. 42)*(Eds, Brunia, C. H. M., Mulder, G. and Verbaten, M. N.) Elsevier, Amsterdam, pp. 80-96.

Sasaki, K., Gemba, H., Hashimoto, S. and Mizuno, N. (1979) Influences of cerebellar hemispherectomy on slow potentials in the motor cortex preceding self-paced hand movements in the monkey. *Neurosci. Lett., 15,* 23-28.

Sasaki, K., Gemba, H. and Tsujimoto, T. (1989) Suppression of visually initiated hand movement by stimulation of the prefrontal cortex in the monkey. *Brain Res., 495,* 100-7.

Shibasaki, H. (1975) Movement-associated cortical potentials in unilateral cerebral lesions. *J. Neurol., 209,* 189-98.

Shibasaki, H., Barrett, G., Halliday, E. and Halliday, A. M. (1980) Components of the movement-related cortical potential and their scalp topography. *Electroencephalogr. Clin. Neurophysiol., 49,* 213-26.

Shibasaki, H., Barrett, G., Neshige, R., Hirata, I. and Tomoda, H. (1986) Volitional movement is not preceded by cortical slow negativity in cerebellar dentate lesion in man. *Brain Res., 368,* 361-5.

Shibasaki, H., Shima, F. and Kuroiwa, Y. (1978) Clinical studies of the movement-related cortical potential (MP) and the relationship between the dentatorubrothalamic pathway and readiness potential (RP). *J. Neurol., 219,* 15-25.

Singh, J. and Knight, R. T. (1990) Frontal lobe contribution to voluntary movements in humans. *Brain Res., 531,* 45-54.

Singh, J. and Knight, R. T. (1993) Effects of posterior association cortex lesions on brain potentials preceding self-initiated movements. *J. Neurosci., 13,* 1820-9.

Tarkka, I. M. and Hallett, M. (1991) The cortical potential related to sensory feedback from voluntary movements shows somatotopic organization of the supplementary motor area. *Brain Topogr., 3,* 359-63.

Tarkka, I. M., Massaquoi, S. and Hallett, M. (1993) Movement-related cortical potentials in patients with cerebellar degeneration. *Acta Neurol. Scand., 88,* 129-35.

Toro, C., Matsumoto, J., Deuschl, G., Roth, B. J. and Hallett, M. (1993) Source analysis of scalp-recorded movement-related electrical potentials. *Electroencephalogr. Clin. Neurophysiol., 86,* 167-75.

Tsujimoto, T., Gemba, H. and Sasaki, K. (1993) Effect of cooling the dentate nucleus of the cerebellum on hand movements of the monkey. *Brain Res., 629,* 1-9.

Tsujimoto, T., Ogawa, M., Tsukada, H., Kakiuchi, T. and Sasaki, K. (1998) Activation of the ventral and mesial frontal cortex of the monkey by self-initiated movement tasks as revealed by positron emission tomography. *Neurosci. Lett., 258,* 117-20.

Turton, A., Wroe, S., Trepte, N., Fraser, C. and Lemon, R. N. (1996) Contralateral and ipsilateral EMG responses to transcranial magnetic stimulation during recovery of arm and hand function after stroke. *Electroencephalogr. Clin. Neurophysiol., 101,* 316-28.

Weeks, R., Gerloff, C., Dalakas, M. C. and Hallett, M. (1999) PET study of visually and non-visually guided finger movements in patients with severe pan-sensory neuropathies and healthy controls. *Exp. Brain Res., 128,* 291-302.

Weiller, C., Chollet, F., Friston, K. J., Wise, R. J. and Frackowiak, R. S. J. (1992) Functional reorganization of the brain in recovery from striatocapsular infarction in man. *Ann. Neurol., 31,* 463-72.

Zentner, J., Hufnagel, A., Pechstein, U., Wolf, H. K. and Schramm, J. (1996) Functional results after resective procedures involving the supplementary motor area. *J. Neurosurg., 85,* 542-49.

MOVEMENT and ERD/ERS

Gert Pfurtscheller[1] and Christa Neuper[2]

[1] Department of Medical Informatics
   Institute of Biomedical Engineering
[2] Ludwig Boltzmann Institute of Medical Informatics and Neuroinformatics
   University of Technology, Graz, Austria

INTRODUCTION

One characteristic feature of the brain is its ability to generate rhythmic potentials or oscillatory activity. Already in 1949 Jasper and Penfield discovered this fact and discussed the relationship between alpha and beta rhythms and their functioning in relation to underlying neural networks. The frequency of brain oscillations depends both on membrane properties of single neurons and the organization and interconnectivity of networks to which they belong (Lopes da Silva, 1991). Such a network can either comprise a large number of neurons controlled, for example, by thalamo-cortical feedback loops or only a small number of neurons interconnected, for example, by intra-cortical feedback loops. Coherent activity in large neuronal pools can result in high amplitude and low frequency oscillations (e.g. alpha band rhythms), whereas synchrony in localized neuronal pools can be the source of gamma oscillations (Lopes da Silva and Pfurtscheller, 1999). The dynamic of such a network can result in phasic changes in the synchrony of cell populations due to externally or internally paced events and lead to characteristic EEG patterns. Two such pattern types are observed, the event-related desynchronization, or ERD, in form of an amplitude attenuation and the event-related synchronization, or ERS, in form of an enhancement of specific frequency components (Pfurtscheller and Lopes da Silva, 1999a, 1999b).

The study of voluntary movement can be considered as a good model to investigate the dynamics of brain oscillations since it is controlled by a number of circumscribed and relatively well defined cortical and subcortical areas. Furthermore, it is supposed that a good deal of cortical activity is involved in preparing, producing and controlling motor behavior, which means that motor behavior requires the activation of a large number of cortical and subcortical systems.

Voluntary movement can be divided into self-paced (internally-paced) and externally-paced movement, which in the latter case means that movement is made in response to a certain cue or stimulus that can indicate, for example, the type of movement to be performed, like in the classical CNV-paradigm (Walter et al., 1964). Although, compared to self-paced movement, externally paced movement is a more complex process, not only including motor preparation and execution but also stimulus processing and expectancy, the performance of the motor act might involve similar brain structures.

Two kinds of changes in brain potentials accompanying voluntary movement can be differentiated: alterations of rhythmic components of the EEG and the occurrence of slow cortical potential shifts. Jasper and Penfield (1949) first reported on blocking or desynchronizing central beta rhythms over the hand area by fist clenching, and later Gastaut (1952) and Chatrian et al. (1959) described the blocking of the Rolandic wicket rhythm (mu rhythm) in relation to movement. While the slow cortical negative potential shift preceding voluntary movement, called Bereitschaftspotential (BP), was first described by Kornhuber and Deecke (1964), the term "event-related desynchronization" (ERD) was introduced by Pfurtscheller and Aranibar (1977) who, for the first time, reported on the quantification of the time course of the ERD during voluntary hand movement.

The BP begins as much as 1.0-1.5 s before EMG onset, is symmetrically distributed on both hemispheres and reaches its maximum on the midline close to the vertex (Kornhuber and Deecke, 1964; Shibasaki et al., 1981). Shortly (approximately 0.7 s) before movement-onset the BP becomes preponderant on the contralateral side and ends with the motor potential, which is strictly contralateral and localized over the primary sensorimotor cortex.

In contrast to the BP, the mu ERD already starts around 2 s prior to the onset of a voluntary self-paced finger movement (Pfurtscheller and Berghold, 1989) over the contralateral hemisphere and later on, shortly before movement-onset, also on the ipsilateral side. During execution of movement the ERD becomes almost symmetric on both hemispheres and recovers after movement-offset. The contralateral preponderance of the ERD during voluntary movement depends on handedness, side of the movement and the analyzed frequency band.

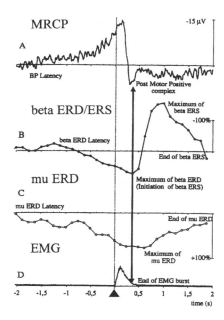

Figure 1. Temporal evolution of (A) movement-related potentials (MRCP), (B) ERD and ERS of beta rhythm, (C) ERD of mu rhythm and (D) rectified EMG for one subject (electrode position C3, right thumb movement). Time is expressed on the horizontal axis from 2 s before to 2 s after movement offset. Modified from Guieu et al. (1999).

After termination of movement, the BP displays a fast decline and is followed by a sequence of positive and negative components similar to the somatosensory evoked potential (SEP, Deecke et al., 1976). The ERD recovers relatively slowly, within a few seconds, in the alpha (mu) band and relatively quickly in the beta band. Characteristic for the beta band is not only the fast recovery but also a rebound in the form of a burst of oscillations (Pfurtscheller, 1981; Pfurtscheller et al., 1996, 1997). A comparison of movement-related potentials and ERD/ERS in a simple finger movement task is given in Fig. 1.

QUANTIFICATION OF ERD/ERS IN TIME AND SPACE

The classical method to compute ERD/ERS time course includes the following steps:
(i)   band pass filtering of all event-(stimulus) triggered EEG trials
(ii)   squaring of amplitude samples to obtain power samples
(iii)   averaging of power samples across all trials
(iv)   averaging over time sample to smooth the data and reduce the variability

This procedure results in a time course of band power including phase-locked (e.g. evoked potentials) and not phase-locked amplitude changes as well. For the discrimination of both, namely to exclude the contribution of event-related potentials, a slightly modified procedure is recommended:
(i)   band pass filtering
(ii)   calculation of the point-to-point intertrial variance
(iii)   averaging over time.

A detailed description of the method is given in Pfurtscheller and Lopes da Silva (1999a, 1999b). To obtain percentage values for ERD/ERS, the power within the frequency band in the period of interest is given by A, whereas that of the preceding baseline or reference period is given by R. ERD or ERS is defined as the percentage of band power decrease or increase, respectively, according to the expression $ERD\% = ((A-R)/R)100$. For spatial mapping of ERD/ERS, different methods are available, such as the calculation of surface Laplacian, cortical imaging and distributed source imaging. References to different deblurring methods, either using a realistic head model or a spherical model, are found in van Burik et al. (1999).

INDEX FINGER, WRIST AND FOOT MOVEMENT

Based on the pioneering findings of Penfield and Boldrey (1937), who first reported on the somatotopic organization of the motor cortex, and on a number of recent fMRI and PET studies reporting specific patterns of cortical rCBF increase during different types of voluntary movements in human subjects (Colebatch et al., 1991; Deiber et al., 1991), we expected to find differences in distribution and/or extent of ERD and ERS depending on the movement type performed.

Taking into account that movements of the individual fingers and the wrist are quite different in respect to moved mass and involved muscle force, differences in the reactivity pattern of sensorimotor rhythms are expected with different types of movement.

In 23-channel EEG recordings with nine healthy, right-handed volunteers (mean age: 24.4 years, SD: 3.3) the following three types of movement were investigated: (i) a brisk extension and flexion of the right index finger, (ii) a brisk pressing of a button on a joystick with the right thumb and (iii) a brisk 45° flexion of the right wrist with brisk return to the resting position. Each session comprised 70-80 movements, whereby the subjects were given a break of about a quarter of an hour between the sessions for relaxation and training of the subsequent task. Details are reported in Pfurtscheller et al. (1998).

The study of ERD/ERS diversity in different kinds of movement revealed two main results: (i) similar pre-movement mu ERD for wrist, index finger and thumb movement (see Fig. 2) and (ii) significantly increased post-movement beta ERS in wrist as compared to finger and thumb movement (see Fig. 3).

According to the finding that the pre-movement ERD is almost the same independent of the type and quality of the movement, activated cortical structures of about the same size can be presumed for movements of a single finger or of the whole hand. The pre-movement mu ERD, starting about 2 seconds prior to voluntary movement-onset, is supposed to reflect a kind of general readiness or the presetting of neurons in different motor and sensory areas, needed to execute a forthcoming movement, while different movement properties such as type (finger vs. wrist), speed (brisk vs. slow) or force are apparently not coded in the relatively long-lasting mu ERD.

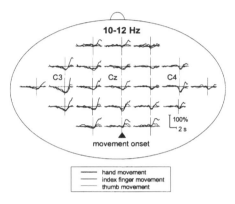

Figure 2. Topographical display (nose on top) of grand average power time courses (N=9) displaying pre-movement ERD in the 10-12 Hz band for wrist, index finger and thumb movement of the right hand. The horizontal lines mark the level of reference power and the vertical line movement-onset. Note the similar pre-movement ERD for the three movement tasks. Modified from Pfurtscheller et al. (1999).

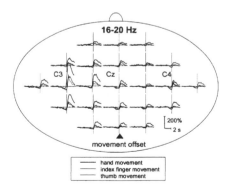

Figure 3. Topographical display of grand average power time courses (N=9) displaying post-movement ERS in the 16-20 Hz band. The vertical lines mark movement-offset. For further explanation see Fig. 2. Modified from Pfurtscheller et al. (1999).

In contrast to the great similarity of mu ERD patterns the post-movement beta oscillations of the three movement tasks are clearly different and largest with wrist movement. While the control of a single finger movement involves a large number of muscle spindles, movement of the whole hand requires, in particular, muscle force and hence, more mass of muscular fibers has to be activated. Button pressing by one finger movement is not only accompanied by cutaneous, but also by proprioceptive afferent activity, whereas wrist movement mainly results in proprioceptive activity from the joints. Activation of a larger muscle mass may require a relatively larger population of cortical neurons, indicating that our finding of larger beta oscillations with wrist as compared to finger or thumb movement can be understood as the change of a larger population of motor cortex neurons from an activated state during the motor act to a state of cortical deactivation or cortical "idling" after termination of movement.

In previous neuromagnetic recordings, moreover, enhanced beta oscillations were found not only following movement of fingers, but also of toes and mouth (Salmelin et al., 1995). In a voluntary foot movement experiment we investigated whether EEG recordings can also detect changes of alpha and beta band components (ERD/ERS) related to self-paced movement of a part of the lower limb and whether a topographical differentiation with regard to the motor homunculus is possible (see Neuper and Pfurtscheller, 1996).

Nine healthy right-handed subjects (mean age: 25.3 years, SD: 6.1), who also showed a preference for the right foot, participated in the experiment. During the experiment the subjects were asked to keep their eyes closed and to perform brisk self-paced dorsiflexions of the right and left foot in intervals of about 10-15 s. Each foot was rested on a sensor on a platform, which indicated the lifting of the foot and served as a trigger for data processing. Both left- and right-sided foot movements were performed in two blocks of about 70 movements, whereby the order of the blocks was balanced across subjects. Details are published elsewhere (Neuper and Pfurtscheller, 1996).

Individual reactive movement-sensitive mu as well as beta frequency components were selected using the respective power spectra from electrode position Cz for both left and right foot movement. While the mean reactive frequency band (for determination of the reactive frequency band see Pfurtscheller and Lopes da Silva, 1999b) displaying largest ERD during initiation and execution of movement was found between 7-11 Hz, post-movement synchronization was most pronounced in a beta range from 12-32 Hz.

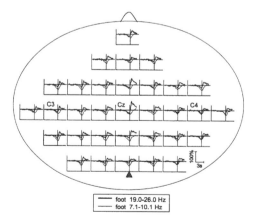

Figure 4. Topographical display of grand average power time courses (n = 9) of alpha band (thin line) and beta band (thick line) power during left foot movement. Movement-offset is indicated by a vertical line. Modified from Neuper and Pfurtscheller (1996).

As can be seen in Fig. 4, a different temporal behavior of mu and beta band reactivity was observed. While the beta frequency components recovered very quickly after termination of movement and reached the maximum rebound level on average 0.76 s (SD:

± 0.32) after movement-offset, alpha frequency components recovered only slowly and showed hardly any synchronization. Furthermore, analogous to the previously reported experiments topographical differences were found with respect to specificity of mu and beta frequency components. The alpha band ERD was found to be widespread and showed no movement-specific desynchronization on a certain electrode or cortical region.

Similar experiments reporting a relatively widespread 10-Hz desynchronization over the sensorimotor areas during self-paced dorsiflexion movement of the foot are based on both subdural recordings in patients and scalp recordings in normal subjects (Arroyo et al., 1993; Toro et al., 1994). The post-movement beta oscillations, in contrast, were focused and especially enhanced along the midline electrodes close to the foot representation area.

DISTINCTION BETWEEN DIFFERENT MU RHYTHMS

It is of interest that during finger and wrist movements frequency components in the upper alpha band (10-13 Hz) are mainly affected, whereas foot and arm movements result in a desynchronization of lower (8-10 Hz) alpha frequency components (Pfurtscheller et al., 1999). Another difference between the alpha band desynchronization during wrist and finger movements on the one side, and foot and arm movements on the other side, is that the former is clearly focused to the respective cortical representation area. Foot and arm movements, in contrast, affect a broader frequency band and result in a relatively widespread ERD, covering hand, arm and foot representation areas.

Fig. 5 compares reactivity of upper and lower alpha band rhythms during self-paced movements of the right index finger (upper panel) and the right foot (lower panel). These results (reported in detail in Pfurtscheller et al., 2000a) further support the notion that different types of sensorimotor rhythms in the alpha band can be differentiated: (i) The classical mu rhythm with frequency components in the upper alpha band originates in the hand area and desynchronizes during finger or wrist movement. (ii) Another class of mu rhythms also originating in the sensorimotor strip with components in the lower alpha band, desynchronizes with arm, foot and hand movement.

In subdural recordings, moreover, the existence of not only a hand area specific mu rhythm, but also of specifically reacting foot and face area mu rhythms, each blocking when the corresponding area is activated, has been proven (Arroyo et al., 1993). Therefore, we can suggest that there exist several types of mu rhythms that are specifically blocked with hand, leg or face/tongue movement. With scalp electrodes, however, only the hand area mu rhythm can be reliably measured.

The desynchronization of lower mu components found over frontal, central and parietal areas may indicate the existence of a distributed neural network in motor and somatosensory cortices, which is activated by different types of motor behavior, but not necessarily critically to support a specific movement. This system may act as a relatively non-specific activation or presetting of somatosensory and motor neurons in cortical areas prior to a specific motor act. Another aspect of interpretation could be a neurophysiological mechanism that serves as general motor attention to all cortical areas involved in a motor task including, apart from primary sensorimotor, also premotor and parietal areas. Lower alpha desynchronization occurs also in response to a variety of non task-specific factors which may be best summed up under the term "attention". This ERD is topographically widespread over the scalp and probably reflects general task demands and attentional processes (Klimesch, 1999). It is not unlikely that similar neurophysiological mechanisms operate in the lower alpha band during cognitive and motor processing.

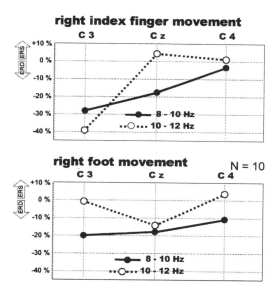

Figure 5. Relationship between movement type (hand vs. foot), band power changes in the 8-10 Hz and 10-12 Hz bands and electrode locations (electrode positions C3, Cz and C4) measured in the last 500 ms prior to movement-onset (N=10). Note the about similar behavior of lower mu components and the different behavior of upper mu components. Modified from Pfurtscheller et al. (2000a).

SELF-PACED MOVEMENT VERSUS CUE-PACED MOVEMENT

Execution of self-paced hand movement in intervals of approximately 10 s is accompanied by a contralateral dominant mu ERD in both the upper and the lower alpha band. Self-paced foot movement executed one after the other in intervals of a few seconds results in a widespread ERD in the 6–10 Hz range (Neuper and Pfurtscheller, 1996). In this case no characteristic change over the hand area in the upper alpha band was found (Pfurtscheller et al., 2000). However, when the same movement types, either a hand or a foot movement, have to be made cue-dependent, the intrinsic activity of the hand area shows an antagonistic behavior in the 10–13 Hz band: hand movement results in a decrease of rhythmic activity, or ERD, similar to the findings with consecutive self-paced hand movements, whereas foot movement is accompanied by a power increase, or ERS, in the 10–13 Hz frequency range (Fig. 6).

One explanation for this phenomenon may be that prior to the cue-stimulus, indicating the type of movement (hand or foot) to be executed, various sensorimotor areas, including those representing the hand and the foot, are pre-activated or primed and therefore transferred to a state of increased excitability. Characteristic for such a cortical state are slow negative potential shifts, such as the contingent negative variation (CNV; Walter et al., 1964), and concurrently, a desynchronized EEG pattern. When the cue-stimulus indicates 'foot movement', the hand area pre-activation has to be 'inhibited', whereas the foot area pre-activation is facilitated. This 'inhibition' of the hand area network is time-locked to induced oscillations in the upper alpha band and topographically restricted to electrodes overlying the hand area. It is of interest that this antagonistic behavior of alpha band components (ERD/ERS) is a dominant feature of the upper alpha band and not seen with lower frequency components.

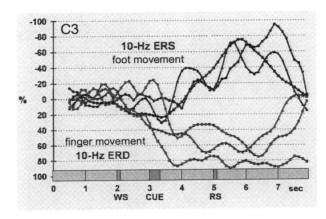

Figure 6. ERD/ERS time courses (10-12 Hz) of 3 subjects display percentage power decrease (10-Hz ERD) during finger movement and power increase (10-Hz ERS) during foot movement over the primary hand area (electrode C3). Modified from Pfurtscheller and Neuper (1994).

MOTOR IMAGERY

Motor imagery can be defined as an imagined rehearsal of a motor act, without any simultaneous sensory input or any overt output in form of muscular movements. It is broadly accepted that mental movements involve similar brain regions/functions which are involved in programming and preparing actual actions (Jeannerod, 1995). The main difference between performance and imagery is that in the latter case execution would be blocked at some cortico-spinal level (Decety, 1996). This explains, for example, that mental rehearsal may have positive effects on motor skill learning and on motor performance.

During movement imagery tasks, an increase of the regional cerebral blood flow (rCBF) has been located in the premotor and frontal regions, in the supplementary motor areas (Roland et al., 1980), and in prefrontal and parietal areas (Deiber et al., 1998). It is of interest, that recent fMRI studies detected some activation in the primary motor cortex during motor imagery, though to a lesser extent than during actual motor performance (e.g. Hallett et al., 1994). Increased motor cortex excitability during motor imagery has also been demonstrated by transcranial magnetic stimulation (e.g. Stephan and Frackowiak, 1996).

A blocking of the central mu rhythm with motor imagery was reported in early clinical EEG studies (Chatrian et al., 1959; Gastaut et al., 1965). Similar cortical activity over the contralateral hand area during execution and imagination of hand movement has further been found with DC potential measurements (Beisteiner et al., 1995) and based on dipole source analysis of electric and magnetic fields (Lang et al., 1996).

Furthermore, recent results obtained in high-resolution EEG experiments strongly indicate activity of primary motor structures not only during actual execution, but also during mental simulation of movement (Pfurtscheller and Neuper, 1997; Neuper and Pfurtscheller, 1999). The experimental procedure of the motor imagery experiments was as follows: The subjects were sitting in a darkened and electrically shielded cubicle and facing a computer monitor placed approximately 2 m in front at eye level. Each trial started with a 2-second period during which a fixation cross was presented at the center of the monitor. Then, a warning tone ("beep") was presented, and one second later the fixation cross at the center of the monitor was overlaid with an arrow for 1.25 s; the arrow was pointing either to the left or to the right ("cue").

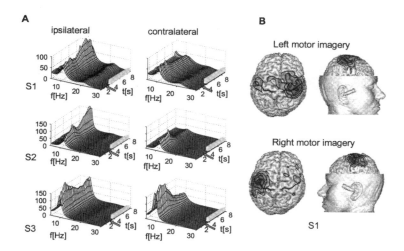

Figure 7. (A) Dynamic spectral arrays of 3 subjects (S1-S3) during imagination of hand movement, computed for EEG signals recorded from the ipsi- and contralateral hemisphere. The spectra are shown at a rate of 4/s from 1-8 s and for 5-30 Hz. Cue presentation, starting at t=3.0 s, is indicated by an arrow. The gray bar along the time axis indicates the imagination period. (B) ERD maps calculated for the cortical surface of a realistic head model of one subject (for details about computation see van Burik et al., (1999)) The model was constructed from 200 T1-weighted transversal MR images. The 3D electrode positions, measured with a magnetic digitizer, were transformed into the MRI co-ordinate system. The spline surface Laplacian method was applied to the bandpass filtered (9-13 Hz) single-trial EEG data and the distribution of the alpha band ERD was calculated for left and right motor imagery. The spline Laplace maps are shown at 625 ms after the presentation of the cue (arrow in left or right direction). Isocontour levels are shown in steps of 20%. The central sulcus is marked by a think black line. Modified from Neuper et al. (1999).

During the execution condition, the subjects were required to perform a brisk dorsiflexion movement of the left or right hand, as indicated by the direction of the arrow. During the imagination condition, the subjects were asked to imagine such a movement while remaining relaxed and avoiding any motions. EEG, sampled at 256 Hz, was recorded from 60 Ag/AgCl electrodes placed over frontal, central and parietal areas equally spaced approximately 2.5 cm apart. (For further details see Neuper and Pfurtscheller, 1999).

Independent of the task, imagination vs. overt execution of the movement, the most prominent EEG changes were localized over the primary sensorimotor cortex. In particular, we found a similar ERD over the contralateral hand area during imagination as is usually found during planning or preparation of a real movement. An example of ERD distribution during motor imagery, mapped on the reconstructed cortical surface of one representative subject is presented in Fig. 7B. As can be seen, a circumscribed ERD with a focus close to the primary hand area can be distinguished during imagination of right-hand as well as left-hand movement. During overt execution of the movement the initially contralateral ERD develops a bilateral distribution, whereas during mental simulation this ERD remained mostly limited to the contralateral hemisphere. Examples of spectral arrays of EEG activity recorded from the contra vs. ipsilateral hand area and calculated for the time course of the trial are shown in Fig. 7A. All 3 selected subjects showed a dominant reactive spectral peak around 10 Hz during the reference period at the beginning of the trial. This mu rhythm recorded over the hand area is desynchronized with imaginary movement of the contralateral hand. The same derivation, however, can also show synchronized activity when the subject imagines a movement with the ipsilateral hand.

FOCAL ERD/SURROUND ERS

The term "focal ERD/surround ERS" was introduced by Lopes da Silva (Suffczynski et al., 1999) to describe the observation that blocking or desynchronization of rhythmic activity in the alpha band does not occur in isolation but can be accompanied by an amplitude increase or synchronization in neighboring cortical areas that correspond to the same or to another modality of information processing.

An intermodal interaction in form of a central desynchronization and a parieto-occipital synchronization as typical for voluntary finger movement in the alpha band can be seen in raw EEG data (Fig. 8) and also visualized in form of maps (Fig. 9A). The opposite phenomenon, the enhancement of central mu rhythm and blocking of occipital alpha rhythm during visual stimulation, was reported by Koshino and Niedermeyer (1975) and Kreitmann and Shaw (1965).

A further example of focal ERD/surround ERS, obtained with a dense EEG electrode montage, demonstrating hand area ERD and foot area ERS during hand movement and hand area ERS and foot area ERD during foot movement, is illustrated in Fig. 9B.

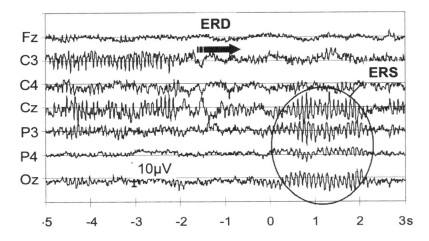

Figure 8. Examples of ongoing EEG data recorded during right finger movement. Movement-onset at t=0 s. Note the EEG desynchronization at electrode location C3 starting about 1.5 s prior to movement-onset and the enhanced alpha band activity over the posterior region (ERS) during movement. Modified from Pfurtscheller and Lopes da Silva (1999b).

The focal mu desynchronization in the 10-12 Hz band as documented in Fig. 9B may reflect a mechanism responsible for selective attention focused to a motor subsystem. This effect of selective attention may be accentuated when other cortical areas, not directly involved in the specific motor task are "inhibited". Increased alpha synchronization over posterior regions (Fig. 9A) may reflect a reduced capacity of information processing in underlying neuronal networks (Pfurtscheller, 1992). In contrast to this fact, the desynchronized mu rhythm during the motor task indicates facilitation of information processing through hand area networks.

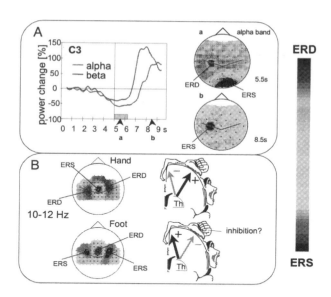

Figure 9. (A) Grand average (N=9) ERD/ERS curves calculated in the alpha and beta bands in a right hand movement task (left side). Grand average maps calculated for a 125 ms interval during movement (a) and after movement-offset in the recovery period (b) (right side). (B) Maps displaying ERD and ERS for an interval of 125 ms during voluntary movement of the hand (left, upper panel) and movement of the foot (left, lower panel). The motor homunculus with a possible mechanism of cortical activation/deactivation gated by thalamic structures is shown on the right. Modified from Pfurtscheller and Lopes da Silva (1999b).

A similar antagonistic behavior with desynchronization of central mu rhythm and synchronization of parieto-occipital alpha rhythms during repetitive brief finger movement was reported by Gerloff et al. (1998). A task-related power increase found in the 9-11 Hz band can be interpreted as possibly reflecting an "inhibitory state" of occipital and parieto-occipital regions, since no visual feedback was presented.

INDUCED BETA OSCILLATIONS AND INTERACTIONS BETWEEN SENSORIMOTOR AREAS

Self-paced limb movements are accompanied by bursts of oscillations in the beta band. Examples of raw EEG data from a foot movement experiment are depicted in Fig. 10, ERD/ERS curves displaying beta power increase after hand and foot movement are shown in Fig. 3 and Fig. 4, respectively. One important feature of these beta oscillations is its strict somatotopical organization in MEG (Salmelin et al., 1995) and EEG (Neuper and Pfurtscheller, 1996; Pfurtscheller et al., 1997).

Another interesting observation is the 'cross-talk' between distinct sensorimotor areas meaning that, for example, finger movement not only induces beta oscillations close to the hand representation area, but also oscillations at electrode position Cz, close to the mid-central foot representation area (Pfurtscheller et al., 2000b). Analogously, foot movement is accompanied by beta oscillations with a maximum near to the vertex (electrode position Cz, example: see Fig. 10), but in addition also by induced beta oscillations close to the hand representation area. Characteristic for this 'cross-talk' phenomenon is that the largest beta oscillations are always found over the movement corresponding representation area and, which is of special interest, the frequencies of the induced hand area oscillations are usually lower than those of the mid-centrally induced oscillations (Neuper and Pfurtscheller, 2001; submitted). The mean frequencies ($\pm$ SD) of hand area oscillations are 17.4 $\pm$ 1.8 Hz and of 'foot area' oscillations 21.5 $\pm$ 3.3 Hz. Group data of voluntary finger and foot movement are displayed in Fig. 11, for the average bands 15-20 Hz and 20-25 Hz.

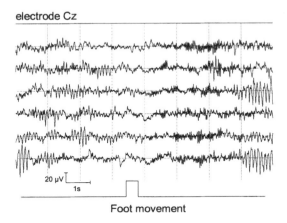

Figure 10. Examples of raw EEG data recorded during self-paced foot movement, recorded from electrode position Cz. Note the induced beta oscillations after movement-offset.

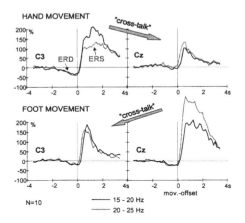

Figure 11. Grand average ERD/ERS time courses calculated for the subject-specific frequency bands obtained over the hand and foot areas. The time courses show finger and foot movement separately for electrode positions C3 and Cz. The average hand area-specific frequency band was about 15-20 Hz (full line curves), and the average foot area-specific frequency band was about 20-25 Hz (dashed line curves). The vertical line indicates the trigger (movement-offset), and the horizontal line marks the baseline. Modified from Neuper and Pfurtscheller (2001).

The novel aspect is that the same event, namely finger (foot) movement, creates induced beta oscillations of different frequencies over neighboring cortical areas representing the hand and perhaps the foot. The primary foot representation area is located on the mesial brain surface close to the SMA (Tanji, 1996) and the potentials generated within this area are not easy accessible to scalp electrodes. Beside the strict localization of induced beta oscillations close to the hand and foot representation areas there is another aspect of interest, namely the preferred oscillatory frequency in each area, always being slightly lower in the hand area. The 'cross-talk' provides the non-linear coupling between beta generating networks with different 'resonance-like' frequencies.

One explanation for the area specific 'resonance' frequency could be the topology and functional interconnectivity of each network. According to the motor homunculus the hand area is of larger size and higher functional importance than e.g. the foot area. Singer (1993) reported frequency differences depending on the size of synchronously activated cell assemblies, with a lower peak frequency in larger networks (such as hand) as compared with smaller networks (such as foot). We speculate therefore that the lower frequency of beta oscillations recorded over the hand area as compared to the mid-central foot area may be indicative of a neural network with a larger density of neurons which form larger functional interconnected assemblies.

VOLUNTARY MOVEMENT VS. SOMATOSENSORY STIMULATION

It is known that similar to voluntary hand movement, stimulation of the median nerve can also elicit short-lasting bursts of beta oscillations that are localized predominantly in the contralateral hand sensorimotor cortex (Salenius et al., 1997). In general, electrical median nerve stimulation above the sensory threshold can generate two types of cortical responses, namely a phase-locked (evoked) response localized in the postcentral somatosensory cortex (Baumgartner et al., 1991) and a non-phase-locked (induced) response in the form of beta oscillations in the precentral motor cortex. The latter appear within 1 s after the delivery of the stimulus and follow a shortlasting ERD immediately after stimulation. One has to keep in mind, however, that stimulation intensity for stimulating the median nerve at the wrist is usually adjusted to produce a visible thumb twitch. Hence, the question remained, whether similar beta oscillations over the motor cortex can be induced by simple mechanical stimulation of the fingertip, without any visible muscular activity.

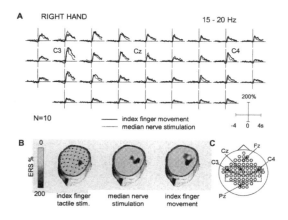

Figure 12. (A) Topographical display (nose on top) of grand average ERD/ERS curves comparing movement vs. stimulation of the hand (mean frequency band: 15-20 Hz). The horizontal line marks the level of reference power and the vertical line movement-offset/stimulation. Band power increase (beta ERS) is indicated by upward deflection, band power decrease (beta ERD) by downward deflection. (B) Topographical maps of one subject, each representing an interval of 125 ms, showing the distribution of beta ERS (16–20 Hz) following tactile stimulation of the index fingertip (electrode positions are marked), median nerve stimulation and index finger movement. "Black" indicates beta power increase (ERS). (C) Electrode montage. Modified from Neuper and Pfurtscheller (2001).

Fig. 12A provides the topographical display of grand average beta band ERD/ERS time courses comparing voluntary index finger movement and median nerve stimulation. In addition, examples of individual data can be seen in Figure 12B. Here topographic maps of one subject illustrate the beta ERS distribution obtained in three different conditions, mechanical finger stimulation, electrical nerve stimulation, and voluntary finger movement. In general, a similar beta ERS distribution was found during active movement and passive stimulation showing in all conditions the focus of beta ERS over the contralateral primary sensorimotor cortex. Since the application of the tactile stimuli did not result in a visible finger movement, it seems unlikely that reafferent input from the kinesthetic receptors would be responsible for the occurrence of the beta ERS.

The observed similarity between beta ERS distribution following unilateral finger movement and somatosensory stimulation indicates that the underlying neuronal populations overlap widely. This result is in line with converging evidence that the sources of beta ERS related to both movement and stimulation of the hand are mainly concentrated in the hand sensorimotor cortex (Salenius et al., 1997). In intracranial recordings, the central beta rhythm that originated in the motor cortex was not only affected by initiation of a voluntary movement, but also by somatosensory stimulation (Jasper and Penfield, 1949).

CONCLUSION

Execution and imagination of movement involve various cortical and subcortical structures and affect different neural networks responsible for the generation of oscillatory activity.

In summary, the following findings on brain oscillations can be emphasized:

- There exists a great variety of mu rhythms, some display a widespread ERD, some a somatotopic organization of the ERD.
- The blocking or ERD of mu rhythms can be accompanied either by an intramodal (e.g. hand vs. foot) or an intermodal enhancement (ERS) of oscillations in the upper alpha band.
- Termination of movement can result in short bursts of central beta oscillations (beta ERS) localized over the corresponding cortical representation areas.
- A beta ERS with a similar spatiotemporal pattern can also be found with active movement, motor imagery and passive somatosensory stimulation. It demonstrates that induced beta oscillation does not exclusively depend on afferent information.
- There is strong evidence that beta oscillations (beta ERS) originating over the hand representation area have a slightly lower frequency compared to oscillations found midcentrally, i.e. close to the foot representation area. This might be interpreted as showing that different neuronal networks or circuitries have their own characteristic "resonance"-frequency.
- Note that during the occurrence of post movement beta ERS the corticospinal excitation is reduced (Chen et al., 1998). This may be the first evidence that the beta ERS coincides with a deactivated cortical structure.

ACKNOWLEDGEMENT

This study was partially supported by the "Fonds zur Förderung der wissenschaftlichen Forschung" in Austria, projects P12407-MED and P14831-PSY.

REFERENCES

Arroyo, S., Lesser, R.P., Gordon, B., Uematsu, S., Jackson, D., Webber, R., 1993, Functional significance of the mu rhythm of human cortex: an electrophysiological study with subdural electrodes, Electroencephalogr. clin. Neurophysiol. 87:76-87.

Baumgartner, C., Doppelbauer, A., Deecke, L., Barth, D.S., Zeitlhofer, J., Lindinger, G., Sutherling, W.W., 1991, Neuromagnetic investigation of somatotopy of human hand somatosensory cortex, Exp. Brain Res. 87:641-648.

Beisteiner, R., Höllinger, P., Lindinger, G., Lang, W. and Berthoz, A., 1995, Mental representations of movements. Brain potentials associated with imagination of hand movements, Electroenceph. clin. Neurophysiol. 96:183-192.

Chatrian, G.E., Petersen, M.C., Lazarte, J.A., 1959, The blocking of the rolandic wicket rhythm and some central changes related to movement, Electroenceph. clin. Neurophysiol. 11:497-510.

Chen, R., Yassen, Z., Cohen, L.G. and Hallett, M., 1999, The time course of corticospinal excitability in reaction time and self-paced movements, Ann. Neurol. 44:317-325.

Colebatch, J.B., Deiber, M.-P., Passingham, R.E., Friston, K.J., Frackowiak, R.S.J., 1991, Regional cerebral blood flow during voluntary arm movements in human subjects, J. Neurophysiol. 65:1392-1401.

Decety, J., 1996, The neurophysiological basis of motor imagery, Behav. Brain Res. 77:45-52.

Deecke, L., Grözinger, B., Kornhuber, H.H., 1976, Voluntary finger movement in man: cerebral potentials and theory, Biol. Cybern. 23:99-119.

Deiber, M.-P., Passingham, R.E., Colebatch, J.G., Friston, K.J., Nixon, P.D. and Frackowiak, R.S.J., 1991, Cortical areas and the selection of movement: a study with positron emission tomography, Exp. Brain Res. 84:393-402.

Deiber, M.P., Ibanez, V., Honda, M., Sadato, N., Raman, R. and Hallett, M., 1998, Cerebral processes related to visuomotor imagery and generation of simple finger movements studied with positron emission tomography, Neuroimage 7(2):73-85.

Gastaut, H., 1952, Etude electrocorticographique de la reactivite des rhythmes rolandiques, Rev. Neurol. 87:176-182.

Gastaut, H., Naquet, R. and Gastaut, Y., 1965, A study of the mu rhythm in subjects lacking one or more limbs, Electroenceph. clin. Neurophysiol. 18:720-721.

Gerloff, C., Hadley, J., Richard, J. Uenishi, N., Honda, M., Hallett, M., 1998, Functional coupling and regional activation of human cortical motor areas during simple, internally paced and externally paced finger movements, Brain 121:1513-1531.

Guieu, J.D., Bourriez, J.L., Derambure, P., Defebvre, L. and Cassim, F., 1999, Temporal and spatial aspects of event-related desynchronization and movement-related cortical potentials, in: Event-Related Desynchronization, Handbook of Electroenceph. and Clin. Neurophysiol., Revised Edition, Vol. 6, Pfurtscheller, G., Lopes da Silva, F.H., eds., Elsevier, Amsterdam, pp. 279-290.

Hallett, M., Fieldman, J., Cohen, L.G., Sadato, N. and Pacual-Leone, A., 1994, Involvement of primary motor cortex in motor imagery and mental practice, Behav. Brain Sci. 17:210.

Hjorth, B., 1975, An on-line transformation of EEG scalp potentials into orthogonal source derivations, Electroenceph. clin. Neurophysiol. 39:526-530.

Jasper, H.H., Penfield, W., 1949, Electrocorticograms in man: effect of the voluntary movement upon the electrical activity of the precentral gyrus, Arch. Psychiat. Z. Neurol. 183:163-174.

Jeannerod, M., 1995, Mental imagery in the motor context, Neuropsychologia 33(11):1419-1432.

Klimesch, W., 1999, Event-related band power changes and memory performance. Event-related desynchronization and related oscillatory phenomena of the brain, in: Handbook of Electroencephalography and Clin. Neurophysiology, Revised Edition, Vol. 6, Pfurtscheller, G., Lopes da Silva, F.H., eds., Elsevier, Amsterdam, pp. 151-178.

Kornhuber, H.H., Deecke, L., 1964, Hirnpotentialänderungen beim Menschen vor und nach Willkürbewegungen, dargestellt mit Magnetbandspeicherung und Rückwärtsanalyse, Plügers Arch. Ges. Physiol. 281:52.

Koshino, Y., Niedermeyer, E., 1975, Enhancement of rolandic mu-rhythm by pattern vision, Electroenceph. clin. Neurophysiol. 28:535-538.

Kreitmann N., Shaw J.C., 1965, Experimental enhancement of alpha activity, Electroenceph. clin. Neurophysiol. 18:147-155.

Kuhlman, W.N., 1978, Functional topography of the human mu rhythm, Electroencephalogr. clin. Neurophysiol. 44:83-93.

Lang, W., Cheyne, D., Hollinger, P., Gerschlager, W. and Lindinger, G., 1996, Electric and magnetic fields of the brain accompanying internal simulation of movement, Brain Res. Cogn. Brain Res. 3:125-129.

Lopes da Silva, F.H., 1991, Neural mechanisms underlying brain waves: from neural membranes to networks, Electroenceph. clin. Neurophysiol. 79:81-93.

Lopes da Silva, FH., Pfurtscheller, G., 1999, Basic concepts on EEG synchronization and desynchronization. Event-related desynchronization and related oscillatory phenomena of the brain, in: Handbook of Electroencephalography and Clin. Neurophysiology, Revised Edition, Vol. 6, Pfurtscheller, G., Lopes da Silva, FH., eds., Elsevier, Amsterdam, pp. 3-11.

Neuper, C., Pfurtscheller, G., 1996, Post-movement synchronization of beta rhythms in the EEG over the cortical foot area in man, Neurosci. Lett. 216:17-20.

Neuper, C. and Pfurtscheller, G., 1999, Motor imagery and ERD, in: Event-Related Desynchronization. Handbook of Electroenceph. and Clin. Neurophysiol., Revised Edition, Vol. 6, Pfurtscheller, G., Lopes da Silva, F.H., eds., Elsevier, Amsterdam, pp. 303-325.

Neuper, C. and Pfurtscheller, G., 2001, Evidence for distinct beta resonance frequencies related to specific sensorimotor cortical areas. Clinical Neurophysiology, in revision.

Penfield, W., Boldrey, E., 1937, Somatic motor and sensory representation in the cerebral cortex of man as studied by electrical stimulation, Brain 60:389-443.

Pfurtscheller, G., 1981, Central beta rhythm during sensory motor activities in man, Electroenceph. clin. Neurophysiol. 51:253-264.

Pfurtscheller, G., 1992, Event-related synchronization (ERS): an electrophysiological correlate of cortical areas at rest, Electroenceph. clin. Neurophysiol. 83:62-69.

Pfurtscheller, G., Aranibar, A., 1977, Event-related cortical desynchronization detected by power measurements of scalp EEG, Electroenceph. clin. Neurophysiol. 42:817-826.

Pfurtscheller, G., Berghold, A., 1989, Patterns of cortical activation during planing of voluntary movement, Electroenceph. clin. Neurophysiol. 72:250-258.

Pfurtscheller, G., Neuper, C., 1997, Motor imagery activates primary sensorimotor area in humans, Neurosci. Lett. 239:65-68.

Pfurtscheller, G., Neuper, C., 1994, Event-related synchronization of mu rhythm in the EEG over the cortical hand area in man, Neurosci. Lett. 174:93-96.

Pfurtscheller, G., Stančák, A., Neuper, C., 1996, Post-movement beta synchronization. A correlate of an idling motor area, Electroenceph. clin. Neurophysiol. 98:281-293.

Pfurtscheller, G., Stančák, A., Edlinger, G., 1997, On the existance of different types of central beta rhythms below 30 Hz, Electroenceph. clin. Neurophysiol. 102:316-325.

Pfurtscheller, G., Zalaudek, K., Neuper, C., 1998, Event-related beta synchronization after wrist, finger and thumb movement, Electroenceph. clin. Neurophysiol. 109:154-160.

Pfurtscheller, G. and Lopes da Silva, F.H, 1999a, Event-Related Desynchronization. Handbook of Electroenceph. and Clin. Neurophysiol., Revised Edition, Vol. 6. Elsevier, Amsterdam.

Pfurtscheller, G. and Lopes da Silva, F.H., 1999b, Event-related EEG/MEG synchronization and desynchronization: Basic principles, Clin. Neurophysiol. 110:1842-1857.

Pfurtscheller, G., Pichler-Zalaudek, K., Neuper, C., 1999, ERD and ERS in voluntary movement of different limbs, in: Handbook of Electroencephalography and Clin. Neurophysiology, Revised Edition, Vol. 6, Pfurtscheller, G., Lopes da Silva, F.H., eds., Elsevier, Amsterdam, pp. 245-268.

Pfurtscheller, G., Neuper, C., Krausz, G., 2000a, Functional dissociation of lower and upper frequency mu rhythms in relation to voluntary limb movement, Clin. Neurophysiol. 111:1873-1879.

Pfurtscheller, G., Neuper, C., Pichler-Zalaudek, K., Edlinger, G. and Lopes da Silva, F.H., 2000b, Do brain oscillations of different frequencies indicate interaction between cortical areas in humans, Neurosci. Lett. 286:66-68.

Roland, P.E., Larsen, B., Lassen, N.A., Skinhoj, E., 1980, Supplementary motor area and other cortical areas in organization of voluntary movements in man, J. Neurophysiol. 43:118-136.

Salenius, S., Schnitzler, A., Salmelin, R., Jousmäki, V. and Hari, R., 1997, Modulation of human cortical rolandic rhythms during natural sensorimotor tasks, Neuroimage 5:221-228.

Salmelin, R., Hämäläinen, M., Kajola, M., Hari, R., 1995, Functional segregation of movement-related rhythmic activity in the human brain, Neuroimage 2:237-243.

Singer, W., 1993, Synchronization of cortical activity and its putative role in information processing and learning, Ann. Rev. Physiol. 55:349-374.

Shibasaki, H., Barrett, G., Haliday, E., Halliday, A.M., 1981, Cortical potentials associated with voluntary foot movements in man, Electroenceph. clin. Neurophysiol. 52:507-516.

Stephan, K.M., Frackowiak, R.S.J., 1996, Motor imagery – anatomical representation and electrophysiological characteristics, Neurochem. Res. 21(9):1105-1116.

Suffczynski, P., Pijn, J.M.P., Pfurtscheller, G., Lopes da Silva, F.H., 1999, Event-related dynamics of alpha band rhythms: a neuronal network model of focal ERD/surround ERS. Event-related desynchronization and related oscillatory phenomena of the brain, in: Handbook of Electroencephalography and Clin. Neurophysiology, Revised Edition, Vol. 6, Pfurtscheller, G., Lopes da Silva, F.H., eds., Elsevier, Amsterdam, pp. 67-85.

Tanji, J., 1996, New concepts of the supplementary motor area, Curr. Opin. Neurobiol. 6:782-787.

Toro, C., Deuschl, G., Thatcher, R., Sato, S., Kufta, C., Hallett, M., 1994, Event-related desynchronization and movement-related cortical potentials on the ECoG and EEG, Electroenceph. clin. Neurophysiol. 93:380-389.

Van Burik, M., Edlinger, G., Pfurtscheller, G., 1999, Spatial Mapping of ERD/ERS, in: Event-Related Desynchronization. Handbook of Electroenceph. and Clin. Neurophysiol, Revised Edition, Vol. 6, Pfurtscheller, G., Lopes da Silva, F.H., eds., Elsevier, Amsterdam, pp. 107-118.

Walter, W.G., Cooper, R., Aldridge, V.J., McCallum, W.C., Winter, A.L., 1964, Contingent negative variation: An electrical sign of sensori-motor association and expectancy in the human brain, Nature 203:380-384.

CNV AND SPN: INDICES OF ANTICIPATORY BEHAVIOR

Cornelis H.M.Brunia

Department of Psychology
Tilburg University

INTRODUCTION

In the title of the first report about the Contingent Negative Variation (CNV) two things are suggested: The CNV is a sign of "sensori-motor association" and the CNV is an index for "expectancy" (Walter et al., 1964). In other words, the CNV reflects processes, which are, at first sight, quite different from the processes discussed in the other chapters in this book. Why then is it important to add a "deviant" chapter to this book? There are two reasons. First, the CNV is also a movement-preceding negativity (MPN), just as the Readiness Potential (RP). The RP reflects processes involved in the preparation of *voluntary* movements, and the CNV reflects processes involved in the preparation of *signaled* movements. In other words the RP and the CNV are both reflections of anticipatory behavior, at least as far as the motor system is involved. Moreover it doesn't seem too difficult to put up a case for the view that most of our motor activity is elicited by the presence of some kind of stimuli, rather than being "voluntary". This being a sufficient reason for a chapter on the CNV, there is a second reason. Anticipatory behavior is not restricted to the motor system; it involves attention to the surrounding and to stimuli, which are relevant for our ongoing behavior. This becomes clear in a simple paradigm such as the forewarned reaction time task, used to elicit a CNV. In such a paradigm, a warning signal (WS) alerts the subject to an upcoming imperative signal (RS) to which the subject has to respond, e.g. by pressing a button. The consequence is that the CNV is a MPN, confounded by activity related to anticipatory attention for the response signal (RS). This causes serious difficulty for the interpretation of the CNV. I will discuss later on a paradigm in which separation in time of motor preparation and anticipatory attention is possible. The latter function is reflected in a second category of anticipatory slow waves: the Stimulus Preceding Negativity (SPN).

The CNV is mostly interpreted as an anticipatory slow wave, which ends with the execution of a short-lasting movement such as a button press (Figure 1). Yet movements aimed at a particular target are also accompanied by a slow wave. In other words slow waves as such are not necessarily a reflection of *anticipatory* behavior. They are even not necessarily reflecting ongoing motor behavior. Also the execution of cognitive or mental tasks, with no concomitant overt behavior, can be accompanied by slow waves in the EEG.

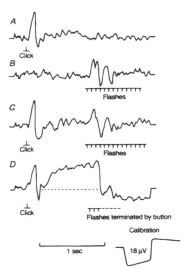

Figure 1. Average of responses to 12 presentations. A. Evoked potentials (EPs) recorded over the frontal cortex in response to clicks; B: EPs in response to flicker; C: EPs in response to clicks, followed by EPs in response to flicker; D: Clicks followed by flicker, terminated by subject's pressing a button as instructed. The contingent negative variation (CNV) appears as a consequence of the instruction. (From Walter et al. 1964, Fig 1. p.381) Reprinted with permission of Nature. Copyright Macmillan Magazine Ltd.).

Since the main topic of this chapter is anticipatory behavior, I will start with a description of the CNV as an anticipatory slow wave. Then I discuss the relationship between RP and CNV. Next I describe some slow wave studies related to task execution. At the end I will argue that recording of the SPN is preferable to that of the CNV, if one is interested in anticipatory attention.

THE CONTINGENT NEGATIVE VARIATION

In the original Walter et al. paper it was mentioned that the WS was a single click, while the imperative or RS was a repetitive series of flashes, interrupted by the RS. The CNV, recorded after the evoked potential following the WS, showed first a sharp increase in negativity until a certain plateau was reached from where the negativity slightly increased to the end of the 1s interval. "This pattern is maintained indefinitively as long as the subject is attentive and presses the button promptly" (Walter et al., 1964, p.381). The latter description underlines the importance of both anticipatory attention and motor preparation.

Very influential papers run the risk that the paradigm used becomes something of a holy commandment. For the CNV it meant that people thought that the foreperiod *had to be* a 1s epoch, and that the RS *had to be* a series of flashes. Neither of these is true of course. We now know that the use of a longer foreperiod has led to the discovery of two different slow waves hidden in the 1s-interval CNV, the so-called early and late waves. (Connor and Lang, 1969). Since these slow waves reflect different processes, their analysis during a 1s interval is hampered, but not impossible (McCarthy and Donchin, 1978). The use of a series of flashes as the RS results in another difficulty for the interpretation of the CNV. Since the button press interrupts the series of stimuli, the subject gets implicit information about response speed. If the number of flashes is small, the subject knows that the response was relatively fast, and if the number of flashes is large the response must have been rather slow. In other words this type of RS provides the subject with a kind of feedback about his or her performance, known as Knowledge of Results (KR). Later, I will show that potentials recorded prior to KR stimuli have a very typical potential distribution, quite different from the movement-preceding negativity. The CNV late wave contains potentials related to the preparation of the response, confounded by those related to attention for the upcoming KR. The simultaneous occurrence of processes related to movement preparation and processes related to anticipatory attention to the RS, is a problem, which is difficult to overcome. Their respective contribution to the CNV cannot be disentangled. So, the simple CNV has proved to be a complex of different slow waves, which may or may not become visible depending upon the paradigm used by the investigator. As a consequence it is very difficult to design a CNV experiment in such a way that the results can be interpreted in an unambiguous way. In the next paragraphs I will discuss the effects of different variables upon the emergence of the CNV (See also McCallum, 1986, 1988).

THE LENGTH OF THE FOREPERIOD

Early studies in which a bilateral EEG was recorded during a 1s foreperiod, reported a symmetrical distribution over the scalp, but it became clear from the very beginning that there were differences along the anterior-posterior axis (Walter, 1967; Cohen, 1969). Largest amplitudes were found over Cz and smaller amplitudes over the frontal and posterior areas. Järviletho and Frühstorfer (1970) distinguished a central premotor wave and a frontally dominant wave related to auditory discrimination. Hillyard (1973) suggested that different task-related potentials might contribute to the CNV. Connor and Lang (1969), using longer foreperiods, reported a negative slow wave with a peak amplitude within one second after the WS and an increasing negativity related to the upcoming RS. Loveless and Sanford (1974) suggested a more explicit relation to the existing reaction time (RT) literature. In that literature it was known that foreperiod length has a predictable effect upon reaction time. If subjects know the length of the foreperiod, they are faster compared to a condition in which they are uncertain about the arrival of the RS (Tychner, W.H., 1954). Loveless and Sanford presented a number of foreperiods with a different length (1,3 or 8 sec) in a regular condition and in an irregular condition. In the regular condition the length of the foreperiod was predictable within a block of trials. In the irregular condition the length changed per trial. In the irregular condition the authors found a slow wave following the WS. This *early wave* was independent of the length of the foreperiod and it was the only electrophysiological measure recorded. In the regular condition the early wave was present as well, but it was followed by a second negative slow wave with increasing amplitude towards the point of presentation of the RS. The authors considered the *late wave* to be related to the preparation of the movement. They selected the trials with a 6s foreperiod from the irregular condition and subtracted the average from the averaged 6s recordings in the regular condition. The subtraction yielded a slow negative wave prior to the presentation of the RS. This led to the conclusion that the late wave *is* a RP: a conclusion that needs further consideration (Loveless, 1979).

ARE THE RP AND THE CNV LATE IDENTICAL?

McAdam et al., (1969) noted that the anticipatory slow wave recorded during larger foreperiods had a similar morphology as the RP. The results of Loveless (1979) suggest an affirmative answer to the question posed above. In contrast, Deecke and Kornhuber (1977, pp. 136-137) pointed to the following differences between RP and CNV:

1) The potential distribution is different along the anterior-posterior axis: the CNV is larger over the frontal areas, and the RP over the parietal areas.
2) In contrast to the RP, the CNV is symmetrical, except if verbal stimuli or responses are involved (Low and Fox, 1977). The RP is symmetrical over the parietal areas, but over the precentral electrode positions amplitudes are larger over the hemisphere contralateral to the movement side.
3) The RP increases gradually, while the CNV increases more suddenly.
4) The RP is smaller than the CNV.

Some years later, Rohrbaugh and Gaillard (1983) advanced the opposite point of view. They presented the following arguments in favour of the RP and the CNV being identical.

1) If *no motor response* is required the CNV is attenuated during a short foreperiod, and the CNV late wave is absent or attenuated during long foreperiods.
2) Task variables that in RT studies are known to influence *motor preparation,* affect the CNV late wave. (Niemi and Näätänen, 1981).
3) Prior to *difficult discriminations* the amplitude of the CNV late wave is not different from those preceding easy ones.
4) If *muscular effort* is required for a response to the RS, amplitudes of the CNV late wave increase, compared to a condition in which this effort is not needed. Kutas and Donchin (1977) have described a similar result for the RP.
5) *Speed instructions* enhance the amplitude of the CNV. Comparing CNV recordings prior to fast and slow responses from the same series of trials, the largest amplitudes are found prior to fast responses (Brunia and Vingerhoets, 1980, see also Figure 2).
6) RP and CNV late wave show *the same left-right asymmetry* preceding finger movements (Rohrbaugh et al., 1976) and foot movements (Brunia and Vingerhoets, 1980). The paradoxical potential distribution prior to foot movements, described for the first time by Brunia (1980), is present in both the RP and the CNV late wave.

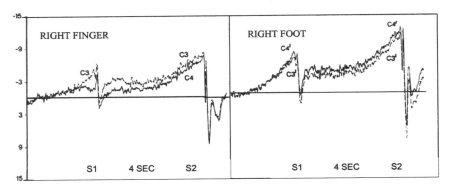

Figure 2. RP and CNV recorded in one trial. A voluntary movement with the right index finger (left) or the right foot (right) initiated the trial. Four seconds later the RS was presented, upon which the same response had to be given. Electrode positions: left panel C3 and C4, right panel C3' and C4'. The CNV late wave is larger than the RP, suggesting that there is a sustained negativity present during the whole length of the foreperiod. The paradoxical distribution prior to foot movements is present in RP and CNV. Modified from Brunia and Vingerhoets, 1981.

The arguments of Deecke and Kornhuber (1977) concerned the size, morphology and the distribution of these two potentials. The length of the foreperiod was not so much an issue for them. Yet if the comparison between the RP and the CNV is made in a foreperiod of 4s, the argument about the size becomes much more convincing (See Figure 2). Another consequence is, however, that the presumed symmetry is in fact gone. Since in most CNV studies some response has to be given upon the arrival of the RS, the results of Low and Fox (1977) should be considered the rule, rather than an exception. The use of longer foreperiods has provided supportive evidence for the notion that the late wave indeed is response-related. The arguments brought forward by Rohrbaugh and Gaillard (1983) all point in the same direction. They are supported by the fact that during the foreperiod, and

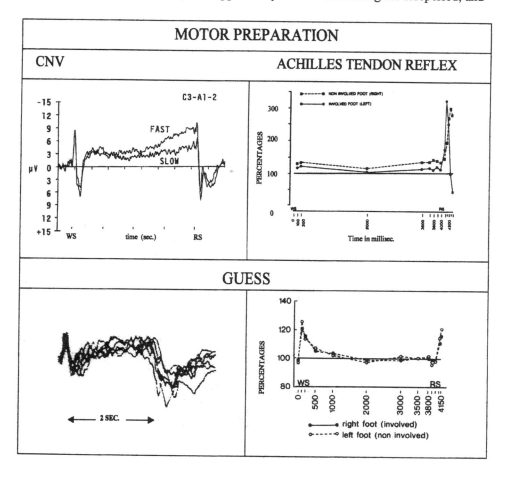

Figure 3. Above: *Motor preparation during a 4s foreperiod.* Left: The late wave shows larger amplitudes prior to faster responses (from Brunia and Vingerhoets, 1980). Right: there is an increase in excitability of spinal motoneurons during the whole foreperiod, as suggested by larger amplitudes of Achilles tendon reflexes in muscles not involved in the response. The reflexes were evoked at moments unpredictable to the subjects (from Brunia and Scheirs, 1982). Below: *Guessing the properties of S2.* Movements to indicate the guess were made prior to S1. S2 confirmed or disconfirmed the guess. Left: A non-motor CNV was recorded at Cz (from Donchin et al., 1972). Spinal reflexes were evoked under almost the same condition. Apart from the initial increase in amplitude 100 ms after WS, there is no further change during the last part of the foreperiod, indicating the absence of any sign of spinal motoric activation Note the difference in scale between both right panels (from Scheirs and Brunia, 1985).

especially during the last part of it, an increase in surface EMG of the agonist has been found in both monkey (Wiesendanger et al., 1987) and man (Brunia and Vingerhoets, 1980). This points to an increase in excitability of the spinal agonist motoneurons. In another study Brunia, Scheirs and Haagh (1982) evoked monosynaptic Achilles tendon reflexes during a 4s foreperiod at different points in time, unpredictable for the subjects (Figure 3). Muscles not involved in the response, showed larger reflex amplitudes than during the inter-trial interval. Thus motoneurons not involved in the response have also an increased excitability, especially during the second part of the foreperiod. This suggests a general motor readiness, which is simultaneously present with the CNV, and particularly with the late wave. Although these results underline the motor-relatedness of the CNV late wave, there remain difficulties for those who consider the CNV late wave and the RP identical. If one looks again at Figure 2, it is clear that the size argument of Deecke and Kornhuber (1977) still holds. In other words it remains to be explained what that "extra" negativity represents.

The size argument only holds if the same simple movement is executed twice, once in the voluntary movement condition and once in the simple, fully informed stimulus-induced condition. As soon as S2 contains more complex information the preceding CNV late wave amplitude changes, and the comparison with the RP no longer makes sense. In the studies of Ikeda et al. (1997) and of Hamano et al. (1997), the CNV was recorded in a choice Go-NoGo paradigm, in which S2 indicated the response to be given. Go and NoGo stimuli were equally distributed in a random fashion. In such a situation motor preparation can't be optimal because of the uncertainty about the content of S2. This could be the reason for the smaller CNV late wave amplitudes found by these authors, compared to the results of Brunia and Vingerhoets (1981) and of Kristeva et al. (1987). In the section on motor preparation, it will be shown that larger late wave amplitudes are found if subjects can prepare better the future response, that is, if they are fully informed about its parameters. This was the case in the studies of Brunia and Vingerhoets (1981) and of Kristeva et al. (1987), and not in those of Ikeda et al. (1997) and of Hamano et al. (1997). From the other side we will see in that same section also that uncertainty itself might go along with an increase in negativity. Thus the size issue is certainly not settled yet.

Apart from the difference in size of the CNV late wave and the RP, there is reason to doubt the conclusion that the CNV late wave is only movement-related. CNVs can be recorded in short and long foreperiods, even if no motor response is required (Donchin et al., 1972; Simons et al., 1978). In one of the conditions of the study of Donchin et al. (1972) subjects had to predict with a button press before the WS whether the S2 ("RS") would be a star or a circle. Thus no movement was to be made at the arrival of S2. The authors found a clear CNV in a 2s foreperiod, suggesting the existence of a negative slow wave, not related to any motor activity (Figure 3). Using the same paradigm, but in a 4s inter-stimulus interval, Scheirs and Brunia (1985) again studied the amplitude of Achilles tendon reflexes, in the same way as described above. Interestingly, they only found the increase in amplitude 100ms after the WS, which is always present, but no further changes from baseline during the whole foreperiod. (See Figure 3). This suggests that 1) if the CNV late wave reflects motor preparation, there is a simultaneous increase in spinal excitability and 2) if the CNV reflects non-motor activity, there is no change in spinal excitability. The question asked at the beginning of this section has to be answered in the negative: RP and CNV late wave are different phenomena.

Ikeda et al. (1997) presented further evidence for this. These authors compared the CNV and RP in patients with Parkinson's disease and progressive supranuclear palsy. They distinguished two groups, with mild and severe symptoms, respectively. Patients with mild symptoms showed a clear CNV and RP, those with severe symptoms had a normal RP but a smaller or absent CNV late wave, suggesting that the two potentials stem from different brain areas. Hamano et al. (1997) confirmed this with subdural recordings of RP and CNV. The RP was recorded predominantly in the primary motor cortex (MI), the primary somatosensory cortex (SI), and the supplementary sensorimotor area (SSMA). The CNV was not a unitary phenomenon. The authors distinguished two different CNV potentials. Type 1 started immediately after S1, while Type 2 onset was 1 to 0.5 s prior to S2. Type 1

CNV, corresponding to the early and late wave CNVs recorded from the scalp, were found in the prefrontal area and the SSMA. Type 2 CNVs, corresponding to the late wave in scalp recordings, were obtained from the prefrontal, temporal and occipital areas, and in one patient also from MI and SI. In other words, it is clear that the origin of the two potentials is different, as was also suggested by Ikeda et al. (1994). In a patient with a selective impairment of cerebello-thalamic output, they found an absence of the RP but not of the CNV. Taken together, this evidence for a different origin of the two potentials can no longer be denied. This is not to say that the problems relating to the RP and the CNV are solved. The conclusion that the RP is not affected in Parkinson's disease (Ikeda et al., 1997) is, for example, in contrast to other publications (Deecke and Kornhuber, 1977, Dick et al., 1989, Jahanshahi et al., 1995, Praamstra et al., 1996). Another point that needs to be investigated in the near future is the fact that in all subjects, except one, Hamano et al. (1977) failed to find an activation of MI, while patients still made the response required. This is compatible with the results of Shibasaki et al. (1978) and Ikeda et al. (1994), who found an absence of the RP in patients, who still could make voluntary movements. We recently described also an absence of the RP in a patient with essential tremor, who was able to make voluntary movements. The RP returned under the influence of electro-stimulation in the thalamic ventralis lateralis posterior nucleus (Brunia et al., 2000).

It should be kept in mind that the CNV late wave is for a large part, but not exclusively, related to motor preparation. Attention for a forthcoming stimulus plays a role as well, and the question to be addressed is how to separate the activity related to anticipatory attention from that related to motor preparation. Before considering this, I will first try to answer the question why there is an extra negativity in the CNV compared to the RP.

TIME ESTIMATION / TIMING

Preparation for a movement can only be studied if a WS informs a subject about the possible arrival of an RS, which has to be responded to. The RT depends, among other things, on two factors: the probability that the RS will be presented and the moment the RS can be expected to appear. If the subject knows *when* to give *what* response, an optimal preparation is possible, resulting in a short RT. Time uncertainty exists when a number of different foreperiod lengths are used during one block of trials. Remember that Loveless and Sanford (1974) found attenuation or an absence of the CNV late wave under such circumstances. This is probably due to a decreasing accuracy in time estimation (Requin et al., 1991). Time uncertainty can be counteracted to an extent by the presence of a timer. Thus, Besrest and Requin (1973) found, in the presence of a clock, shorter RTs and larger CNV amplitudes than when no timer was available. This suggests that part of the negativity, recorded during the foreperiod can be related to processes underlying time estimation. Supportive evidence for this point of view is provided by animal studies (Macar and Vitton, 1979) and experiments in humans (Macar and Vitton, 1982; Macar and Besson, 1985). It could also be one of the causes of the difference in amplitude between the RP and the CNV late wave in Figure 2. Thus apart from the early wave and the late wave, there is a third negativity, which is related to time estimation and which is present during the whole length of the foreperiod. If the WS indicates that after some known or unknown time a compatible or incompatible response has to be given upon the presentation of the RS, the different behavioral elements have to be integrated in the correct temporal order. This is what we call the programming of the response. Apart from the temporal ordering of behaviour, time also plays a role in the form of the duration of the interval between two stimuli. If two stimuli are presented with a short or a long interval in between, subjects can distinguish between them, and the distinction can be made difficult if the interval is only very short. Apart from the perceptual influence of duration, duration has also an effect on motor processes. One can be asked to press a button at the onset and the offset of an estimated interval, or to press a button continuously during an estimated interval. In these cases, duration is part of the response and can be programmed in advance (Macar et al., 1990; Vidal et al., 1995). In an fMRI study, Rao et al. (1997) asked subjects who were performing a paced tapping task to continue rhythmic tapping without further sensory guidance. It is plausible that here, as in the studies of Macar et al. (1990) and of Vidal et al. (1997), the executed movement pattern is based upon the presence of a template of the relevant duration in a working memory. Fuster (1997) has suggested the existence of two types of prefrontal memory cells, one sensory-coupled and one motor-coupled. The activity of the latter might be related to the duration experiments described above. Niki and Watanabe (1971) have reported delay-related firing of single units in the prefrontal cortex of the order of some seconds, which is the same interval used in most CNV-studies. It should be noted however that it is not clear whether this firing is related to a clock function or to working memory. For the time being the "extra" negativity in the CNV late wave (Figure 2) is interpreted as the manifestation of the temporal integration of different behavioral elements, while the early and late waves reflect the orientation to the WS and motor preparation, respectively. The latter process will be discussed next.

MOTOR PREPARATION

Although the BP and the CNV late wave are different, they are both involved in motor preparation. In the classic CNV the subject knows what to do and when. Thus in a simple RT paradigm the response can be fully prepared. This is not the case in a choice RT task, in which one or several parameters of the response may be unknown. The subject might be uncertain about the kind of response, the response side, the duration, the force level to be used, the distance to be overcome etc. When several parameters are unknown, the experimenter may decide to provide partial information by means of a cue, thus decreasing the uncertainty of the subject. These factors affect the CNV late wave, but the investigation of the influence of these different response parameters does not solve the problem of the contamination of the late wave by sensory-related activity. The development of a technique to remove all sensory-related activity from the CNV late wave has led to a very "handsome" measure for the onset of lateral cortical processes related to response preparation (Coles et al., 1988; De Jong et al, 1988). As a caveat it should be remembered that bilateral cortical activity and parallel bilateral spinal (and brain stem) activity are present during the whole foreperiod. Thus, there is a lot going on, before the asymmetrical cortical activity shows up as the "Lateralized Readiness Potential" (LRP). The LRP remains a *stimulus-evoked* response in the motor system and is as such different from the RP, which is related to self-paced movements. Its discovery is, however, a logical step in our efforts to split the motor-related activity from the stimulus-related activity. As a counter part we will see later that the use of another paradigm has made it possible to separate in time the movement-preceding negativity from a stimulus preceding negativity (SPN; Brunia, 1988). In a certain sense one could argue that this is the end of the CNV late wave as a tool for investigating preparatory processes, since these can now be better studied separately in the sensory domain and in the motor domain. This conclusion might be premature, though, since in a study on motor programming, Ulrich et al. (1998) found a dissociation between the CNV and the LRP. Thus for the motor domain it seems to be worthwhile to continue the simultaneous investigation of both potentials, because they reflect non-lateralized and lateralized aspects of motor preparation, respectively. Therefore I will continue describing the relation between motor preparation and the CNV late wave, keeping the SPN for a later section. The LRP has been discussed in a separate chapter of this book.

We have seen above that uncertainty about the response is manifested in the CNV. What happens when a cue resolves part of the uncertainty? The answer is not unanimous, as we will see. I will first discuss an experiment in which the uncertainty is not movement-related. Ruchkin et al. (1986) recorded a terminal CNV or Expectancy (E)-wave in the absence of a motor response, comparing the time course of the E-wave in a 1s inter-stimulus interval (ISI) and a 3s ISI. Stimuli were flashes that might or might not be accompanied by a sound. Subjects predicted whether both stimuli would be the same or different with respect to the presence of the sound. The amplitude of the E-wave was the same for both ISIs, but the time course was different. This apparently depended upon the timing of S2. The amplitudes were largest over the posterior half of the scalp and decreased immediately after S2, suggesting that it reflected uncertainty about the outcome of the prediction. In this experiment subjects were waiting for a stimulus to be presented. They attended to the locus where the stimulus would be presented and were uncertain about its information content. In other words the topic of the study was anticipatory behavior in the sensory domain. This behavior is accompanied by a negativity preceding S2. In a following experiment the locus of the uncertainty was manipulated during a 3s ISI, in which uncertainty could be resolved by either S1 or S2. There were three blocks of trials. In one block subjects had to predict whether the successive stimuli would be the "same" or "different", without knowing anything about them. In a second block subjects knew that there was always a sound in S1. They had to predict whether there would be a sound in S2. In the third block subjects knew that there was a sound in S2 and they had to predict whether S1 had a sound as well. The E-wave was smallest as S1 resolved uncertainty and S2 provided no more information. If both stimuli were necessary to reach a conclusion, the E-wave was largest. If S2 was necessary to resolve uncertainty, the E-wave was about as large as when both stimuli were needed. Interestingly prior to S1 there was also some negativity, showing the largest amplitude if S1 contained the crucial information. I will return to such SPN later. The conclusion of these experiments is that if uncertainty about the outcome of a future stimulus exists, that stimulus will be preceded by a negative slow wave. If S2 had a known content, the E-wave had its smallest amplitude, suggesting that the greater the uncertainty, the larger the negativity. It should be noted, finally, that even if there is no information in S2 there is still anticipatory negativity, especially over the posterior and frontal areas.

The next experiment is that of Vidal et al. (1995). In self-initiated trials subjects waited some seconds for a visual cue, telling them to prepare a long or a short interval between two button presses, or giving no information (neutral condition). The response had to be initiated after the RS, which was presented 2s after cue onset and which informed the subjects about the correct response. Information was valid in 81% of the precued trials. The different conditions were presented in mixed blocks. The question was whether duration could be programmed in advance and whether this was reflected in the CNV. The authors recorded Laplacian derivations over FCz, C4' and P4, supposedly reflecting activity in, respectively, the SMA, MI and the superior parietal area. They found larger amplitudes in the precued conditions as compared to the neutral condition over the SMA and over MI. The timing was different though. After the offset of the cue, the slow wave over the SMA was larger in the precued condition than in the neutral condition. At the end of the preparatory interval the difference was no longer significant. In contrast, the difference became significant over MI during the last 300ms, supporting the notion that the SMA, being activated first, is a "supra-motor" area instead of a "supplementary" one. The number of electrodes and their position doesn't allow distinctions between the pre-SMA and the SMA proper, as suggested by Rizolatti et al., (1996). It does suggest though that the mesial motor structures are activated earlier than the lateral ones. Concerning the late wave, the authors found larger amplitudes for the short and then the long interval (in that order) and the smallest amplitudes in the neutral condition. The parietal derivation didn't show any difference. The results were considered to be in agreement with an earlier study in which *duration* of a movement had to be programmed instead of the *interval between two short movements* (Macar et al., 1990). This experiment provides an example of *smaller* amplitudes when uncertainty exists about a response. It is in line with an experiment of MacKay and Bonnet (1990), who also reported an increase in CNV amplitude with increasing advance information about the future movement.

In contrast to these studies, van Boxtel and Brunia (1994) reported larger CNV amplitudes with increasing uncertainty. These authors investigated the motor and non-motor components in a CNV recorded during a 4s foreperiod. The task was to produce a squeeze with either hand at 15% of maximal force level, either in a fast or a slow fashion. Two seconds after an acoustic WS, a visual cue (S1) was presented to the subjects followed again two seconds later by a RS (S2). There were two experimental conditions: Precued or Choice. In the precued condition the speed instruction was given by S1 and repeated by S2. In the choice condition the cue information had no relation to the information in S2. The WS started half of the trials, and a voluntary movement the other half. The main result was a larger amplitude of the late wave in the choice condition, as compared to the precued condition. In other words the amplitude of the late wave was larger if subjects were uncertain about the speed of contraction required. Obviously the motor task was quite different from that of Vidal et al. (1995). Yet it is puzzling that in our study the uncertainty goes along with larger amplitudes, while Vidal et al. (1995) and MacKay and Bonnet (1990) found smaller amplitudes.

Ulrich et al. (1998) studied the effect of advance information about a future movement in an experiment in which CNV, LRP and RT were recorded. A visual WS consisted of 4 open squares placed in the corners of an imaginary rectangle on a black background. It was followed 450ms later by a cue stimulus, changing one or two of the four open squares to (a) filled one(s). A square on the left indicated a left-sided movement and a square on the right side indicated a right-sided movement. An upper square was used for flexion movements and a lower square for extension movements. The size of the square indicated the force to be used. Two left squares indicated a left-sided movement; two right-sided squares indicated a right-sided movement. In these cases the choice between flexion and extension could only be made at RS. Two upper squares indicated a flexion movement; two lower squares indicated an extension movement. Response side remained unknown until RS presentation. A full precue about movement side, direction and force resulted in shortest RTs, and largest CNV amplitudes. Partial precues, providing information about movement side and force, or about movement side and direction, resulted in longer RTs and smaller CNV amplitudes. Only precuing movement side caused still longer RTs and smaller amplitudes of the CNV late wave. The LRP showed a different picture. The amplitude of the LRP was largest for the full precue, in accordance with the CNV late wave. The remaining precues did not result in a different size of the LRP. The conclusion of the authors was that the CNV late wave is related to the central assembly of a motor program, and the LRP to the implementation of the program at more peripheral levels. Such a two-phase process was suggested to be in line with Rosenbaum's ideas about the organization of motor preparation (Rosenbaum, 1985). The results of the study of Ulrich et al. (1998) show that for a psychophysiological analysis of motor preparation the LRP cannot be considered a sufficient tool. This is presumably caused by the fact that motor preparation is for a large part hidden in non-lateralized processes.

In line with the study of Vidal et al. (1995), Ulrich et al. (1998) reported larger amplitudes when subjects were fully informed by a cue. They too were puzzled by the contrasting results of van Boxtel and Brunia (1994). They offered two explanations. One is a procedural difference: In the study of van Boxtel and Brunia the different conditions were presented in blocks, while in all other experiments randomly mixed blocks were used. Ulrich et al. (1998) suggested that subjects in the van Boxtel and Brunia study didn't recognize the speed advantage from the partial information and as a consequence didn't prepare well. In the study of Ulrich et al. the interval between cue and RS was 1500ms. Since in the study of van Boxtel and Brunia (1994) an interval of 3000ms was used, Ulrich et al. (1998) argued that in the latter case motor preparation was weaker, in line with results from the RT literature (Niemi and Näätänen, 1981). Another point that should be stressed here is that the task in the study of van Boxtel and Brunia was very demanding, which might have been of influence too.

If the negativity in the study of van Boxtel and Brunia (1994) is indeed more related to stimulus processes as suggested by Ulrich et al., it is interesting to look back to the results of Ruchkin et al, (1987) suggesting that the larger the uncertainty the larger the CNV amplitudes, at least in the sensory domain. The implication would be that preparation is differently organized in the sensory domain and in the motor domain. In motor preparation, the more is known about the response to be given, the larger the amplitude of the preceding slow wave. In the sensory domain the anticipatory negativity seems to run in parallel with the uncertainty about S2. What is required is an experiment with the SPN, in which degrees of uncertainty are manipulated. As far as I know this hasn't been done yet. We have however results of an unpublished non-motor experiment, in which following a WS subjects were confronted with three stimuli in a row, with an interval of 2s between each of them. The stimuli were numbers, and the subject knew that if a "9" would be presented three times, a large bonus would be earned. If the first number was a "9" there was of course uncertainty about S2 and there was a negative slow wave prior to it. If S2 was a "9" the uncertainty about S3 remained, reflected in a large anticipatory negative slow wave. If S2 was not a "9", they were certain that S3 could never be accompanied by the bonus, and that certainty went along with a decrease in amplitude of the negative slow wave preceding S3. This suggests indeed that the larger the uncertainty, the larger the amplitude of the anticipatory slow wave. Uncertainty is not the only factor, though. Another explanation would be that, S2 being a "9", the subject was maximally motivated to attend to S3. So I consider the issue of uncertainty not solved, at least no in the sensory domain. Concerning studies on programming of a movement, the conclusion seems warranted that the more information is available about the upcoming movement, and thus, the better the possibility to prepare it, the larger the CNV late wave. Therefore Cui et al. (2000) found larger CNV amplitudes prior to complex movements than prior to simple movements. Apart from the preparatory activity, which is in line with the original interpretation of Grey Walter, there are a large number of studies, in which slow potentials are investigated in relation to ongoing task performance. These will be discussed next.

NEGATIVE SLOW WAVES DURING TASK EXECUTION

Since the major part of the CNV late wave is related to motoric aspects of behavior, it is no wonder that people have not only investigated brain activity related to the *preparation* of a movement, but also negativity related to movement *execution*. In an overview of several of their experiments, Grünewald and Grünewald-Zuberbier (1983,1984) reported *during* upper limb movements aimed at a target, a continuation of the contralateral larger negativity, found in the RP *prior to* the movement. Unlike the CNV paradigm, the target was present from the outset of a trial. The authors suggested that the result was typical for phasic aiming movements, because no such negativity was present during hold contractions. They related their findings to the results of Roland et al. (1980), who reported a 50 % less increase in regional blood flow in the contralateral motor cortex during sustained isometric finger contractions, compared to single finger flexions. It cannot be ruled out that reafferent somatosensory input via Ia, Ib and skin-afferents might cause (part of) the negativity. Yet Grünewald and Grünewald-Zuberbier (1983) pointed to the results of Vallbo (1974), who reported continuous firing of muscle spindle afferents during a sustained contraction, i.e. under conditions in which the asymmetric negativity was found to be absent. Muscle spindles of the agonist are silent during shortening of the muscle, while the impulse rate of tendon afferents may increase (Najem et al.1990). Since shortening of the agonist goes along with lengthening of the antagonist, it remains possible that part of the negativity during movement execution might be related to reafferent input from the periphery. A central process, related to attention for the ongoing movement execution remains another possibility. Grünewald and Grünewald-Zuberbier (1983) suggest that the bilateral component of the execution related negativity is susceptible to attention for and concentration upon task execution, but not the asymmetric part of it.

In another paradigm Lang et al. (1984) recorded movement execution-related activity during stimulus-guided hand movements. Stimuli in either the visual or the somato-sensory mode were presented in the left hemifield or on the surface of the left hand. Tracking movements had to be made with the right hand. Movements were accompanied by negative slow waves, which were interrupted by each change in direction of the guiding stimuli. Slow potentials were interpreted as reflecting "directed attention". Their potential distribution suggested partial modality specificity. The continuously present stimuli were an insufficient condition for the emergence of the slow waves: the slow waves appeared only if the stimuli were to be monitored in order to be tracked correctly. Subsequently Foit et al. (1982) suggested the term "movement-monitoring negativity", to indicate any negativity related to processes underlying the execution of a movement. Finally the term "performance-related negativity" was proposed, because negativity was also present during the execution of tasks in which the motor aspect was trivial. As a formal description of the existence of any negativity present during task execution, the term might be acceptable, as long as one realizes that nothing is explained by it.

The last two decades have seen an enormous increase in studies in which cognitive processes have been investigated with the help of ERPs. The discovery of the N400 (Kutas and Hillyard, 1980) has opened up a large area of research in language processing. Similar developments are seen in the field of memory (Rugg, 1995). Regardless of whether the ERPs reflect "specific" functions, they only appear *after* presentation of the crucial stimulus. The investigation of slow potentials recorded *during* cognitive task performance, remains an interesting tool however, to clarify covert cognitive processes. While "specific" ERPs seem to reflect an outcome, slow potentials reflect the on-line processing during the functions one is interested in. Detailed consideration of these studies is beyond the scope of this chapter. However, in the rest of this section I will briefly describe a few relevant cognitive studies in which slow potentials have been recorded, related to memory functions. In these studies, an area-specific activation reflects which cortical regions are involved in the task to be executed (See also: Birbaumer et al., 1992).

Some ERP memory studies employ variations on the delayed match-to-sample task: S1 has to be kept in mind in order to allow a decision about whether S2 matches S1 or not (Ruchkin et al., 1995). The slow waves recorded during the ISI are presumably related to one or more of the operations executed in a *working memory*, i.e. storage, maintenance and retrieval. The distribution of the slow waves is related to the type of material and amplitudes are related to information load (Ruchkin et al., 1995). Thus, Barret and Rugg (1989b, 1990) found larger amplitudes over the right hemisphere during analysis of visual features, and over the left hemisphere during phonological processing. Slow waves recorded between 2 successive stimuli can also be interpreted in terms of anticipatory processes. Therefore Ruchkin et al. (1995) presented two visual processing tasks to their subjects, requiring a similar preparation for S2. For one of the tasks additional memory rehearsal operations were needed. Subjects had to memorize the visuospatial configuration of either three or five unconnected asterisks, and to decide whether S2 showed the same pattern or not. The response had to be postponed until a third stimulus was presented. In the "warning task" S1 was either a triangular pattern, consisting of three unconnected digits

(3s) or a cross, consisting of five unconnected digits (5s). S2 was a random pattern of three or five unconnected asterisks. S3 was again a set of 3 or 5 question marks, arranged in an identical way as the asterisks in S2, or in a different way. Subjects had to indicate whether the patterns of S2 and S3 differed or not. The crucial comparison concerned the slow waves recorded between S1 and S2, reflecting either "preparation alone" or "preparation + memory" functions. In the memory task there was a negative slow wave over the right parietal and central electrode positions, which varied with memory load. In the warning task the slow wave had lower amplitudes, was restricted to the posterior cortex and had no relation with memory load. The authors suggested that the negativity in the memory task was related to storage and retention and that the differing potential distribution in the two conditions reflected the activity of different neural generators. The authors confirmed the findings of an earlier experiment, in which a similar increase in negativity, parallel to an increased memory load was reported (Ruchkin et al, 1992).

Rösler and coworkers investigated *long-term memory* functions with slow potentials in a series of experiments (Rösler et al., 1995a,b; Heil et al., 1996,1997) in which retrieval of verbal, spatial and color information was investigated. They were interested in the question of whether distinct cortical areas are activated when differing types of information are retrieved from long-term memory. In a first session, Rösler et al. (1995) asked their subjects to learn a number of associations, either between nouns and nouns (verbal condition), between pictures and spatial positions (spatial condition) or between pictures and color patches (color condition). The learning procedure was the same for the three conditions and so was the associative structure. Slow potentials were recorded during retrieval, which followed one day later. Maximal negativity was found over the left frontal cortex in the verbal condition, over the parietal cortex in the spatial condition and over the right occipital and temporal cortex in the color condition. In other words the cortical areas known, from other studies, to be involved in the perceptual processing of the respective information, are also active during retrieval. These results were interpreted as evidence for a retroactivation of specialized cortical modules during intentional memory recall. In a follow-up study Heil et al. (1997) demonstrated that the cortical areas activated during the retrieval were indeed the areas, which were activated by the learning of the associations. In essence, the same procedure was followed as in the former study, but now EEG was recorded both during learning of the associations and during the retrieval the day after. In this way a comparison was possible between the slow potentials recorded when subjects anticipated either the spatial or the verbal information in the learning phase with when they were involved in retrieval of that information. The slow waves showed the same topographic pattern during the learning phase and during cued recall, suggesting that the same cortical area, which was activated during learning, was re-activated during recall. The conclusion was that memory representations are reactivated in cortical cell assemblies specialized for particular codes.

CLINICAL STUDIES

In some neurological diseases the pathophysiology is relatively well defined, so that the investigation of the CNV can lead to a better understanding of the pathways via which the potential is generated. This is the case, as we have seen above, in studies of Parkinsonian patients in which CNV and RP are compared. Ikeda et al. (1994, 1997) and Hamano et al. (1997) found dissociation between the two potentials and proposed a reasonable hypothesis about the possible underlying mechanisms. Pulvermüller et al. (1996), recording smaller amplitudes of the CNV in Parkinson patients compared to controls, interpreted their results in terms of a diminished input from the striato-thalamo-cortical loop, known to be affected in this disease.

It is also known that in monkeys and in cerebellar patients in whom the pathway from the dentate nucleus via thalamus to MI is disturbed, no RP emerges (Sasaki et al., 1979;

Shibasaki et al., 1978, 1986), while the CNV may still be present (Ikeda et al., 1994). Wessel et al., (1994) investigated patients with cerebellar atrophy, in whom the CT scan revealed mainly cortical lesions and no brain stem involvement. In contrast to a simple button press, they used sequential and goal-directed movements, expected to be more sensitive to cerebellar degeneration. They recorded smaller RP amplitudes, but the RP was not completely absent. For the latter result a disturbance of the efferent pathways via the dentate nucleus is probably necessary. The authors did not record CNVs in this study, but Verleger et al. (1999) recently did. In the latter study the authors investigated twelve patients, three of whom had exclusively cerebellar symptoms, while the other nine all had extra-cerebellar symptoms. This is in sharp contrast to Wessel et al. (1994), who explicitly excluded any brain stem involvement in their group of patients. The patients of Verleger et al. (1999) showed "drastically" smaller CNV amplitudes. Since most of the patients of Verleger et al. (1999) had extra-cerebellar symptoms, their results suggest that the CNV is affected by (extra)cerebellar pathways, which differ from the output channel of the dentate nucleus. Although their use of more complex motor tasks is methodologically favourable for the investigation of the role of the cerebellum, it is necessary to asses a group of patients with "pure" cerebellar lesions, with simultaneous recording of the RP and the CNV, if one is interested in deciphering the origin of both slow waves (Figure 2; Brunia and Vingerhoets, 1981).

Studies in psychiatric patients are much more difficult to interpret because the underlying pathophysiology is less well known or not known. In earlier studies psychopaths are, for example, said to have the same CNVs as normal control groups (Syndulko et al., 1975; Fenton et al., 1978), whereas it seems a safe guess that there was a behavioural difference between these groups. A number of studies in patients with schizophrenia reported smaller CNV amplitudes, especially at Cz (van den Bosch, 1983; Rockstroh et al., 1994; Wagner et al., 1996; Klein et al. 1997, Verleger et al. 1999). Since the underlying pathophysiological mechanism in schizophrenia is largely unknown, the research questions are necessarily too unspecific, and the results are rather disappointing. In my opinion the chance that something fundamental will be discovered about the CNV in schizophrenic patients is at this moment very small, and so is the likelihood of fully elucidating the underlying mechanisms of schizophrenia. This is less so in the case of phobia. In this field there is sufficient theorizing about neural networks underlying emotional activation (Adolphs et al., 1995) to suggest reasonable hypotheses about the anticipation of fear (See also: Öhman et al., 2000). Extensive animal research has provided a good insight into the pathways involved in the modification of the startle reflex (Davis et al., 1999), while about that same reflex in human research a sound database is obtained over the last ten years (Lang et al., 1990; Dawson et al., 1999).

ANTICIPATION OF EMOTIONAL STIMULI

Reports about the existence of a CNV in the absence of a motor response can be found in the literature from the first years after its discovery (Donchin et al., 1972). In an overview of slow wave studies on the anticipation of emotional stimuli, Simons (1986) suggests that slow waves recorded prior to significant events and in the absence of a need to respond, are characterized by simultaneous changes in autonomic nervous system (ANS) variables. Heart rate (HR) deceleration is a very frequently recorded phenomenon. At the time Simons wrote his review, only few relevant papers were available in the literature. Best known are the studies of Simons et al. (1979) and Klorman and Ryan (1980). Both found CNV late wave prior to emotional stimuli. Trying to accommodate the contrasting results in the literature, Simons suggested that there might be a difference in the anticipation of cognitive and emotional information. In the latter case, an anticipatory slow wave would be the rule in combination with HR deceleration, while the anticipation of cognitive information would be associated with HR deceleration only. This suggestion was based upon a study of Lacey and Lacey (1974). These authors found HR deceleration accompanied by a BP prior to a button press, and HR deceleration without any slow wave prior to a feedback stimulus. Both Simons (1986) and Lacey and Lacey (1973, 1974) considered this to be characteristic for cognitive anticipation. However, in contrast to this opinion, Damen and Brunia (1987) found in an almost similar experiment as that of the Lacey's, a HR deceleration combined with a slow wave. This was the very beginning of our series of SPN studies, which will be discussed below. In fact we never doubted the potentially motivational function of the feedback stimuli, but their pure informative effect on performance should not be underestimated either. It will be difficult to disentangle these.

Since the review of Simons (1986) other studies on slow wave studies have become available in the literature, such as the work of Lumsden et al. (1986) and of Regan and Howard (1995). Howard et al. (1982) considered the size of the CNV as directly related to expectancy for reward or non-punishment, and inversely related to expectancy for punishment and non-reward. Given the paucity of experiments in the emotional domain, a brief reconsideration of our as yet unpublished study described above may be appropriate. In this experiment subjects could earn or lose a large amount of money. Three stimuli were presented successively after the WS. The stimuli were numbers between 0 and 9. If the number was three times a "9" we called that BINGO. In the "loss" condition it meant that the subjects lost a large amount of money; in the "profit" condition they earned a lot of money. We found the largest amplitudes prior to S3, if S1 and S2 were two times a "9". There was no difference in size between the two conditions, though, which is of course in contrast to the model of Howard et al. (1982). Given the paradigm used in this experiment, I consider the SPN amplitude in this case to reflect emotional anticipation more than attention.

STIMULUS PRECEDING NEGATIVITY

In the introduction I indicated that response preparation and attention for the upcoming RS occur simultaneously during a warned RT task. This results in an unavoidable contamination of the accompanying slow waves. In order to study anticipatory attention separately from motor preparation, we have used over the last decade a time estimation paradigm (Damen and Brunia 1988, Brunia and Damen, 1989). In most experiments a warning stimulus had to be followed by a button press, after a fixed time interval of some seconds. Two or three seconds after the movement a Knowledge of Results (KR) stimulus was provided to inform the subject about whether the response was in time, too early or too late. This was done by means of three visual symbols: a plus sign indicated that the subject was too late, a vertical line implied that the response was in the correct time window, and a horizontal line indicated that the response was too early. Auditory KR stimuli have been used too. Experiments in which this paradigm has been used resulted in the emergence of two different slow waves: a Movement Preceding Negativity (MPN) and a Stimulus Preceding Negativity (SPN). From Figure 4 it becomes clear that they have a different scalp distribution, suggesting different underlying sources. The RP is larger over the hemisphere contralateral to the finger movement, and has its maximum amplitude over the motor cortex. The SPN has a right-hemisphere preponderance, and shows an RP-like steeply increasing negativity over the parietal cortex and a sustained negativity over the frontal cortex (Brunia and Damen, 1989).

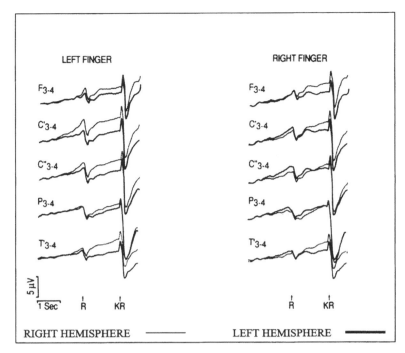

Figure 4. In a time estimation task subjects had to press a button with either the right or the left hand in intervals of 20-22 s. Two seconds after each button press a Knowledge of Results (KR) stimulus was presented on a computer screen. The KR stimulus indicated whether the preceding interval was too short, correct or too long. Preceding the movement a RP was recorded and prior to the KR stimulus an SPN was recorded. RP and SPN have a different potential distribution. (After Brunia, 1988).

To investigate whether attention for the KR stimulus was the crucial factor in the emergence of the SPN, Chwilla and Brunia (1991) recorded the SPN under three different conditions: true feedback, false feedback and no feedback. In the first condition subjects were correctly informed about their performance; in the second condition the same symbols were used, but in a random order, and in the third condition there was no feedback. The conditions were presented in blocks and only in the true feedback condition an SPN was recorded. Since the subjects knew that the feedback was false in the second condition, there was no need for them to attend the screen where the symbols were presented. We offered that as an explanation for the absence of the SPN in this condition. In an earlier paper Brunia (1988) pointed to the different role of RS in a warned RT task and the KR stimulus in a time estimation task. An RS in a CNV paradigm indicates *that* a response should be produced, or *what* response should be produced, while a KR stimulus indicates *how well a past performance* has been. In other words the first is directed to the future, while the second is directed to the past. Therefore Damen and Brunia (1994) investigated next, whether the SPN prior to an instruction stimulus is different from the SPN prior to a KR stimulus. The SPNs indeed showed a different size and potential distribution. The SPN prior to the KR stimuli showed the well-known right hemisphere preponderance with largest amplitudes over the parietal cortex. The sustained negativity over the frontal electrodes was replicated too. In contrast, prior to the instruction stimuli only a slight bilateral negativity was found over the parietal cortex, which was marginally significant. In a spatio-temporal dipole study, Böcker et al. (1994) suggested that a bilateral fronto-temporal dipole could explain most of the variance in the interval between movement and the presentation of the KR stimulus. It was hypothesized that the Insula Reili might be one of the locations, which is activated when a subject is waiting for feedback about a past performance. Böcker et al. (1994) questioned the difference in shape of the SPN recorded over the frontal and parietal derivations by Brunia and Damen (1988). They suggested that the SPN could be a steady potential over all electrode positions, and that an overlap of this and a recurrence to baseline of the post-movement P3 would cause the increase in negativity over the parietal cortex. This interpretation casts some doubt on a significant role of the parietal cortex in this very task. Thus it is important to investigate which brain areas are activated when a subject is waiting for KR, to ameliorate performance in the next trial. In a subsequent PET study Brunia et al. (2000) presented the first two conditions of the study of Chwilla and Brunia (1991) to their subjects, who were confronted with either true feedback of false feedback in blocks, two of each kind. Three areas in the right hemisphere were found to be activated: the prefrontal cortex (area 45), the Insula Reili, but now temporo-parietally, and the parietal cortex. So it seems that the performance of the time estimation task with KR about that performance, is accompanied by electrophysiological activity in a network in which these three areas play a major role. In our low-density EEG recordings we found in essence the same potential distribution every time. Yet in one of our unpublished experiments we recorded clear modality effects suggesting that activity in the hypothetical network can be initiated via differing sensory channels. These experiments elucidate how anticipatory attention can be investigated without contamination of movement-related activity. Therefore I consider the time estimation paradigm a better tool to study anticipatory attention than the CNV (See also Brunia and van Boxtel, 2002).

So far, we have seen that apart from the pre-KR SPN there is also an SPN preceding instruction stimuli. A major condition for their appearance is the predictability in time of the crucial stimulus. This holds too for cues used in a number of experiments to (partially) inform the subject about what the response should be, after the presentation of the RS. It is therefore not surprising that an SPN can also be recorded prior to their appearance. Thus, Gaillard and van Beijsterveldt (1991) reported a symmetrical SPN prior to a cue, as did van Boxtel and Brunia (1994). Ruchkin et al. (1986) described an SPN in a paradigm in which S2 followed S1 after a fixed interval of 3 s. They found larger amplitudes over the left hemisphere prior to probe stimuli, which indicated to a subject whether a perceptual (pattern recognition) or conceptual (arithmetic) task result did match with a probe stimulus presented at the end of a trial. In summary, we conclude that an SPN has been found prior to the onset of a KR stimulus, prior to the onset of a cue and prior to the onset of an instruction stimulus (for a review see Böcker and van Boxtel, 1997). In all three cases the onset of the stimuli was predictable in time, and there was no confound with motor preparation.

CONCLUSION

The CNV is a complex of slow waves related to the analysis of the WS, to time estimation, to anticipatory attention to the RS and to preparation of the response. In so far as it is a MPN, it remains a useful tool to study motor programming. The LRP by definition only records asymmetrical aspects of response preparation and a large part of the preparatory process seems to be bilateral. On the contrary, if one wants to study anticipatory attention the time estimation paradigm with feedback offers a better opportunity to record an SPN not contaminated by motor activity. The recording of slow waves during cognitive task performance should be used more frequently, since it gives an on-line picture of ongoing mental processes with a high time resolution.

ACKNOWLEDGEMENT

It is a pleasure to thank Han Brunia-Touw for preparation of the figures, and Ingrid Beerens for bibliographic assistance.

REFERENCES

Adolphs, R., Tranel, D., Damasio, H. and Damasio, A.R. (1995). Fear and the human amygdala. *Journal of Neuroscience*, **15**, 5879-5891.

Al-Falahe, N.A., Nagaoka, M. and Vallbo, A.B. (1990). Response profiles of human muscle afferents during active finger movements. *Brain*, **113**, 325-346.

Barrett, S.E. and Rugg, M.D. (1989a). Asymmetries in event-related potentials during rhyme matching: confirmation of the null effects of handedness. *Neuropsychologia*, **27**, 539-548.

Barrett, S.E. and Rugg, M.D. (1989b). Event-related potentials and the semantic matching of faces. *Neuropsychologia*, **27**, 913-922.

Barrett, S.E. and Rugg, M.D. (1990). Event-related potentials and the phonological matching of picture names. *Brain and Language*, **38**, 424-437.

Besrest, A. and Requin, J. (1973). Development of expectancy wave and the time course of preparatory set in a simple reaction time task. In: Kornblum, S. (Ed.): Attention and Performance IV, pp. 209-219. New York: Academic Press.

Böcker, K.B.E., Brunia, C.H.M., and van den Berg-Lenssen, M.M.C. (1994). A spatiotemporal dipole model of the Stimulus Preceding Negativity (SPN) prior to feedback stimuli. *Brain Topography*, **7**, 71-88.

Böcker, K.B.E. & Van Boxtel, G.J.M. 1997. Stimulus-preceding negativity: a class of anticipatory slow potentials. In: van Boxtel, G.J.M and Böcker, K.B.E. (Eds.) Brain and Behavior: Past, Present, and Future, pp. 105-116. Tilburg: Tilburg University Press.

Birbaumer, N., Roberts, L.E., Lutzenberger, W., Rockstroh, B., and Elbert, T. (1992). Area-specific self-regulation of cortical slow potentials on the sagittal midline and its effects on behavior. *Electroencephalography and Clinical Neurophysiology*, **84**, 353-361.

Brunia, C.H.M. (1980). What is wrong with legs in motor preparation? In: Kornhuber, H.H. and Deecke, L. (Eds.) Progress in Brain Research, volume 54, Motivation, Motor and Sensory Processes of the Brain, pp. 232-236. Amsterdam: Elsevier.

Brunia, C.H.M. (1988). Movement and stimulus preceding negativity. *Biological Psychology*, **26**, 165-178.

Brunia, C.H.M. (1993). Stimulus preceding negativity: arguments in favour of non-motoric slow waves. In: McCallum, W.C and Curry, S.H. (Eds.) Slow Potential Changes in the Human Brain, pp. 147-161. New York: Plenum Press.

Brunia, C.H.M. (1999). Neural aspects of anticipatory behavior. *Acta Psychologica*, **101**, 213-242.

Brunia, C.H.M. and Damen, E.J.P.(1988). Distribution of slow potentials related to motor preparation and stimulus anticipation in a time estimation task. *Electroencephalography and Clinical Neurophysiology*, **69**, 234-243.

Brunia, C.H.M. and Vingerhoets, A.J.J.M. (1980), CNV and EMG preceding a plantar flexion of the foot. *Biological Psychology*, **11**, 181-191.

Brunia, C.H.M. and Vingerhoets, A.J.J.M. (1981), Opposite hemisphere differences in movement related potentials preceding foot and finger flexions. *Biological Psychology*, **13**, 261-269.

Brunia, C.H.M., Scheirs, J.G.M. and Haagh, S.A.V.M. (1982). Changes of Achilles tendon reflex amplitudes during a fixed foreperiod of for seconds. *Psychophysiology*, **19**, 63-70.

Brunia, C.H.M., Haagh, S.A.V.M. & Scheirs, J.G.M. (1985). Waiting to respond. Electrophysiological measurements in man during preparation for a voluntary movement. In: Heuer, H, Kleinbeck, U. and Schmidt, K.H. (Eds.) Motor Behavior: Programming, control, and acquisition, pp. 35-78. Berlin: Springer Verlag.

Brunia, C.H.M., de Jong, B.M., van den Berg-Lenssen, M.M.A.C. & Paans, A.M.J. (2000). Visual feedback about time estimation is related to right hemisphere activation measured by PET. *Experimental Brain Research*, **130**, 328-337.

Coles, M.G.H., Gratton, G. and Donchin, E. (1988). Detecting early communication: using measures of movement-related potentials to illuminate human information processing. *Biological Psychology*, **26**, 69-89.

Connor, W.H. & Lang, P.J. (1969). Cortical slow wave and cardiac rate responses in stimulus orientation and reaction time conditions. *Journal of Experimental Psychology*, **82**, 310-320.

Chwilla, D.J., & Brunia, C.H.M. (1991). Event-Related potentials to different feedback stimuli. *Journal of Psychophysiology*, **28**, 123-132.

Cui, R.Q., Egker, A., Huter, D., Lang, W., Lindinger, G. and Deecke, L. (2000) High-resolution spatiotemporal analysis of the contingent negative variation in simple or complex motor tasks and a non-motor task. *Clinical Neurophysiology*, **111**, 1847-1860.

Damen, E.J.P. & Brunia, C.H.M. (1987). Changes in heart rate and slow potentials related to motor preparation and stimulus anticipation in a time estimation task. *Psychophysiology*, **24**, 700-713.

Damen, E.J.P. & Brunia, C.H.M. (1994). Is a stimulus-conveying task relevant information a sufficient condition to elicit stimulus-preceding negativity (SPN)? *Psychophysiology*, **31**, 129-139.

Davis, M., Walker, D.L. and Lee, Y. (1999) Neurophysiology and neuropharmacolgy of startle and its affective modulation. In: Dawson, M.E., Schell, A.M. and Böhmelt, A.H. (Eds.) Startle Modification. pp. 95-113. Cambridge: Cambridge University Press.

Dawson, M.E., Schell, A.M. and Böhmelt, A.H. (1999). Startle Modification. Cambridge: Cambridge University Press.

Deecke, L. and Kornhuber, H. (1977). Cerebral potentials and the initiation of voluntary movement. In: Desmedt, J.E. (Ed.) Attention, Voluntary Contraction and Slow Potential Shifts, pp. 132-150. Basel: Karger.

De Jong, R., Wierda, M., Mulder, G and Mulder, L.J.M. (1988). Use of partial information in responding. *Journal of experimental psychology: Human perception and performance.* **14**, 682-692.

Dick, J.P.R., Rothwell, J.C., Day, B.L., Cantello, R., Buruma, O., Gioux, M., Benecke, R., Bernardelli, A., Thompson, P.D., and Marsden, C.D. (1989). The Bereitschaftspotential is abnormal in Parkinson's disease. *Brain,* **112**, 233-244.

Donchin, E., Gerbrandt, L.A., Leiffer, L. and Tucker, L. (1972). Is the contingent negative variation contingent upon a motor response? *Psychophysiology,* **9**, 178-188.

Foit, A.B., Grözinger, B and Kornhuber, H.H. (1982). Brain potential differences related to programming, monitoring and outcome of aimed and non-aimed fast and slow movements to a visual target: The movement-monitoring potential (MMP) and the task outcome evaluation potential (TEP). *Neuroscience,* **7**, 571.

Fuster JM (1997) The prefrontal cortex: Anatomy, Physiology and Neuropsychology of the Frontal Lobe. 3rd ed. Raven Press: New York.

Gaillard, A.W.K. and van Beijsterveld, C.E.M. (1991). Slow brain potentials elicited by a cue signal. *Journal of Psychophysiology,* **5**, 337-347.

Grünewald, G., & Grünewald-Zuberbier, E. (1983). Cerebral potentials during voluntary ramp movements in   aiming tasks. In: Gaillard, A.W.K. and Ritter, W. (Eds.), Tutorials in ERP research: Endogenous components, pp. 311-327. Amsterdam: Elsevier.

Grünewald, G., Grünewald-Zuberbier, E., Hömberg, V., Schuhmacher, H. (1984). Hemispheric asymmetry of feedback-related slow negative potential shifts in a positioning movement task. In: Karrer, R. and Tueting, P., (Eds.) Brain and Information: Event-related potentials, pp. 470-476. New York: New York Academy of Sciences.

Hamano, T., Lüders, H.O., Ikeda, A., Collura, T.F., Comair, Y.G. and Shibasaki, H. (1997). The cortical generators of the contingent negative variation in humans: a study with subdural electrodes. *Electroencephalography and Clinical neurophysiology,* **104**, 257-268.

Heil, M., Rösler, F. and Henninghausen, E. (1996). Topographically distinct cortical activation in episodic long-term memory: The retrieval of spatial versus verbal material. *Memory and Cognition,* **24**, 777-795.

Heil, M., Rösler, F. and Henninghausen, E. (1997). Topography of brain electrical activity dissociates the retrieval of spatial versus verbal information from episodic long-term memory in humans. *Neuroscience Letters,* **222**, 45-48.

Hillyard S.A. (1973). The CNV and human behavior. *Electroencephalography and Clinical Neurophysiology.* Supplement **33**, 161-171.

Hultin, L., Rossini, P. Romani, G.L., Högstedt, P, Tecchio, F. and Pizella, V. (1996). Neuromagnetic localization of the late component of the contingent negative variation. *Electroencephalography and Clinical Neurophysiology,* **98**, 435-438.

Ikeda, A., Shibasaki, H., Nagamine, T., Terada, K., Kaji, R., Fukuyama, H. and Kimura, J. (1994). Dissociation between contingent negative variation and Bereitschaftspotential in a patient with cerebellar efferent lesion. *Electroencephalography and Clinical Neurophysiology,* **90**, 359-364.

Ikeda, A., Shibasaki, H., Kaji, R., Terada, K., Nagamine, T., Honda, M. and Kimura, J (1997). Dissociation between contingent negative variation and Bereitschaftspotential in a patient with Parkinsonism. *Electroencephalography and Clinical Neurophysiology,* **102**, 142-151.

Jahanshahi, M., Jenkins, I.H., Brown, R.G., Marsden, C.D., Passingham, R.E. and Brooks, D.J. (1995). Self-initiated versus externally triggered movements. I. An investigation using measurement of blood flow with PET and movement-related potentials in normal and Parkinson's disease subjects. *Brain,* **118**, 913-933.

Jarvilehto, T. and Frühstorfer, H. (1970). Differentation between slow cortical potentials associated with motor and mental acts. *Experimental Brain Research,* **11**, 309-317.

Kitamura, J-I, Shibasaki, H. and Kondo, T. (1993). A cortical slow potential is larger before an isolated movement of a single finger than simultaneous movement of two fingers. *Electroencephalography and clinical Neurophysiology,* **86**, 252-258.

Klein, C., Berg, P., Cohen, R., Elbert, T. and Rockstroh, B. (1993). Topography of CNV and PINV in schizophrenic patients and healthy subjects during a delayed matching-to-sample task. *Journal of Psychophysiology,* **11**, 322-334.

Kristeva, R., Jankov, E. and Gantchev, G. (1987). Differences in slow potentials in Bereitschaftspotential and Contingent Negative Variation. In: Johnson Jr., R., Rohrbaugh J.W. and Parasuraman R., (Eds.) Current Trends in Event-related Potential Research. EEG supplement 40, 41-46.

Kutas, M. and Donchin, E. (1977). The effects of handedness, responding hand, response force, and asymmetry of readiness potential. In: Desmedt J.E. (Ed.) Attention, Voluntary Contraction and Slow Potential Shifts, pp 189-210. Basel: Karger.

Kutas, M. and Hillyard, S.A. (1980). Reading senseless sentences: Brain potentials reflect semantic incongruity. *Science,* **207**, 203-205.

Lacey, J.I. and Lacey, B.C. (1973). Experimental association and dissociation of phasic bradycardia and vertex-negative wave: A psychophysiological study of attention and response intention. In: McCallum, W.C. and Knott, J.R. (Eds.) Event-related Slow Potentials of the Brain. Their relations to behaviour. pp. 87-94. Amsterdam: Elsevier.

Lacey, J.I. and Lacey, B.C. (1974). Some autonomic-central nervous system interrelationships. In: Black, P. (Ed.) Physiological correlates of emotion. pp. 205-227. New York: Academic Press.

Lang, P.J., Bradley, M.M. & Cuthbert, B.N. (1990). Emotion, attention and the startle reflex. *Psychological Review,* **7**, 377-395.

Lang, W., Lang, M., Heise, B. Deecke, L. & Kornhuber, H.H. (1984). Brain potentials related to voluntary hand tracking, motivation and attention. *Human Neurobiology,* **3**, 235-240.

Loveless, N.E. (1979). Event-related slow potentials of the brain as expressions of orienting function. In H.D. Kimmel, E.H van Olst & J.F. Orlebeke (Eds.) The Orienting Reflex in Humans, pp. 77-100. Hillsdale, New Jersey: Lawrence Erlbaum Associates, Publishers.

226   *Cornelis Brunia*

Loveless, N.E. and Sanford, A.J. (1974). Slow potential correlates of preparatory set. *Biological Psychology*, 1, 303-314.
Low, M.D. and McSherry, J.W. (1968). Further observations of psychological factors involved in CNV genesis. *Electroencephalography and Clinical Neurophysiology*, 25, 203-207.
Kornhuber, H.H. & Deecke, L. (1965). Hirnpotentialänderungen bei Willkürbewegungen und passiven Bewegungen des Menschen: Bereitschaftspotential und reafferente Potentiale. *Pflügers Archiv*, 284, 1-17.
Macar, F. and Besson, M. (1985). Contingent negative variation in processes of expectancy, motor preparation and time estimation. *Biological Psychology*, 21, 293-307.
Macar, F. and Besson, M. (1985). Contingent negative variation in processes of expectancy, motor preparation and time estimation. *Biological Psychology*, 21, 293-307.
Macar, F. and Vitton, N., 1979. Contingent negative variation and accuracy of time estimation: a study in cats. *Electroencephalography and Clinical Neurophysiology*, 47, 213-218.
Macar, F. and Vitton, N., 1982. An early resolution of contingent negative variation in time discrimination. *Electroencephalography and Clinical Neurophysiology*, 54, 426-435.
Macar, F., Vidal, F. and Bonnet, M., 1990. Laplacian derivations of CNV in time Programming. In: Brunia, C.H.M., Gaillard A.W.K and Kok, A. (Eds.), Psychophysiological Brain Research, Volume I, pp. 69-77. Tilburg: Tilburg University Press.
McAdam, D.W., Knott, J.R. and Rebert, C.S. (1969). Cortical slow potential changes in man related to inter-stimulus interval. *Psychophysiology*, 5, 349-358.
McCallum, W.C. (1988). How many separate processes constitute the CNV? In: McCallum, W.C., Zappoli, R. and Denoth, F. Cerebral Psychophysiology: Studies in event-related potentials. EEG supplement 38. pp.192-196. Amsterdam: Elsevier Science Publishers.
McCallum, W.C. (1988). Potentials related to expectancy, preparation and motor activity. In Picton T.W. (Ed.), EEG Handbook: Vol. 3. Human Event-Related Potentials, pp. 427-533. Amsterdam: Elsevier.
McCarthy, G. and Donchin, E. (1978). Brain potentials associated with structural and functional visual matching. Neuropsychologia, 16, 571-585.
MacKay, D.M. and Bonnet, M. (1990). CNV, stretch reflex and reaction time correlates of preparation for movement direction and force. *Electroencephalography and Clinical Neurophysiology*, 56, 696-698.
McCallum, W.C. (1988). Potentials related to expectancy, preparation and motor activity. In: Picton, T.W. (Ed.) Human event-related potentials. EEG handbook (Revised series, Volume 3), pp. 427-534. Amsterdam: Elsevier Science Publishers.
Niki, H. and Watanabe, M. (1979). Prefrontal and cingulate unit activity during timing behavior in the monkey. *Brain Research*, 171, 213-224.
Niemi, P. and Näätänen, R. (1981). Foreperiod and simple reaction time. *Psychological Bulletin*, 89, 133-162.
Öhman, A., Flykt, A. and Lundqvist, D. (2000). In: Lane R.D. and Nadel L. (Eds.) The Cognitive Neuro- science of Emotion. pp. 296-327. New York: Oxford University Press.
Passingham, R.E. (1987). Two cortical systems for directing movement. In: Motor areas of the cerebral cortex. Ciba Foundation Symposium, 132, pp.151-161. Chichester: John Wiley.
Passingham, R.E. (1993). The frontal Lobes and Voluntary Action. Oxford: Oxford University Press.
Praamstra, P., Meyer, A.S., Cools, A.R., Horstink, M.W.I.M. and Stegeman, D.F. (1996). Movement preparation in Parkinson's disease: time course and distribution of movement-related potentials in a movement-precuing task. *Brain*, 119, 1689-1704.
Pulvermüler, F., Lutzenberger, W, Müller, V., Mohr, B., Dichgans, J. and Birbaumer, N. (1996). P3 and the contingent negative variation in Parkinson's disease. *Electroencephalography and Clinical Neuro-physiology*, 98, 456-467.
Rao, S.M., Harrington, D.L., Haaland, K.Y., Bobholz, J.A, Cox, R.W. and Binder, J.R. (1997). Distributed neural systems underlying the timing of movements. *Neuroimage*, 5, 13.
Rebert, C.S. (1977). Intracerebral slow potential changes in monkeys during the foreperiod of reaction time. In: Desmedt, J. E (Ed.) Attention, Voluntary Contraction and Slow Potential Shifts, pp 242-253. Basel: Karger.
Requin, J., Brener, J. and Ring, C. (1991). Preparation for action. In: Jennings, J.R. and Coles, M.G.H. (Eds.) Handbook of Cognitive Psychophysiology. pp. 357-448.Chichester: John Wiley and Sons.
Rizolatti, G., Luppino, G. and Matelli, M. (1996). The classic supplementary motor area is formed by two independent areas. In: Lüders, H.O. Advances in Neurology, Vol.70, Supplementary Sensorimotor Area pp. 45-56. Philadelphia: Lippincott-Raven Publishers.
Rockstroh, B., Elbert, T. and Lutzenberger, W. (1989). Slow potentials of the brain and behavior: is there a non-motor CNV? *Psychophysiology*, 4A, S1.
Rohrbaugh, J. & Gaillard, A.W.K. (1983). Sensory and motor aspects of the Contingent Negative Variation. In: Gaillard A.W.K and Ritter W. (Eds.) Tutorials in Event-related Potentials Research: Endogenous Components, pp. 269-310. Amsterdam: North-Holland.
Roland, P.E., Skinhoj, E., Larsen, B and Lassen, N.A. (1980). The role of different cortical areas in the organization of voluntary movements in man. A regional cerebral blood flow study. In Ingvar, D.H. and Lassen, N.A. (Eds.) Cerebral Function, Metabolism and Circulation. Acta Neurologica Scandinavica, 56, 542-543.
Rosenbaum, D.A. (1985). Motor programming: A review and scheduling theory. In: Heuer, H, Kleinbeck, U. and Schmidt, K.H. (Eds.) Motor Behavior: Programming, control, and acquisition, pp.1-35. Berlin: Springer Verlag.
Rösler, F., Heil, M. and Henninghausen, E. (1995). Distinct cortical activation patterns during long-term memory retrieval of verbal, spatial and color information. *Journal of Cognitive Neuroscience*, 7, 51-65.
Rösler, F., Heil, M. and Henninghausen, E. (1995). Exploring memory functions by means of brain electrical topography: a review. *Brain Topography*, 7, 301-313.
Rösler, F., Pechmann, T., Streb, J., Röder and Henninghausen, E. (1998). Parsing sentences in a language with varying word order: word-by-word variations of processing demands are revealed by event-related brain potentials. *Journal of Memory and Language*, 38, 150-176.

Ruchkin, D.S., Sutton, S. Mahaffey, D. & Glaser, J. (1986). Terminal CNV in the absence of motor response. *Electroencephalography and Clinical Neurophysiology*, **63**, 445-463.

Ruchkin, D.S., Johnson Jr., R., Canoune, H and Ritter, W. (1992) Distinctions and similarities among working memory processes: an event-related potential study. *Cognitive Brain Research*, **1**, 53-66.

Ruchkin, D.S., Canoune H.L., Johnson Jr., R., and Ritter, W. (1995). Working memory and preparation elicit different patterns of slow wave event-related potentials. *Psychophysiology*, **32**, 399-410.

Rugg, M.D. (1995). ERP studies of memory. In Rugg, M.D. and Coles, M.G.H. (Eds.) Electrophysiology of Mind, pp. 132-170. Oxford: Oxford University Press.

Sasaki, K., Gemba, H., Hashimoto, S. and Mizuno, N. (1979). Influences of cerebellar hemispherectomy on slow potentials in the motor cortex preceding self-paced hand movements in the monkey. *Neuroscience Letters*, **15**, 23-28.

Sasaki, K. and Gemba, H. (1991). Cortical potentials associated with voluntary movements in monkeys. In: Brunia, C.H.M, Mulder G. and Verbaten, M.N. (Eds.) Event-related Brain Research, pp. 80-96. Amsterdam: Elsevier.

Scheirs, J.G.M. and Brunia, C.H.M. (1985). Achilles tendon reflexes and surface EMG activity during anticipation of a significant event and preparation for a voluntary movement. *Journal of Motor Behavior*, **17**, 96-109.

Shibasaki, H., Shima, F. and Kuroiwa, Y. (1978). Clinical studies of the movement-related cortical potential (MP) and the relationship between the dentato-rubro-thalamic pathway and the readiness potential (RP). *Journal of Neurology*, **219**, 15-25.

Shibasaki, H., Barrett, G., Neshige, R., Hirata, I. & Tomoda, H. (1986). Volitional movement is not preceded by cortical slow negativity in cerebellar dentate lesion in man. *Brain Research*, **368**, 361-365.

Simons, R.F. (1988). Event-related slow brain potentials: a perspective from ANS psychophysiology. In: Ackles, P.I. Jennings J.R. and. Coles, M.G.H (Eds.) Advances in Psychophysiology, Volume 3, pp. 223-267. Greenwich, Connecticut: JAI Press.

Simons, R.F., Öhman, A. and Lang, P.J. (1979). Anticipation and response set: cortical, cardiac and electro-dermal correlates. *Psychophysiology*, **16**, 222-233.

Tychner, W.H. (1954). Recent studies of simple reaction time. *Psychological Bulletin*, 51, 128-149.

Ulrich, R., Leuthold, H. and Sommer, W. (1998). Motor programming of response force and movement direction. *Psychophysiology*, **35**, 721-728.

Vallbo, A.B. (1974). Human muscle spindle discharge during isometric voluntary contractions. Amplitude relations between spindle frequency and torque. *Acta Physiologica Scandinavica*, **90**, 319-336

Van den Bosch, R.J. (1983). Contingent negative variation: components and scalp distribution in psychiatric patients. *Biological Psychiatry*, **19**, 963-972.

Van Boxtel, G.and Brunia, C.H.M. (1994). Motor and non-motor aspects of slow brain potentials. *Biological Psychology*, **38**, 35-51.

Van Boxtel, G.and Brunia, C.H.M. (1994). Motor and non-motor components of the contingent negative variation. *International Journal of Psychophysiology*, **17**, 269-279.

Verleger, R., Wascher, E., Wauschkuhn, B., Jaskowski, P., Allouni, B., Trillenberg, P. and Wessel, K. (1999). Consequences of altered cerebellar input for the cortical regulation of motor coordination, as reflected in EEG potentials. *Experimental Brain Research*, **127**, 409-422.

Vidal, F., Bonnet, M. and Macar, F. (1995). Programming of duration of a motor sequence: role of the primary and supplementary motor areas in man. *Experimental Brain Research*, **106**, 339-350.

Wagner, M., Rendtorff, N., Kathmann, N. and Engel, R.R. (1996). CNV, PINV and probe-evoked potentials in schizophrenics. *Electroencephalography and Clinical Neurophysiology*, **98**, 130-143.

Walter, W.G., Cooper, R., Aldridge, V.J., McCallum, W.C.& Winter, A.L. 1964. Contingent Negative Variation: an electric sign of sensori-motor association and expectancy in the human brain. *Nature*, **203**, 380-384.

Wiesendanger, M., Hummelsheim, H., Bianchetti, M, Chen, D.F., Hyland, B., Maier, V. and Wiesendanger, V. (1987). Input and output organization of the supplementary motor area. In: Motor areas of the cerebral cortex. CIBA Foundation Symposium 132, pp. 40-53. Chichester, John Wiley and Sons.

THE LATERALIZED READINESS POTENTIAL

Martin Eimer[1] and Michael G. H. Coles[2]

[1]Department of Psychology
Birkbeck College
University of London
London, England

[2]Department of Psychology
University of Illinois
Champaign, Illinois, USA

INTRODUCTION

The pioneering work of Kornhuber and Deecke on the Bereitschaftspotential (Kornhuber & Deecke, 1965) suggested that movement-related brain activity could be recorded from the scalps of human subjects. Not only did this open up the possibility of studying brain-movement relationships using non-invasive procedures, it also pointed to a method by which measures of event-related brain potentials (ERPs) could be used to study cognitive processes. The benefits of the discovery of the Bereitschaftspotential for the analysis of cognitive function were not perhaps as obvious as those that related to the discovery of other ERP components during the 1960s, such as the P300 (Sutton, Braren, Zubin, & John, 1965) and the CNV (Walter, Cooper, Aldridge, McCallum, & Winter, 1964). Nevertheless, it has become clear in the last 35 years that the impact of its discovery has been just as profound. This is because knowledge of covert movement-related processes has proved to be extraordinarily important in exploring human cognitive function.

Models of the structure and dynamics of cognition propose the existence of cognitive processes which function to produce a given behavioural output, usually as a consequence of some input. The problem confronted by the cognitive psychologist is how to evaluate claims about the nature and mode of operation of these processes when the data available are limited to parameters of the output, such as speed and accuracy. It is the problem of making inferences about covert processes from observations of overt behaviour. In this regard, what was so striking about the studies of Kornhuber and Deecke, and those by Vaughan and his colleagues (Vaughan, Costa, & Ritter, 1968) which followed shortly thereafter, was that brain activity could be detected prior to an overt movement, and that the nature of this activity seemed to depend on the nature of the impending movement. The Bereitschaftspotential begins several

The Bereitschaftspotential
Edited by Jahanshahi and Hallett, Kluwer Academic/Plenum Publishers, New York, 2003

hundred ms before the movement and, as the time for movement approaches, the scalp distribution depends on what movement is about to be executed.

It was the latter observation that gave rise to the measure called the lateralized readiness potential (LRP). With the studies of Kornhuber and Deecke (1965) as a starting point, Kutas and Donchin (1980) had examined movement-preceding brain activity for a variety of conditions which differed in terms when subjects knew which of two manual responses they would have to make. Kutas and Donchin found that the time at which the brain activity became asymmetrical was closely related to the time at which subjects knew whether a right or left-hand response would be required, and concluded that the asymmetry reflected preparation to execute specific motor acts.

On the basis of these results, groups in Groningen (e.g. Smid, Mulder, & Mulder, 1987) and Illinois (e.g. Coles & Gratton, 1986) simultaneously and independently reasoned that, under certain circumstances, the presence of asymmetries could be used to *infer* the presence of preferential preparatory activity. They proposed procedures to derive a measure of asymmetric, movement-related brain activity and these procedures yielded the measure now referred to as the LRP (although it was originally referred to as "corrected motor asymmetry" by the Groningen group).

THE LATERALIZED READINESS POTENTIAL (LRP)

Deriving the LRP

Both Groningen and Illinois groups recognised that asymmetries in electrical brain activity can be observed for a variety of reasons, only some of which involve preparation for movement. The problem of isolating the movement-related contribution to the asymmetries was solved in a similar way by both groups and, as a result, similar computational formulae were proposed for deriving the LRP. The main difference between the procedures involved the use of two-stage subtraction procedure (the double subtraction method: see De Jong, Wierda, Mulder, & Mulder, 1988) or subtraction-averaging sequence (the averaging method: see Gratton et al., 1988, Coles, 1989).

We present here the double subtraction method as described by Eimer (1998: see Figure 1). The situation is a choice reaction time task in which the left- and right-hand is used to respond to an imperative stimulus. ERP data are available for an epoch that begins 100 ms before the imperative stimulus for 600 ms post-stimulus. The first stage in the analysis is to sort trials into two groups: those for which the left hand response is correct and those for which the right hand response is correct. Then, ERP data from two lateral electrodes placed over the left and right motor cortices (C3' and C4', respectively) are averaged separately for the two groups of trials. In the upper panel of Figure 1, these averages are characterised by: (a) a general positive response in all waveforms; (b) an asymmetry between the averages for C3' and C4' which begins between 150 and 200 ms post stimulus; (c) for left-hand responses the voltage at C4' is more negative than that at C3'; while (d) the converse is the case for right hand responses. These lateralized negativities are isolated in the next step of the procedure which involves subtracting the voltage recorded at C4' from that recorded at C3' (see lower left panel of Figure 1). These difference waveforms reflect the asymmetrical activity associated with left and right-hand movements. The third step in the procedure involves a second subtraction: the asymmetry waveform for right-hand responses is subtracted from the asymmetry waveform for left-hand responses (see lower right panel of Figure 1). If lateralized negativities are larger over the motor cortex contralateral to the correct response, the waveform will be positive; if the lateralized negativities are larger over the motor cortex ipsilateral to the correct response, the waveform will be negative. Thus, positive values represent activation of the correct response, and negative values represent activation of the incorrect response. Finally, it is important to note the fate of asymmetrical activity that it not related to the side of movement. During the last 50 ms of the epoch, the same asymmetrical activity is present in the waveforms for both left and right-hand responses. In each case, the voltage at C3' is less positive than that at C4'. This common asymmetrical activity disappears in the double subtraction procedure. The value of the LRP for the last 50ms of the epoch is close to zero.

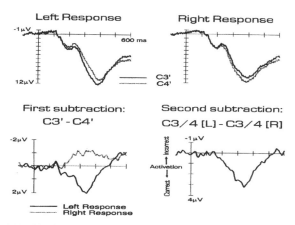

Figure 1. Derivation of the LRP with the double subtraction method on the basis of ERP waveforms elicited at electrodes C3' (left hemisphere) and C4' (right hemisphere). Top: Grand-averaged ERP waveforms from ten participants elicited at C3' (solid lines) and C4' (dashed lines) in response to stimuli requiring a left-hand response (left side) or a right-hand response (right side). Bottom left: Difference waveforms resulting from subtracting the ERPs obtained at C4' from the ERPs obtained at C3' separately for left-hand responses (solid line) and right-hand responses (dashed line). Bottom right: LRP waveform resulting from subtracting the C3'-C4' difference waveform for right-hand responses from the C3'-C4' difference waveform for left-hand responses. A downward-going (positive) deflection indicates an activation of the correct response, an upward-going (negative) deflection indicates an activation of the incorrect response.

Although the double subtraction and averaging procedures are computationally equivalent, there is an important difference that can lead to confusion. As we have noted, positive LRP values derived from the double-subtraction procedure reflect preferential activation of the correct response; negative values reflect preferential activation of the incorrect response. In contrast, positive LRP values derived from the averaging procedure reflect preferential activation of the incorrect response; while negative values reflect preferential activation of the correct response. In the sections that follow, we describe LRP data derived from both these procedures. Thus, it is important to keep this difference in mind.

Validating the LRP

Although the procedures used to derive the LRP provide some face validity for the claim that it reflects preferential preparatory activity, two kinds of additional evidence are available. First, there is evidence that variables that are related to aspects of response preparation influence parameters of the LRP. Second, there is evidence that the LRP indeed reflects the activity of brain systems involved in motor output.

Preparation, Movement Parameters and the LRP. If the LRP reflects the central activation of a specific response, it should be observed not only immediately before and during the execution of a response, but already in the interval when this response is being prepared. For example, LRPs should be elicited in the foreperiod of a warned RT task, when a precue presented prior to a response-relevant target predicts the likely side of the upcoming response. In an experiment by Gratton, Bosco, Kramer, Coles, Wickens, & Donchin (1990), cues indicated with 80% probability which response (left vs. right hand) was likely to be required to a target stimulus that was presented 1000 ms after the cue. On neutral cue trials, cues were uninformative about the upcoming response. The LRPs measured in the cue-target interval in response to correct (valid), incorrect (invalid), and neutral cues are shown in Figure 2. When the cue was valid, the correct response was partially activated, whereas the incorrect response was partially activated on invalid trials. No LRP was elicited in the cue-target interval in neutral trials. In other words, participants in this experiment prepared the response predicted by the precue. As a result, responses were fast and accurate when cues were valid, and slow as well as error-prone when cues were invalid. Thus, the LRP can be used to measure covert response preparation processes in the foreperiod preceding the presentation of a target that are responsible for these costs and benefits observed for overt performance measures in response to this target (see also Gehring, Gratton, Coles, & Donchin, 1992, for similar results).

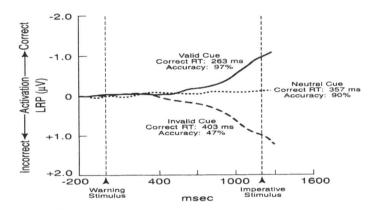

Figure 2. Effects of movement precue validity on the LRP measured in the interval between cue and target stimulus. Data from Gratton et al., 1990.

The observation that LRPs are elicited in the period between a warning stimulus and an imperative stimulus when warning stimuli specify the likely side of an upcoming response demonstrates that the LRP is linked to the preparation of unimanual responses. However, response preparation is a sequential process, which includes not only the selection of global parameters like response side, but also the programming of more specific response properties such as movement direction and response force, and ultimately the selection and activation of specific motor units. The LRP may be sensitive to some, but not all of the processes involved in the programming of responses. To further validate the LRP as a measure of response preparation, we need to separate those aspects of response preparation which have systematic effects on LRP waveforms from other aspects which leave the LRP unaffected. If the LRP was exclusively linked to the abstract specification of response side, it should not be influenced by the preparation of other, more specific movement parameters. In contrast, if the LRP was also sensitive to processes involved in the programming of movement direction or amplitude, or even to the efferent activation of peripheral muscle groups, experimental manipulations of these processes should result in systematic effects on LRP measures.

The relationship of the LRP to different parameters of a movement can be investigated by modifying the response precueing procedures described earlier. Instead of measuring LRPs in response to warning stimuli that only specify the side of an upcoming response, LRPs can also be recorded to other types of movement precues, which either convey full response information, or specify some parameters of an upcoming movement, while leaving other parameters unspecified. Behavioural experiments using this movement precueing technique have generally found that reaction times decrease as a function of the amount of information provided by a precue, suggesting that different movement parameters can be programmed in parallel. If the LRP reflected this programming process, LRP amplitudes measured in the interval between movement precue and imperative stimulus should increase with the number of response parameters specified by a movement cue.

Several experiments have adopted the movement precueing technique to investigate which aspects of movement preparation are reflected by the LRP. One negative result obtained with movement precueing was entirely predictable. LRPs were found to be entirely absent during response preparation when cue stimuli indicated the direction of a movement (finger extension vs. finger flexion), but left response hand unspecified (Leuthold, Sommer, & Ulrich, 1996). This is hardly surprising, since any programming of movement direction should occur simultaneously for both hands when response side is unknown. Because bilateral response preparation will not be accompanied by a lateralization of the RP, no LRP can be observed. In order to investigate the sensitivity of the LRP to different types of movement precues, response side needs to be among the response parameters specified by these cues.

Two studies investigated whether the LRP is related to the selection of peripheral muscle groups. If there was such a link, LRP amplitudes should be influenced by variations in the required force of a subsequent response. In an experiment by Sommer, Leuthold, & Ulrich (1994), neither response force nor rate of force production had any effect on LRP amplitudes measured during response preparation, thus indicating that the LRP is not linked to the recruitment of peripheral force units, but reflects more central stages of motor programming. In a more recent study (Ray, Slobounov, Mordkoff, Johnston, & Simon, 2000), LRP amplitudes were affected by differences in the rate of force production, suggesting that more peripheral motor factors may contribute to the LRP. The factors responsible for this inconsistency between the findings of Sommer et al. (1994) and Ray et al. (2000) remain to be determined.

A number of studies have found larger LRP amplitudes when response-related information is increased beyond the specification of response hand. LRPs were larger in response to cues that specified complex movements (a sequence of successive keystrokes performed with one hand) than when cues signalled that a single keystroke had to be performed in response to an upcoming target (Hackley & Miller, 1995). LRPs were larger in response to precues that fully specified an upcoming response in terms of movement direction (extension vs. flexion) and response hand (Leuthold, Sommer, & Ulrich, 1996), or in terms of movement direction, movement force, and response hand (Ulrich, Leuthold, & Sommer, 1998) than for precues that determined response hand, but left other movement parameters unspecified. The fact that LRP amplitudes are influenced by the amount of information conveyed by a movement precue demonstrates that the LRP does not exclusively reflect the selection of response side at an early, global level of motor programming. However, LRP amplitudes do not simply increase as a function of the number of response parameters determined by a movement cue. When increasing the amount of advance information beyond the specification of response hand, corresponding increases in LRP amplitudes were only found when a response was fully specified by the cue, but not when one response parameter remained uncertain (Ulrich, Leuthold, & Sommer, 1998). This suggests that the response preparation processes reflected in the LRP may consist of two stages – the selection of response side, which is unaffected by the presence of additional response-related information, and the preparation of additional response parameters, which can only proceed when full response information is available.

In summary, the results from studies measuring LRPs during response preparation suggest that the LRP reflects the selection of response side, which can take place even when all other features of a response remain unspecified. In contrast, the advance preparation of response parameters other than response side affects the LRP only under conditions where all properties of an upcoming response are known and this response can therefore be fully prepared.

The Neural Basis of the LRP. Suggestive evidence concerning the neural basis of the LRP is provided by studies of its scalp distribution. First, as we have noted, the original Kornhuber & Deecke (1965) study showed that the negative potential that precedes a hand movement becomes larger at electrode sites contralateral to the side of the movement as the time for the movement approaches. Given that the motor system is organised contralaterally, this result certainly points to a source for the LRP in the motor system. Second, the locus of the maximum negativity depends on the kind of movement that is to be made in a manner that is consistent with the known somatotopic organisation of motor cortex (Vaughan, Costa, & Ritter, 1968).

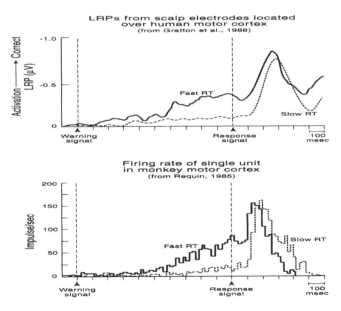

Figure 3. Upper panel: LRPs recorded from scalp electrodes located above motor cortex on trials with fast and slow reaction times. Data from Gratton et al. (1988). Lower panel: Firing rate of cell in monkey motor cortex on trials with fast and slow reaction times. Data from Requin (1985).

While this kind of distributional evidence implicates activity in motor-related brain areas as being responsible for the LRP we record at the scalp, it is not conclusive. Because of volume conduction, intracranially generated brain activity propagates through the brain and skull to the scalp. As a result, the locus of maximum voltage on the scalp may not correspond to the site of generation. More definitive evidence could be provided by an analysis of the relationship between scalp-recorded activity and intracranial activity. However, there have been no studies in which the measures of the LRP and intracranial activity have been obtained simultaneously. Nevertheless, there are several examples of parallel relationships between behavioural signs of preparation and both unit activity and the LRP. As an example, consider the relationship between the speed of a motor response to an imperative stimulus and the preparatory activities that occur during the time preceding the presentation of the stimulus. In a study with humans, Gratton, Coles, Sirevaag, Eriksen, & Donchin (1988) showed that the LRP preceding fast responses was more negative than that preceding slow responses (see upper panel in Figure 3). In a similar study with monkeys, Requin and his colleagues (Requin, 1985) recorded single unit activity from neurons in layers 4 and 5 of motor cortex and found units at the border of Areas 4 and 6 whose activity varied as a function of response speed. In particular, the units fired more in the interval preceding fast responses than when reaction times were long (see lower panel of Figure 3). Thus, there is a remarkable parallel between the LRP data recorded from the scalps of human subjects and the unit data derived directly from the motor cortex of monkeys.

While the relationship between the LRP and activity in motor cortex is reasonably well-established, such a relationship does not necessarily imply that the LRP exclusively reflects response-related processes. This is because there are neurons in motor cortex whose activity seems to be related more to sensory than to motor processes (see Requin & Riehle, 1985). This suggests that there may be circumstances under which the LRP may reflect sensory, as well as motor, processes. This is a point to which we will return below.

Interpretative issues

As with other ERP components, different kinds of inferences can be made on the basis of LRP recordings (Fabiani, Gratton, & Coles, 2000; Rugg & Coles, 1995). These inferences are based on different measurement operations, and assumptions about the meaning of variation in the amplitude and latency of the LRP. The four main inferences are:

a. that deviation of the LRP from zero indicates that one or other of the two responses has been preferentially primed. The keyword here is "preferentially": if both responses are primed equally, the LRP will remain at zero. This is the simplest, and least controversial inference in part because it depends solely the logic behind the measurement operation used to derive the LRP and on a simple statistical test of the difference between the value of the LRP and zero.

b. that the magnitude of the LRP indicates by how much one or other of the responses has been preferentially primed. This inference requires an assumption about the meaning of the magnitude of the LRP. Evidence in support of this assumption was provided by Gratton et al. (1988) who showed that the level of the LRP at the time of the response was relatively constant for responses with different reaction times. This observation of an apparent fixed LRP threshold for the execution of responses is consistent with the idea that the larger the LRP, the more pronounced the preferential priming. Note, however, that the concept of a fixed threshold is not universally accepted (Band & Van Boxtel, 1999).

c. that the latency of the LRP indicates the time at which preferential preparation occurs. This inference is based on analytic procedures for determining the LRP onset or the time at which the LRP begins to deviate from zero. Such procedures have to deal with the problem that the LRP is based on the average of many trials. As a result, the latency at which the average waveform deviates from zero will be a function of the latencies of those trials with the earliest onset. (For a review of procedures designed to address this issue see Miller, Patterson, & Ulrich, 1998; Schwarzenau, Falkenstein, Hoormann, & Hohnsbein, 1998).

   d. that LRP onset latencies computed on the basis of stimulus-locked or of response-locked LRP waveforms provide different, and complementary information about the timing of cognitive and response processes. In stimulus-locked LRPs, latency measures reflect the time between stimulus onset and LRP onset. In response-locked LRPs, these latency measures refer to the distance in time between the onset of the LRP deflection and the start of an overt response. Stimulus-locked LRP latency is determined by cognitive processes that occur prior to the selective activation of a response, whereas response-locked LRP latency reflects the duration of processes that take place between LRP onset and response execution. To find out whether experimental factors affect processes prior to or after the start of selective response activation, LRP onset latencies can be determined separately for stimulus-locked and response-locked LRPs (see Osman & Moore, 1993; Leuthold, Sommer, & Ulrich, 1996; and Miller & Ulrich, 1998, for applications of this procedure).

USING THE LRP

Subliminal Perception and Implicit learning

   Few topics in experimental psychology are more controversial than the status of conscious awareness, and the question of how consciousness affects cognitive processing and the control of behaviour. Numerous experimental studies have investigated whether stimuli below the threshold of conscious awareness have systematic effects on behaviour (subliminal perception), or whether knowledge can be acquired without concurrent awareness (implicit learning). LRP measures can help to shed new light on these issues, because they can provide insight into covert response activation processes that may occur without explicit instructions and intentions, or even in the absence of conscious awareness.

   Briefly presented and subsequently masked stimuli can have systematic effects on behaviour even though these stimuli are not accessible to conscious awareness (Fehrer & Raab, 1962). Such effects are often explained by assuming that subliminal stimuli affect performance via direct sensory-motor links. These links can give sensory stimuli immediate access to response-related processing stages, even though masking interrupts the perceptual analysis, and prevents the conscious identification of these stimuli. If there were such direct sensory-motor links, stimuli that are inaccessible to conscious awareness should sometimes be able to activate a response. By measuring LRPs in response to these stimuli, evidence for the existence of such subliminal motor activation processes has indeed been obtained. Leuthold & Kopp (1998) presented brief prime stimuli that were masked by the arrival of subsequent targets. Response hand was determined by target location, and primes appeared at the same location as targets (congruent trials) or at different locations (incongruent trials). Although the masking procedure prevented any conscious awareness of the primes' location, responses were faster in congruent trials than in incongruent trials, and LRPs revealed that the primes triggered an early activation of their corresponding response. In incongruent trials, an early incorrect response activation was observed, whereas an early correct response activation was found for congruent trials. This pattern of results thus provides strong support for the hypothesis that there are direct sensory-motor links, and that these links may be responsible for subliminal perception effects observed in earlier studies.

Additional LRP studies have found that subliminal motor activation processes may be more complex than suggested by the results obtained by Leuthold & Kopp (1998), because they include both facilitatory as well as inhibitory components. Eimer & Schlaghecken (1998; see also Eimer, 1999) measured LRPs in experiments where a left-pointing or right-pointing arrow prime (16 ms duration) was immediately followed by a mask (100 ms duration) and then by a left-pointing or right-pointing arrow target (100 ms duration), which required a left-hand or right-hand response. Primes and targets were mapped to the same response on congruent trials, and to opposite responses on incongruent trials. The major difference between this experiment and the Leuthold & Kopp (1998) study was that masks and targets were now separate, successively presented items, which implied that the interval between primes and targets was increased. Although forced choice tests confirmed that the primes were not consciously perceived, LRP waveforms again revealed an initial partial activation of the response mapped to the prime stimulus, providing further evidence for the existence of direct sensory-motor links (see Figure 4, black arrow). Unexpectedly, and in marked contrast to Leuthold & Kopp (1998), responses to targets were delayed in congruent relative to incongruent trials. LRP waveforms revealed that the initial response activation triggered by the primes was only present for a brief period, and was then replaced by LRP deflections in the opposite direction (Figure 4, white arrow). This pattern of results suggests that subliminal motor priming includes both an early facilitatory phase, where masked primes activate their corresponding response, as well as a subsequent inhibitory phase, where this initial activation is suppressed. The inhibitory component can only be observed when prime-target intervals are sufficiently long, but not when targets are presented in close temporal contiguity with a prime. In the latter case, responses to targets can be selected and activated during the initial facilitatory phase, resulting in performance benefits for congruent trials, and the absence of any LRP evidence for response inhibition (see Eimer, 1999, for a more detailed discussion).

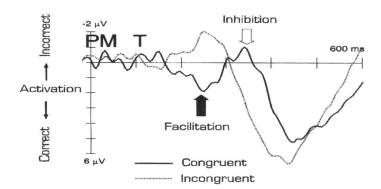

Figure 4. LRP waveforms observed in a 'subliminal priming' experiment where a prime (P) was presented for 16 ms, was immediately followed by a masking stimulus (M, 100 ms), and then by a response-relevant target (T). Primes and targets were mapped to identical responses in congruent trials, and to opposite responses in incongruent trials. LRPs revealed an early facilitation of the response mapped to the prime (black arrow: correct response activation for congruent trials, incorrect activation for incongruent trials). This activation pattern was then replaced by an LRP deflection of opposite polarity (white arrow), indicating response inhibition. Data from Eimer (1999).

In addition to using LRPs to demonstrate subliminal motor activation as evidence for the existence of direct sensory-motor links, LRP measures have also been used to show that subliminal stimuli can be subject to semantic categorisation processes, and that results of this semantic analysis will influence motor processes. Dehaene et al. (1998) presented numerals between 1 and 9 (either digits or written words) as masked primes and targets, and asked participants to classify target numerals as larger or smaller than 5. Masked primes and targets were mapped to the same response on congruent trials, and to opposite responses on incongruent trials. Although primes could not be consciously identified, performance was better on congruent trials, and LRP waveforms revealed an early partial activation of the correct response on congruent trials, and incorrect response activation on incongruent trials. This result indicates that subliminal primes were categorised at the semantic level in accordance with task instructions, and that the result of this categorisation was then transmitted to response-related processing stages.

LRP measures have also been used to study mechanisms involved in implicit and explicit sequence learning. Learning is termed 'implicit' when it occurs without intention to learn and without awareness of the learned material. This type of learning is often studied in serial reaction time (SRT) tasks where imperative stimuli follow a repetitive sequence, although participants are not informed about the existence of this sequence (Nissen & Bullemer, 1987). Extended practice with a particular stimulus sequence results in a decrease of reaction times, and in performance costs when this sequence is suddenly replaced by a random sequence. Because these effects are observed even when participants do not notice the presence of a sequence, and are unable to recall or reproduce parts of it, knowledge about sequential structures seems to be acquired at least in part implicitly. It is not yet clear what kind of knowledge is acquired during SRT tasks. Participants may learn about a sequence of stimulus events, a sequence of motor responses, or acquire knowledge about both stimulus and response sequences. LRP measures have been used to investigate whether motor learning contributes to the behavioural effects observed in SRT tasks. In an experiment where a 10-item stimulus sequence was presented repeatedly, training drastically reduced LRP onset latencies, up to the point where LRP onset coincided with stimulus onset (Eimer, Goschke, Schlaghecken, & Stürmer, 1996). This effect was not only observed for participants reporting some explicit sequence knowledge, but also for 'implicit' participants who were completely unaware of the presence of a sequence. On some trials, a deviant stimulus was presented, which required a response on the side contralateral to the response that would normally have occurred at that serial position. LRPs revealed a partial activation of the expected, but now incorrect response (see Rüsseler & Rösler, 2000, for similar results). This effect was larger for 'explicit' than for 'implicit' participants. Overall, these findings demonstrate that the anticipatory preparation of expected motor responses contributes to serial learning in SRT tasks, that such preparatory motor processes start before the presentation of the next stimulus, and that motor learning may be involved in implicit sequence learning.

S-R Compatibility and the Simon effect

One of the most robust effects in choice reaction experiments is the observation that unimanual responses are faster when stimulus and response locations correspond than when stimuli and responses are located at non-corresponding positions. Such spatial stimulus-response compatibility effects can even be observed when stimulus locations are entirely irrelevant for response selection (Wallace, 1971). The influence of irrelevant stimulus locations on reaction times, commonly called the Simon effect, is explained by a tendency to respond toward the source of stimulation (Simon, 1969). The onset of a stimulus on the left or right side is assumed to trigger an automatic (unintentional and unavoidable) activation of an ipsilateral response (Kornblum, Hasbroucq, & Osman, 1990). This automatic response activation results in performance benefits for compatible trials where stimulus and response positions correspond, and in performance costs for incompatible trials where they are on opposite sides. The hypothesis that lateralised stimuli tend to activate spatially corresponding responses has found strong support from several studies which recorded LRPs in compatible and incompatible trials under conditions where stimulus position was irrelevant for response selection (De Jong, Liang, & Lauber, 1994; Valle-Inclan, 1996, Valle-Inclan & Redondo, 1998). While LRP onset latencies were identical in compatible and incompatible trials, an initial activation of the incorrect response was observed for incompatible trials, whereas only correct response activation was observed for compatible trials. This finding clearly demonstrates that lateralized stimuli tend to activate spatially corresponding responses, even when stimulus location is irrelevant. Eimer (1995) observed similar results for centrally presented left-pointing and right-pointing arrows that preceded an imperative stimulus, and were uninformative with respect to response position. LRPs recorded in response to left-pointing arrows indicated a partial activation of a left response, whereas right-hand responses were partially activated by right-pointing arrows (but see Verleger, Vollmer, Wauschkuhn, van der Lubbe, & Wascher, 2000, for a different interpretation of these effects in terms of an encoding of the arrows' spatial properties for action). It is important to note that while these LRP results show that lateralized stimuli tend to activate spatially corresponding responses, they do not provide any clear-cut evidence in favour of the hypothesis that this type of response activation is 'automatic'. In fact, results from several recent LRP studies (Valle-Inclan & Redondo, 1998; Valle-Inclan, Hackley, & de Labra, in press) suggest that response activation processes in S-R compatibility tasks are strongly affected by strategic control processes, and may therefore be much less automatic than previously thought.

Models of Human Information Processing

Cognitive psychologists are interested in the problem of specifying the nature of the processes that intervene between stimulus and response. One approach to this problem, referred to as "mental chronometry" (Posner, 1978), is to evaluate the effects of various experimental manipulations on measures of response speed. While this approach has its origins in the work of Donders (1868/1969), it is still a dominant force today.

Interpretation of measures of response speed depends on various assumptions about the nature of the interaction among different processing units (or stages). A particularly important distinction can be drawn between those who propose that the stages communicate discretely, with transmission of information between units occurring in an all-or-none fashion (e.g. Sternberg, 1969), and those who propose that the stages communicate continuously (e.g. Eriksen & Schultz, 1979; McClelland, 1979). There are also hybrid models that propose that information is transmitted in discrete "chunks", with each transmission containing only part of the total information (e.g. Miller, 1988).

Resolution of the debate about the nature of transmission has proved to be difficult particularly because of the need to rely on a measure of overt behaviour (response speed) to make inferences about the nature of covert processes. Beginning in the 1980s, several different research groups reasoned that, since the LRP is a measure of covert processes, it might be used to resolve the debate (see Coles, Smid, Scheffers, & Otten, 1995, for review). A key factor behind this reasoning was that the idea that the LRP could provide a window on covert preparatory processes that are not associated directly with any overt behaviour. Two paradigms have been dominant: the conflict paradigm, and the go/no-go paradigm. In the *conflict* paradigm, the subject is required to respond to a visual display using either the left or right-hand. Response choice is guided by a target stimulus (a letter or an arrow) presented at the centre of the display. The target is flanked by letters or arrows that are sometimes the same as the target (compatible or congruent condition) or the flanking stimuli are associated with a different response than the target (incompatible or incongruent condition). The critical question is what happens in the incompatible condition when there is conflict between the target and flanking stimuli. If information about the visual display is transmitted through the system discretely, then the flanking stimuli should have no influence on the activity of the response system. Responses may be delayed while the perceptual system distinguishes between target and flankers, but once the target is identified this information should be passed on in an all-or-none fashion and only the response associated with the target should be prepared and executed. On the other hand, if partial information transmission occurs, then information about the flankers may reach the response system and the incorrect response may be activated, although the correct response may still be executed. Thus, the presence or absence of covert, incorrect response preparation is critical to the differentiation between the two views of the nature of information transmission.

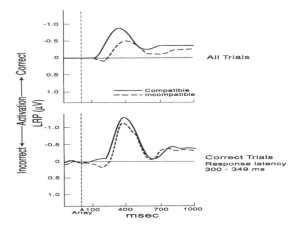

Figure 5. Lateralized readiness potential data for S-R compatible and S-R incompatible trials. For the data shown in the upper panel, all trials regardless of response accuracy or latency were included in the averages. Data shown in the lower panel are based on correct responses with a latency of 300-349 ms. For correct responses, either there was no EMG activity from muscles associated with the incorrect response or EMG activity associated with the correct response occurred first. Data from Gratton et al. (1988).

Figure 5 shows LRP data obtained by Gratton et al. (1988) for compatible and incompatible conditions of an experiment using the conflict paradigm. The important feature of these waveforms is the "dip" in the LRP that occurs on incompatible trials after the onset of the visual stimulus (array). Since there were no overt signs of an incorrect response on these trials (see lower panel of Figure 5), the dip implies that the incorrect response was covertly prepared on these incompatible trials, even though only the correct response was actually executed. Such covert incorrect response preparation is inconsistent with an all-or-none view of information transmission, and consistent with either the continuous or partial transmission views. Similar results have been obtained by several other groups, including Smid et al. (1991) and Praamstra, Plat, Meyer, & Horstink (1999). The latter group used the size of the "dip" to infer that

Parkinson patients were less able than controls to suppress activation of the motor cortex controlling the wrong response hand.

In the *go/no-go* paradigm, the subject is also required to respond to a visual stimulus with the left or right-hand, but only on a subset of the trials (usually 50%). The choice of response hand is determined by one stimulus attribute, while the decision to respond is determined by another attribute. The critical question is what happens in the no-go condition. If information about the attribute associated with response choice (left or right hand) is available before information about the response decision (whether or not to respond), and if this information is transmitted to the response system, then covert preparation of the appropriate response should occur. On the other hand, if transmission occurs in an all-or-none fashion when both stimulus attributes have been processed, there should be no covert response preparation in the no-go condition. Thus, the presence or absence of covert response preparation in the no-go condition is diagnostic of the existence of partial or all-or-none information transmission.

Miller and Hackley (1992) used a variant of the go/no-go paradigm in which the letters S or T were presented in either large or small sizes. The size discrimination was deliberately made more difficult than the letter discrimination. Response choice was mapped to letter identity, while response decision was mapped to letter size. Thus, if information about the letter was transmitted to the response system before information about letter size was available, covert preparation could begin before a decision was made as to whether a response was, in fact, required. Data obtained by Miller and Hackley revealed the presence of an LRP on no-go trials (see Figure 6). The onset of this no-go LRP occurred at the same time as the onset of the LRP on go trials at about 150 ms after the stimulus. However, while the two LRPs showed a similar initial increase in amplitude, the LRP for no-go trials peaked after about 275 ms. In contrast, the LRP on go-trials continued to increase, reaching a maximum at around the time of the response. Note that the LRP occurred on no-go trials even though the subjects correctly withheld their response on these trials.

Similar data have been obtained by Smid et al. (1992). These investigators used letters of different colours as stimuli and showed that information about the colour feature could be used to activate responses before information about letter identity was used to enable the response decision. Osman et al. (1992) used the spatial position of a visual stimulus (to the left or right of fixation) to indicate the required response hand, while stimulus identity (letter or digit) was mapped to the response decision. An LRP on no-go trials indicated that information about the spatial location of the stimulus could be used to prepare responses before information about stimulus identity was available.

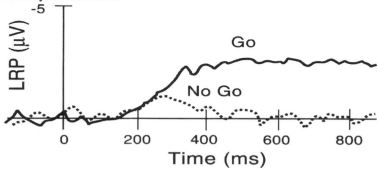

Figure 6. Lateralized readiness potentials elicited on go and no-go trials. Data from Miller & Hackley (1992).

Together, LRP data from both the conflict and the no-go paradigms suggest that responses are prepared covertly on the basis of a partial analysis of the stimulus. Some information about the stimulus is passed to the response system before all information is available. Thus, the LRP data favour the continuous, or partially continuous view of the nature of information transmission over the discrete, all-or-none view.

Using the no-go LRP. The presence and temporal properties of the no-go LRP have been used to make important inferences about the sequence of processing in two domains: speech production and person recognition.

A critical issue in the study of speech production concerns the temporal order of semantic and phonological retrieval processes. As we are about to speak, do we retrieve the semantic properties of a word before we have access to the phonology? Van Turrenout Hagoort, and Brown (1997) answered this question using the go/no-go paradigm imbedded in a picture naming task. Subjects had to prepare to name a picture, and, at the same time, prepare to make a motor response on the basis of the picture's features. In one condition, the response hand was specified by the semantic properties of the picture (is it animate?), while response decision (go or no-go) was specified by the phonological properties of the name of the picture (does the name start with a /s/ or a /b/?). Under these conditions, an LRP was evident on no-go trials, indicating that semantic information was available before phonological information. Importantly, when the assignment of features was reversed, and response hand was mapped to phonology, no LRP was evident on no-go trials. Thus, the data support the inference that we know about the semantics of a word before we know its phonology.

In a similar study, Van Turrenout, Haggort, and Brown (1998) showed that syntactic information is retrieved before phonological information. When syntactic features were mapped to response hand, and phonology (initial phoneme) to response decision, a no-go LRP was observed. These authors also evaluated the temporal properties of the no-go LRP by measuring the interval between LRP onset and its peak. LRP onset occurs when syntactic processing influences the motor system and peaks when sufficient phonological information is available for the decision not to respond to be made. Based on the interval measure, it was concluded that it takes about 40 ms to access phonological information after syntactic information has been retrieved.

The same kind of approach was used by Abdel-Rahman, Sommer, & Schweinberger (2000) to evaluate different models of person recognition following exposure to the person's face. The issue here is why it is more difficult to recall a person's name than other kinds of (semantic) information about that person. According to one view, the serial view, name information can only be recalled after semantic information has been retrieved; according to another view, the parallel view, both kinds of information can be retrieved in parallel, but retrieval of name information usually takes longer. In the Abdel-Rahman et al. study (Experiment 2), semantic attributes of pictures of well-known politicians (their nationality or political party) were mapped to response hand, while phonological attributes of the politicians' names (the initial phoneme) were mapped to response decision. Importantly, information about political party was more difficult to extract than information about nationality. The critical question concerned the presence of a no-go LRP. If information about the features of the person's name is available only after semantic information has been extracted, then there should be a no-go LRP regardless of how long it takes to process the semantic information. Response decision should always follow response choice. On the other hand, if both name and semantic features are processed in parallel, then the presence of a no-go LRP should depend on the relative difficulty of extracting the semantic and name information. Response decision may even precede response choice when semantic information is difficult to retrieve. In fact, while a no-go LRP was observed when response choice depended on the easier semantic attribute (nationality), the no-go LRP was absent when choice depended on the more difficult attribute (political party). Thus, the LRP data supported the parallel view of person recognition.

Stopping

Being able to stop doing something we have started is an important aspect of human executive control. In the laboratory, it has been studied using the stop-signal paradigm (Logan and Cowan, 1984). Subjects are required to perform a reaction time task, but on some trials a stop signal is presented which instructs subjects to withhold their response. The stop-signal is presented at varying delays from the imperative stimulus. As a result, subjects are more or less successful at stopping.

The LRP has been used in this paradigm to evaluate the mechanisms that are responsible for this kind of control. The critical observation concerns the behaviour of the LRP on trials when the subjects succeed in stopping. If there is a threshold level of the LRP at which responses are normally released (c.f. Gratton et al., 1988), then one would expect that the LRP would be below this level on successfully stopped trials. However, De Jong, Coles, Logan, & Gratton (1990) found that the level exceeded that observed on trials when subjects executed a motor response. On the basis of this observation, they reasoned that responses must have been inhibited by a mechanism that is downstream from the system in the motor cortex that generates the LRP.

De Jong, Coles, & Logan (1995) evaluated this idea further in a second series of studies. Based on the work of Bullock and Grossberg (1988), they proposed that two inhibitory mechanisms were available to the subjects in these kinds of tasks. If no further actions were required of the subject after stopping, then a peripheral, inhibit all, mechanism could be used. This mechanism would operate after the central motor command was issued and was presumably used by subjects in the original experiment. Alternatively, subjects could use a more central mechanism, inhibiting the specific response before the central motor command was initiated. This mechanism would be used if some additional action was required after stopping. De Jong et al. predicted that the peripheral mechanism could be used in the standard choice reaction time task, but that the central mechanism would have to be used if stopping had to be immediately followed by another motor response. The critical test of these predictions was provided by the LRP data. If the peripheral mechanism was used, then, as before, the level of the LRP on successfully stopped trials should exceed that usually associated with a motor response; but if the central mechanism was used, then the LRP should be consistently below this threshold level. In fact, the data supported the predictions and, more generally, provided evidence for the existence of two different kinds of inhibitory mechanisms for stopping motor actions.

CAVEATS

Misuse

We have argued that, under appropriate circumstances, the LRP can be used to assess the *preferential preparatory activity* associated with left or right (usually hand) responses. Because of the logic behind the derivation, particularly the need to eliminate lateralized activity that is not associated with movement, it is critical that experimental conditions are not selectively associated with a particular response side. It must be the case that left and right sides are used equally frequently as responses for each experimental condition. Thus, studies that purport to measure the LRP associated with a particular response side fail to appreciate the logic of the derivation.

A second kind of misuse is evident when investigators employ a task involving more than just left- and right-hand responses. Such a task was used by Goodin, Aminoff and Ortiz (1993) who required subjects to respond with either left- or right-hands or with both hands following presentation of one of three tone stimuli which varied in pitch. As Gehring and Coles (1994) pointed out, interpretation of the LRP in this task is impossible. This is because preferential activation of the three possible responses cannot be equally reflected in an unique pattern of the LRP. In particular, a flat LRP might reflect no activation of either response or equal activation of both responses, as would be expected in the bimanual response condition. In other words, the LRP cannot be diagnostic of the differential activation of the three possible responses.

Event-related Lateralization and Artefacts

It is often assumed that applying the double subtraction or averaging method for deriving LRPs from averaged ERP waveforms guarantees that all response-unrelated brain activity is eliminated, and that the resulting LRP waveforms will always be a valid index of response-related processing. Unfortunately, there are circumstances when these assumptions do not hold. In experiments where responses are mapped to visual stimuli presented in the left or right visual field, and response side is determined by stimulus position, lateralizations of the RP related to unimanual response preparation will be confounded with lateralized brain activity related to sensory-perceptual asymmetries. Visual stimuli presented to the left and right of fixation elicit systematically different visual ERPs over ipsilateral and contralateral posterior regions. To illustrate this fact, Figure 7 (left) shows ERPs elicited in response to visual stimuli in the left or right visual field at occipital electrodes ipsilateral and contralateral to the visual field of stimulation. The P1 component peaks about 20 ms earlier over the contralateral occipital cortex, and the amplitude of the contralateral N1 component is considerably larger than the ipsilateral N1. These asymmetries reflect the functional organisation of the visual pathways, as information from the left and right hemifield is received by the contralateral occipital cortex.

Such sensory asymmetries can be propagated by volume conduction to more anterior locations and thus will be picked up by electrodes located over motor cortex. This fact compromises the interpretation of the LRP in experiments where stimulus and response sides are confounded. Figure 7 (right) illustrates this problem. It shows double subtraction waveforms for C3'/C4' (solid line) and for lateral occipital electrodes (dashed line) obtained under conditions where response side (left hand vs. right hand) was determined by the side of a visual stimulus (left vs. right of fixation). Early modulations in the occipital difference wave result from latency differences of the P1 component and from amplitude differences of the N1 component elicited at contralateral and ipsilateral occipital sites (Figure 7, left). The LRP recorded within this time interval shows smaller, but closely parallel deflections. Because these deflections are most likely caused by occipital asymmetries that are propagated to lateral central electrodes, they can not be interpreted as evidence for response preparation processes.

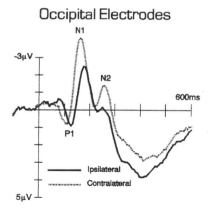

Occipital Electrodes

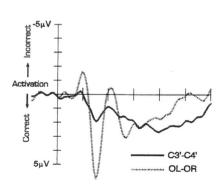

Double Subtraction Waveforms

Figure 7. Left panel: Grand-averaged event-related brain potentials (ERPs) recorded at lateral occipital electrodes (OL, OR) in response to visual stimuli presented ipsilateral (solid line) or contralateral (dashed line) to the recording electrode, and averaged across left and right electrodes. Right panel: LRP waveforms (solid line) and difference waveforms computed with the double subtraction method for lateral occipital electrodes (dashed line) elicited in trials where participants responded with the left hand to left visual stimuli, and with the right hand to right stimuli.

One way to overcome the problem of lateralized sensory asymmetries is to use stimuli and responses that are lateralized along the vertical axis (De Jong et al., 1994; Valle-Inclán, 1996). However, because even ERPs generated in response to visual stimuli presented in the upper and lower visual fields may show distinct lateralization patterns, the assignment of the left and right hand to upper and lower responses needs to be carefully counterbalanced within an experiment.

Another confound between motor and non-motor lateralizations can be encountered when targets are presented together with irrelevant non-target items in bilateral or multi-stimulus arrays. A response-relevant target stimulus in the left or right visual field gives rise to an enlarged posterior negativity over the contralateral hemisphere (Luck & Hillyard, 1994) This 'N2pc' component is assumed to reflect the attentional selection of target stimuli and/or the selective inhibition of distractors. In experiments where the location of targets among non-targets determines the response, the presence of the N2pc can compromise the interpretation of the LRP just as much as the presence of sensory asymmetries elicited by unilateral stimuli. Because LRPs observed over motor cortex are affected by the propagation of posterior attentional asymmetries, they cannot be unequivocally interpreted as an index for preferential response preparation processes (see Eimer, 1998, for further discussion).

Based on the evidence reviewed in this section, it is important to keep in mind that the application of the double subtraction or the averaging methods does not always guarantee that resulting LRP waveforms are a pure measure of motor-related brain activity. In addition to lateralizations caused by sensory asymmetries or by attentional selection processes, other non-motor event-related lateralizations (ERLs) have recently been described, reflecting processes such as the processing of direction information and the orienting of visual attention (Wauschkuhn et al., 1998). Whenever systematic ERLs are observed over non-motor sites, and when such effects are temporally correlated with LRPs recorded over motor cortex, one should be extremely cautious in interpreting LRP measures as indicators of response-related processes.

CONCLUDING REMARKS

More than three decades after Kornhuber and Deecke introduced the Bereitschaftspotential as an electrophysiological measure of response processes, the impact of their discovery on the experimental study of cognitive processes can hardly be overestimated. As demonstrated in the other contributions to this volume, the BP has become an invaluable tool in the study of basic properties of response processes, but also in basic and applied neuropsychological and psychiatric research. In our contribution, we have reviewed the history, the derivation, and current applications of the lateralized part of the Bereitschaftspotential – the lateralized readiness potential (LRP). Introduced in the 1980s as an index of preferential response preparation, the LRP has been widely used as an online measure of response presetting. Because both the BP and the LRP can be measured prior to, and even in the absence of, overt responses, they provide us with a window on covert cognitive processes. The success of the LRP as an electrophysiological tool in the study of cognitive processes is largely due to the fact that LRP measures are immediate reflections of response presetting processes in the motor system. Because of this direct link, the interpretation of LRP amplitudes and latencies is usually much less controversial than the interpretation of ERP components such as the P300. However, as discussed above, the LRP will only be a valid index of preferential response preparation when experimental circumstances are appropriate.

While research on the LRP started with the validation of this measure as an index of preferential preparatory activity in the motor system, application of the LRP is no longer confined to the study of response-related processes. As shown in our review, LRP measures have been successfully applied in many different fields of cognitive psychology and cognitive neuroscience, and have contributed to new insights in areas as diverse as subliminal perception, face recognition, speech production, executive control, or sequence learning.

ACKNOWLEDGMENTS

We dedicate this chapter to the memory of our friend and colleague, Bert Mulder.

REFERENCES

Abdel-Rahman, R., Sommer, W. and Schweinberger, S. R. (1999) Parallel, or sequential access to semantics and proper names of famous faces? *Journal of Cognitive Neuroscience*, Suppl., 84.

Band, G.P.H. and Van Boxtel, G.J.M. (1999) Inhibitory motor control in stop paradigms: Review and reinterpretation of neural mechanisms. *Acta Psychologica* **101**, 179-211.

Bullock, D. and Grossberg, S. (1988) Neural dynamics of planned arm movements: Emerging invariants and speed-accuracy properties during trajectory formation. *Psychological Review* **95**, 49-90.

Coles, M. G. H. (1989) Modern mind-brain reading: Psychophysiology, physiology & cognition. *Psychophysiology* **26**, 251-269

Coles, M. G. H. & Gratton, G. (1986) Cognitive psychophysiology and the study of states and processes. In: G. R. J. Hockey, A. W. K. Gaillard and M. G. H. Coles, (Eds.) *Energetics and human information processing*, pp. 409-424. Dordrecht, The Netherlands: Martinus Nijhof.

Coles, M. G. H., Smid, H. G. O. M., Scheffers, M. K. and Otten, L. J. (1995) Mental chronometry and the study of human information processing. In: M. D. Rugg and M. G. H. Coles, (Eds.) *Electrophysiology of mind: Event-related brain potentials and cognition*, pp. 86-131. Oxford, UK: Oxford University Press

Dehaene, S., Naccache, L., Le Clec'H, G., Koechlin, E., Mueller, M., Dehaene-Lambertz, G., van de Moortele, P.-F. and Le Bihan, D. (1999) Imaging unconscious semantic priming. *Nature* **395**, 597-600.

De Jong, R., Coles, M. G. H., Gratton, G. and Logan, G. L. (1990) In search of the point of no return: The control of response processes. *Journal of Experimental Psychology: Human Perception and Performance* **16**, 164-182.

De Jong, R., Coles, M. G. H. and Logan, G.. L. (1995) Strategies and mechanisms in nonselective and selective inhibitory motor control. *Journal of Experimental Psychology: Human Perception and Performance* **21**, 498-511.

De Jong, R., Liang, C.C. and Lauber, E. (1994) Conditional and unconditional automaticity: A dual-process model of effects of spatial stimulus-response compatibility. *Journal of Experimental Psychology: Human Perception and Performance* **20**,

731-750.

De Jong, R., Wierda, M., Mulder, G. and Mulder, L. J. M. (1988) Use of partial information in responding. *Journal of Experimental Psychology: Human Perception and Performance* 14, 682-692.

Donders, F. C. (1868/1969) On the speed of mental processes. Translation by W. G. Koster. In W. G. Koster, (Ed.) *Attention and Performance II*, pp. 412-431. Amsterdam: North-Holland.

Eimer, M. (1995) Stimulus-response compatibility and automatic response activation: Evidence from psychophysiological studies. *Journal of Experimental Psychology: Human Perception and Performance* 21, 837-854.

Eimer, M. (1998) The Lateralized Readiness Potential as an on-line measure of selective response activation. *Behavior Research Methods, Instruments, and Computers* 30, 146-156.

Eimer, M. (1999) Facilitatory and inhibitory effects of masked prime stimuli on motor activation and behavioral performance. *Acta Psychologica* 101, 293-313.

Eimer, M., Goschke, T., Schlaghecken, F. and Stürmer, B. (1996) Explicit and implicit learning of event sequences: Evidence from event-related brain potentials. *Journal of Experimental Psychology: Learning, Memory, and Cognition* 22, 1-18.

Eimer, M. and Schlaghecken, F. (1998) Effects of masked stimuli on motor activation: Behavioral and electrophysiological evidence. *Journal of Experimental Psychology: Human Perception and Performance* 24, 1737-1747.

Eriksen, C. W. and Schultz, D. W. (1979) Information processing in visual search: A continuous flow conception and experimental results. *Perception and Psychophysics* 25, 581-591.

Fabiani, M., Gratton, G. and Coles, M. G. H. (2000) Event-related brain potentials: Methods, theory and application. In: J. T. Cacioppo, L. G. Tassinary and G. Berntson, (Eds.) *Handbook of Psychophysiology*, pp. 53-84. Cambridge, UK: Cambridge University Press.

Fehrer, E. and Raab, D. (1962) Reaction time to stimuli masked by metacontrast. *Journal of Experimental Psychology* 63, 143-147.

Gehring, W. J. and Coles, M. G. H. (1994) Readiness potential. *Neurology* 44, 2212-2213.

Gehring, W. J., Gratton, G., Coles, M. G. H. and Donchin, E. (1992) Probability effects on stimulus evaluation and response processes. *Journal of Experimental Psychology: Human Perception and Performance* 18, 198-216.

Goodin, D. S., Aminoff, M. J. and Ortiz, T. A. (1993) Expectancy and response strategy to sensory stimuli. *Neurology* 43, 2139-2142.

Gratton, G., Bosco, C.M., Kramer, A.F., Coles, M.G.H., Wickens, C.D. and Donchin, E. (1990) Event-related brain potentials as indices of information extraction and response priming. *Electroencephalography and Clinical Neurophysiology* 75, 419-432.

Gratton, G., Coles, M. G. H., Sirevaag, E. J., Eriksen, C. W. and Donchin, E. (1988) Pre- and post-stimulus activation of response channels: A psychophysiological analysis. *Journal of Experimental Psychology: Human Perception and Performance* 14, 331-344.

Hackley, S.A. and Miller,J. (1995) Response complexity and precue interval effects on the lateralized readiness potential. *Psychophysiology* 32, 230-241.

Kornblum, S., Hasbroucq, T. and Osman, A. (1990) Dimensional overlap: Cognitive basis for stimulus-response compatibility - a model and taxonomy. *Psychological Review* 97, 253-270.

Kornhuber, H. H. and Deecke, L. (1965) Hirnpotentialänderungen bei Wilkürbewegungen und passiven Bewegungen des Menschen: Bereitschaftspotential und reafferente Potentiale. *Pflügers Archiv* 284, 1-17.

Kutas, M. and Donchin, E. (1980) Preparation to respond as manifested by movement-related brain potentials. *Brain Research* 202, 95-115.

Leuthold, H. and Kopp, B. (1998) Mechanisms of priming by masked stimuli: Inferences from event-related brain potentials. *Psychological Science* 9, 263-269.

Leuthold, H., Sommer, W. and Ulrich, R. (1996) Partial advance information and response preparation: Inferences from the Lateralized Readiness Potential. *Journal of Experimental Psychology: General* 125, 307-323.

Logan, G. D. and Cowan, W. B. (1984) On the ability to inhibit thought and action: a theory of an act of control. *Psychological Review* 91, 295-327.

Luck, S.J. and Hillyard, S. A. (1994) Electrophysiological correlates of feature analysis during visual search. *Psychophysiology* 31, 291-308.

McClelland, J. L. (1979) On the time relations of mental processes: A framework for analyzing processes in cascade. *Psychological Review* 86, 287-330.

Miller, J. (1988) Discrete and continuous models of human information processing: Theoretical distinctions and empirical results. *Acta Psychologica* 67, 191-257.

Miller, J. O. and Hackley, S. A. (1992) Electrophysiological evidence for temporal overlap among contingent mental processes. *Journal of Experimental Psychology: General* 121, 195-209.

Miller, J.O., Patterson, T. and Ulrich, R. (1998) A jackknife-based method for measuring LRP onset latency differences. *Psychophysiology* 35, 99-115.

Miller, J. and Ulrich, R. (1998) Locus of the effect of the number of alternatives: Evidence from the Lateralized Readiness Potential. *Journal of Experimental Psychology: Human Perception and Performance* 24, 1215-1231.

Nissen, M. J. and Bullemer, P. (1987) Attentional requirements of learning: Evidence from performance measures. *Cognitive Psychology* 19, 1-32.

Osman, A., Bashore, T. R., Coles, M. G. H., Donchin, E. and Meyer, D. E. (1992) On the transmission of partial information: Inferences from movement related brain potentials. *Journal of Experimental Psychology: Human Perception and Performance* 18, 217-232.

Osman, A. and Moore, C.M. (1993) The locus of dual-task interference: Psychological refractory effects on movement-related brain potentials. *Journal of Experimental Psychology: Human Perception and Performance* **19**, 1292-1312.

Posner, M. I. (1978) *Chronometric explorations of mind*. Hillsdale, N.J.: Erlbaum.

Praamstra P., Plat E. M, Meyer A.S and Horstink M. W. I. M. (1999) Motor cortex activation in Parkinson's disease: Dissociation of electrocortical and peripheral measures of response generation. *Movement Disorders* **14**, 790-799.

Ray, W.J., Slobounov, D., Mordkoff, J.T., Johnston, J. and Simon, R.F. (2000) Rate of force development and the lateralized readiness potential. *Psychophysiology* **37**, 757-765.

Requin, J. (1985) Looking forward to moving soon: Ante factum selective processes in motor control. In: M. I. Posner and O. Marin, (Eds.) *Attention and Performance V*, pp. 147-167. New York: Academic Press

Requin, J. and Riehle, A. (1995) Neural correlates of partial information transmission of sensorimotor information in the cerebral cortex. *Acta Psychologica* **90**, 81-95.

Rugg, M. D. and Coles, M. G. H. (1995) The ERP and cognitive psychology: Conceptual issues. In: M. D. Rugg and M. G. H. Coles, (Eds.) *Electrophysiology of mind: Event-related brain potentials and cognition*, pp. 27-39. Oxford, UK: Oxford University Press

Rüsseler, J. and Rösler, F. (2000) Implicit and explicit learning of event sequences: Evidence for distinct coding of perceptual and motor representations. *Acta Psychologica* **104**, 45-67.

Schwarzenau, P., Falkenstein, M., Hoormann, J. and Hohnsbein, J. (1998) A new method for the estimation of the onset of the lateralized readiness potential (LRP*). Behavior Research Methods, Instruments, and Computers* **30**, 146-156.

Simon, J.R. (1969) Reactions towards the source of stimulation. *Journal of Experimental Psychology* **81**, 174-176.

Smid, H. G. O. M., Lamain, W., Hogeboom, M. M., Mulder, G. and Mulder, L. J. M. (1991) Psychophysiological evidence for continuous information transmission between visual search and response processes. *Journal of Experimental Psychology: Human Perception and Performance* **17**, 696-714.

Smid, H. G. O. M., Mulder, G. and Mulder, L. J. M. (1987) The continuous flow model revisited: Preceptual and central motor aspects. In: R. Johnson, Jr., J. W. Rohrbaugh and R. Parasuraman, (Eds.) *Current trends in event-related potential research (EEG Suppl. 40)*, pp. 270-278. Amsterdam: Elsevier.

Smid, H. G. O. M., Mulder, G., Mulder, L. J. M. and Brands, G. J. (1992) A psychophysiological study of the use of partial information in stimulus-response translation. *Journal of Experimental Psychology: Human Perception and Performance* **18**, 1101-1119.

Sommer, W., Leuthold, H. and Ulrich, R. (1994) The lateralized readiness potential preceding brief isometric force pulses of different peak force and rate of force production. *Psychophysiology* **31**, 503-512.

Sternberg, S. (1969) The discovery of processing stages: Extensions of Donders' method. In: W. G. Koster, (Ed.) *Attention and performance II*, pp. 276-315. Amsterdam: North-Holland.

Sutton, S., Braren, M., Zubin, J. and John, E. R. (1965) Evoked potential correlates of stimulus uncertainty. *Science* **150**, 1187-1188.

Ulrich, R., Leuthold, H. and Sommer, W. (1998) Motor programming of response force and movement direction. *Psychophysiology* **35**, 721-728.

Valle-Inclán, F. (1996) The locus of interference in the Simon effect: an ERP study. *Biological Psychology* **43**, 147-162.

Valle-Inclan, F., Hackley, S.A., de Labra, C. (in press) Automatic response activation and attention shifts in the Simon effect. In: W. Prinz & B. Hommel, (Eds.) *Attention and Performance XIX*.

Valle-Inclán, F. and Redondo, M. (1998) On the automaticity of ipsilateral response activation in the Simon effect. *Psychophysiology* **35**, 366-371.

Van Turrenout, M., Hagoort, P. and Brown, C. M. (1997) Electrophysiological evidence on the time course of semantic and phonological processes in speech production. *Journal of Experimental Psychology: Learning, Memory, and Cognition* **23**, 787-806.

Van Turrenout, M., Hagoort, P. and Brown, C. M. (1998) Brain activity during speaking: From syntax to phonology in 40 ms. *Science* **280**, 572-574.

Vaughan, H. G., Jr., Costa, L. D. and Ritter, W. (1968) Topography of the human motor potential. *Electroencephalography and Clinical Neurophysiology* **25**, 1-10.

Verleger, R., Vollmer, C., Wauschkuhn, B., van der Lubbe, R.H.J. and Wascher, E. (2000) Dimensional overlap between arrows as cueing stimuli and responses? Evidence from contra-ipsilateral differences in EEG potentials. *Cognitive Brain Research* **10**, 99-109.

Wallace, R.J. (1971) S-R compatibility and the idea of a response code. *Journal of Experimental Psychology* **88**, 354-360.

Walter, W. G., Cooper, R., Aldridge, V. J., McCallum, W. C. and Winter, A. L. (1964) Contingent negative variation: A electrical sign of sensorimotor association and expectancy in the human brain. *Nature* **203**, 380-384.

Wauschkuhn, B., Verleger, R., Wascher, E., Klostermann, W., Heide, W., Burk, M. and Kömpf, D. (1998) Lateralized cortical activity for shifts of visuospatial attention and initiating saccades. *Journal of Neurophysiology* **80**, 2900-2910.

MOVEMENT SELECTION, PREPARATION, AND THE DECISION TO ACT: NEUROPHYSIOLOGICAL STUDIES IN NONHUMAN PRIMATES

Steven P. Wise

Section on Neurophysiology
Laboratory of Systems Neuroscience
National Institute of Mental Health
Bethesda, Maryland 20892-4401, USA

OVERVIEW

The selection and preparation of movement, as well as the decision to act, depend on the accumulation and maintenance of information within distributed neural networks. These networks perform several related functions, including: (1) identifying sensory inputs that might be targets of movement; (2) evaluating the biological impact of moving to those potential targets; (3) selecting the targets of action from among these possibilities, along with the movements needed to achieve those goals; (4) maintaining those intentions when action must be deferred; and (5) determining when to act. Neurons in the cerebral cortex of monkeys contribute to these networks, and some have discharge patterns resembling the late phase of the Bereitschaftspotential (BP) or the analogous phase of the Contingent Negative Variation (CNV). Many neurons show a steady build-up in activity over a few hundred milliseconds prior to movement, a property I will call *ramp-to-threshold activity* when it contributes to the selection of movement, or *anticipatory activity* when it contributes to *motor readiness*, an increased but nonspecific tendency to act. Perhaps this latter function comes closest to a correlate of the late BP at the single-cell level. Other cortical neurons have sustained changes in discharge rate that resemble the earlier phases of the CNV, termed *delay period-* or *set-related activity*, which can persist for several seconds. Set-related neurons participate in movement preparation, especially during intervals between movement selection and the decision to act.

HISTORICAL INTRODUCTION

Since its discovery by Kornhuber and Deecke (1965), the BP has allowed a glimpse inside the human brain during the choice and initiation of movement. Contemporaneously, Evarts (1965) developed a method for monitoring the activity of individual neurons in a monkey's brain during the choice and initiation of movement. The prospects for relating these two techniques, and thus determining both the source and function of the BP, must have appeared very bright indeed in the 1960s and 1970s.

Evarts and Tanji (1974) made some initial progress when they presented a monkey with either a red light, which instructed the monkey to pull a handle, or a green light, which meant to push the handle. The monkeys had to withhold movement until a subsequent time, when the handle moved to trigger the instructed movement. Cells in the primary motor cortex (M1) began discharging shortly after the colored light appeared and continued to do so until movement. The similarity between this pattern of neuronal activity and both the BP and the CNV was immediately apparent. Indeed, in their first full-length report about this phenomenon, Tanji and Evarts (1976) devoted most of their discussion to these topics:

> There are a number of features in common between these observations on single-unit discharge in monkey motor cortex and observations obtained by other investigators on slow potentials recorded by means of scalp potentials in man. These slow potentials include the readiness potential (Bereitschaftspotential) of Kornhuber and Deecke [1965] and the contingent negative variation (CNV) or expectancy wave of Walter et al. [1964], potentials which occur in the period prior to voluntary movement.

The optimism generated by these and related discoveries found an influential voice in the writings and speeches of Eccles (1982a, b). In writing about intention and the origins of voluntary movement, Eccles concluded that one particular cortical field, the supplementary motor area (SMA), played the "key role in being excited by the mental act of intention and then calling up the appropriate motor programs to give the desired movement". He proposed that each "mental intention would act on the SMA in a specific manner" and that "the SMA has an 'inventory' and the 'addresses' of stored subroutines of all learnt motor programs and so is able to institute the desired movement by its neuronal connectivities". As Eccles (1982a) summarized the evidence in favor of an SMA intention center, he argued that the:

> [e]mpirical evidence was provided by the readiness potential of Kornhuber and associates [1965], which indicated a key role for the supplementary motor area, SMA. It was found by Brinkman and Porter [1979] that in voluntary movement many neurones of the SMA were activated probably up to 200 msec before the pyramidal tract discharge. Then came the . . . investigations of regional cerebral blood flow by Roland and associates [1980] to reveal that there was neuronal activity in the SMA of both sides during a continued series of voluntary movements, and that this even occurred when the movement was being thought of but not executed . . . .

Leading authorities in the field promptly echoed this basic idea (e.g., Porter, 1990), but serious problems had already emerged with each of Eccles' arguments. (1) As discussed in several chapters of this volume, areas other than the SMA appear to contribute to the BP (Neshige et al., 1988), although under certain circumstances SMA may predominate (Ikeda et al., 1992; Jenkins et al., 2000). (2) Virtually all motor areas in the frontal cortex, as well as several areas in the posterior parietal cortex, show increased rates of blood flow during the planning or imagining of movements, not merely the SMA as Roland and his colleagues originally reported (Deiber et al., 1991; Decety et al., 1993; Krams et al., 1998; Jeannerod and Frak, 1999; Lotze et al., 1999; Porro et al., 2000; Ogiso et al., 2000). And (3) the SMA is by no means unique in having neurons activated hundreds of milliseconds prior to movement (e.g., Weinrich and Wise, 1982; Weinrich et al., 1984; Kalaska and Crammond, 1992).

Nevertheless, if one does away with Eccles' focus on the SMA and generalizes his conceptions to a larger, distributed network of frontal and parietal areas, his general ideas about intention have re-emerged in modified form as *accumulator models* of decision and intention. Generally, these models rely on the assumption that cells in certain areas of cortex represent some action or, as Eccles would have it, "the addresses of stored [motor] subroutines." Inputs — processed by the visual, auditory and somatosensory areas of the occipital, temporal and parietal lobes — then excite neurons representing those actions until they accumulate sufficient information to generate a decision, which corresponds to the intention to make a particular movement. These models generally involve movements guided by vision to spatial targets. However, similar models have been developed for movements guided by symbolic visual information (Fagg and Arbib, 1992), and the guidance of action by other sensory modalities probably follows the same principles. Thus, accumulator models are likely to have a high degree of generality. So, even if the SMA does not function as *the* intention center, as Eccles supposed, some of his ideas about the neural basis of intention have proven edifying when applied to accumulator models, as implemented by distributed neural networks.

In order to discuss the neurophysiological literature on these issues, I will need to use an explicit terminology for the various activity patterns encountered in the cortex. Many of these terms, however, have more restricted connotations in behavioral neurophysiology than in everyday discourse. This is not such an unusual problem. If one looks up the meaning of *cortex* in the *Webster Collegiate Dictionary*, the first meaning involves "plant bark . . . used medicinally." Yet this confuses no one. Similarly, one of the terms often used in discussions of motor preparation is *set*, which is defined by more words than any other entry in the *Oxford English Dictionary*. Indeed, its definitions there run to 22.5 pages! This fact has led to some despair, but no matter: any terminology will do if the underlying concepts remain clear and consistent.

NOMENCLATURE

Activity Patterns and Experimental Events

For the most part, the terminology I will use depends on two events: a sensory *cue* that guides the selection of action and a subsequent "go" signal. The former is termed an *instruction stimulus*; the latter, a *trigger stimulus* (Fig. 1B). The time between these events is often termed an *instructed delay period*, or more simply, a *delay period*. In many experiments, the instruction and trigger stimuli are combined into a single input, much as they are in a choice reaction-time task (Fig. 1A). *Activity* is defined as the number of action potentials discharged by a neuron per unit time, typically impulses per second. An increase in activity before some temporally predictable event, including either an instruction stimulus or a movement, is termed *anticipatory activity*. An activity increase immediately after an instruction stimulus is called either *signal-related activity* (Fig. 1B) or *ramp-to-threshold activity* (Fig. 1A), depending on the type of experiment in which it is observed. A sustained activity increase that continues for up to several seconds after an instruction stimulus (Fig. 1B) is termed either *delay-period activity* or *set-related activity*, and an increase in activity immediately prior to (and during) movement is termed *movement-related activity*. The same terms apply to decreases in activity.

The general similarity between anticipatory activity and ramp-to-threshold activity, which both involve steady increases in discharge rate, suggests a similar, neural-integrator mechanism for both. This kind of mechanism contrasts with that thought to underlie sustained, delay-period activity. Theoretical work has shown that delay-period activity can be accounted for by the dynamics of recurrently connected neural networks (Zipser, 1991).

It has been suggested that fixed point-attractors determine the most common states of these networks, with each of those states reflecting the retention of an item of information.

Note that both of the experimental designs depicted in Figure 1 involve sensorially guided movements, as opposed to the self-paced movements associated with the BP. In most of these experiments, sensory cues not only instruct the movement, but also determine the time to initiate it. Thus, the observations that I will summarize from

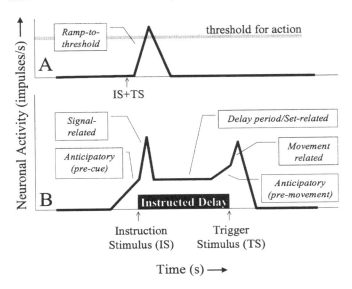

Figure 1. Idealized neuronal activity patterns that occur in many neurophysiological experiments. **A.** Activity during a simple experiment in which a movement target is given (as an instruction stimulus, IS) at the same time as a "go" signal (also known as a trigger stimulus, TS). Usually, the IS and the TS are the same stimulus. Activity in motor networks ramps to the threshold for producing a movement. **B.** Activity for a more complex experiment in which the IS is separated in time from the TS. In that circumstance, several patterns of activity occur, which are shown here as if combined in a single neuron. In the cortex, however, these five activity patterns occur in various combinations for any given neuron. For example, a neuron might have set-related and movement-related activity, but none of the other patterns labeled in part B. In this figure, the instructed delay period is of fixed duration, but it can also vary.

behavioral neurophysiology on nonhuman primates may relate most directly to the CNV rather than to the BP. As we shall see, however, one aspect of these experiments might be more relevant to the BP. After the instruction stimulus appears, monkeys sometimes must self-time the delay period and initiate movement after some minimal time has elapsed (e.g., Kurata and Wise, 1988). The results of those experiments are taken up in the final section of this chapter.

Cortical Areas

To some experts, names such as the motor cortex and the SMA convey a highly specific meaning, connoting strict boundaries and locations; to others, these terms have a more general use, which indicates a general location in the brain. For example, *motor cortex* has a specific sense, which refers to the primary motor cortex (M1), also known as the precentral or Rolandic motor area, the motor strip, and area 4. In its more general sense, the term motor cortex can refer to a group of frontal cortical fields, all of which play a relatively direct role in the cerebral control of movement: relatively direct, that is, by contrast with predominantly sensory areas such as the visual, auditory, and somatosensory cortex. On this view, the motor cortex consists of approximately a dozen distinct cortical fields, many of which are depicted in Figure 2.

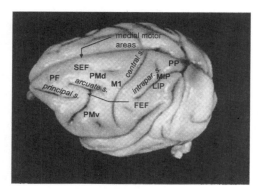

Figure 2. A rhesus monkey brain, showing the approximate locations of certain cortical fields. Abbreviations: FEF, frontal eye field; *intrapar*, intraparietal; LIP, lateral intraparietal area; MIP, medial intraparietal area; PF, prefrontal cortex; PMd, dorsal premotor cortex; PMv, ventral premotor cortex; PP, posterior parietal cortex (which includes LIP, MIP and other areas); *s*, sulcus; SEF, supplementary eye field. Photograph from Welker et al. (2001), http://www.brainmuseum.org.

Included among the motor areas are certain medial premotor areas, such as the SMA, the pre-supplementary motor area (pre-SMA), and a group of motor areas within the cingulate sulcus (not illustrated in Fig. 2). There are also a number of lateral premotor areas, such as the dorsal premotor cortex (PMd) and the ventral premotor cortex (PMv). Some of these areas, such as PMd (Geyer et al., 2000) and PMv (Matelli et al., 1985) might be further subdivided, but that will not be necessary for the purposes of this chapter. In addition, the frontal eye field (FEF) and the supplementary eye field (SEF) appear to play an important role in the control of eye movements. It is important, however, not to overstate the distinction between oculomotor and skeletomotor areas. Experimental work has blurred this distinction for areas such as PMd (Preuss et al., 1996; Fujii et al., 2000), PMv (Fujii et al., 1998), and the SMA (Mitz and Wise, 1987), by showing that they have an oculomotor representation in addition to a skeletomotor one. Interested readers might consult two recent reviews for a detailed discussion of these issues (Tehovnik et al., 2000; Wu et al., 2000).

In addition to these frontal motor areas, parts of the posterior parietal cortex directly interconnect with the premotor areas and play an important role in the cerebral control of movement (see, e.g., Wise et al., 1997; Burnod et al., 1999). Furthermore, the role of prefrontal cortex in the selection and control of motor responses has been increasingly recognized in the past decade (Passingham, 1993).

UNDERSTANDING NEURAL SIGNALS

Nomenclature aside, at least three conceptual problems arise in considering the functional correlates of single-cell activity. The first of these problems involves a shift from the concept of functional centers, as exemplified by Eccles' view of the SMA as an intention center, to one of parallel, distributed neural networks (Zipser and Andersen, 1988). The other two problems are pragmatic. One involves distinguishing sensory from motor signals in the brain. It remains difficult to ascertain when neuronal activity reflects the selection of or preparation for movement, as opposed to the perception or memory of stimuli, places, and events. The other pragmatic problem concerns the large number of spatial variables in neurophysiological experiments. These variables include eye position, the orientation of selective spatial attention, and the target of a reaching movement, among others. These factors typically co-vary. If so, how can experiments control such variables to reach conclusions about the selection and preparation of movement? These three problems will be addressed, in turn.

Neural Centers versus Distributed Networks

Although some textbooks still discuss motor control in terms of neural centers, it has become apparent that, instead, a broadly distributed neural network generates movements. Attempts to show that motor activity begins in some particular center, such as the cerebellum (Thach, 1975), have fallen by the wayside in the light of more recently attained data (Fortier et al., 1993). The once-popular idea that the posterior parietal cortex functions as a center issuing "command signals" (Mountcastle et al., 1975) failed the test of time, as well (see, for example, Kalaska and Crammond, 1992). As noted above, the notion that the SMA was the intention center (Eccles, 1982a, b) fared no better. Even the simpler contention that movement-related activity in the SMA precedes that in M1 has failed to stand up to explicit, critical scrutiny (Chen et al., 1991). Researchers have increasingly recognized the parallel, distributed nature of the neural networks underlying movement generation (Zipser and Andersen, 1988), both in experimental (Alexander and Crutcher, 1990a, b; Crutcher and Alexander, 1990) and theoretical (Fetz and Shupe, 1990; Mel, 1991; Moody and Zipser, 1998) work.

To illustrate this point, let us take as an example the somatotopic organization within M1. Previous ideas about its organization, such as Woolsey's homunculus or the core-surround organization of Murphy and his colleagues (Wong et al., 1977), have proven to be oversimplified (Donoghue and Sanes, 1994). M1's organization respects certain broad somatotopic features, such as the forelimb representation, but little else. Recently, a particularly careful study searched for somatotopic organization within the forelimb representation. It failed to reveal any significant clustering of neurons with movement-related activity related to any given finger (Poliakov and Schieber, 1999). Thus, one can say that M1 lacks an index-finger center. This fact does not, however, prevent relatively independent control of the index finger. Such control arises from the operations of a distributed network, including not only a large portion of the forelimb representation in M1, but also parts of the cerebellum and the basal ganglia.

Sensory versus Motor Signals

In addition to theoretical obstacles, there are pragmatic problems in understanding neuronal activity at the single-cell level. Among these, the differentiation between sensory and motor signals has proven particularly problematic. Two aspects of sensation need to be distinguished for the purposes of this discussion. Some sensory inputs lead relatively directly to movements. This concept is implied in the phrases *sensorimotor integration* and *sensorially guided movement*. Other sensory inputs serve different purposes, including perception, which plays a much less direct role in movement. Clearly, M1, which has corticospinal projections directly to α-motor neurons, plays a relatively direct role in the control of limb movements. Similarly, the FEF (see Fig. 2), which sends a potent projection to the superior colliculus and to other brainstem oculomotor neurons, plays a central role in the control of eye movements. However, neurophysiologists have increasingly found sensory-like signals in these motor networks. Neurons in the intermediate layers of the superior colliculus not only mediate saccadic eye movements, but also have visual responses that cannot be readily accounted for in terms of sensorially guided movement (Horwitz and Newsome, 1999). Similarly, the FEF has neurons that respond to the location of visual stimuli, regardless of whether the monkey uses that information to guide an eye movement (Thompson et al., 1997). As for limb movements, M1 has been reported to show activity reflecting mental rotation (Georgopoulos et al., 1989, 1993; Lurito et al., 1991; Wexler et al., 1998), serial order (Carpenter et al., 1999), vibrotactile signals (Salinas and Romo, 1998) and even "rules" (Zhang et al., 1997). A similar appreciation has developed for the premotor cortex and for the SMA, where nonmotor signals (Weinrich and Wise, 1982; Weinrich et al., 1984; Wise et al., 1986; Boussaoud and Wise, 1993a, b; Graziano and Gross, 1998; Graziano, 1999) and serial-order effects (Clower and Alexander, 1998) have also been reported.

The converse also holds. Even that most sensory of cortical areas, the primary visual cortex, also known as striate cortex, seems to reflect signals other than strictly sensory inputs (Rossi et al., 1996; Sheth et al., 1996; Zipser et al., 1996). It appears that cortical areas cooperate in widely diversified behaviors, rather than restricting their information processing to either sensation or movement.

Confounded Spatial Variables

The discovery of motor preparatory signals in the monkey brain is usually attributed to Evarts and Tanji (1974), based on their recordings from M1. However, the earliest description of preparatory activity may, surprisingly, have come from the prefrontal cortex (Fuster and Alexander, 1971; Kubota and Niki, 1971; Fuster, 1973). Delay-period activity in M1 was interpreted as reflecting motor set, whereas the same kind of activity in prefrontal cortex was interpreted as reflecting sensory short-term memory. In my view, neither of these conclusions should be accepted without explicit experimental examination. The reason for such caution is that a great many confounded variables, especially spatial ones, need to be dissociated experimentally in order to interpret single-cell activity. This principle of experimental design holds for M1, the prefrontal cortex, or any other area. For example, after receiving an instruction stimulus (Fig. 1B), a monkey may be transforming that sensory information into a spatial goal, calling up a spatial limb trajectory, reorienting spatial attention to the target of movement, or shifting gaze to it. Distinguishing among these alternatives has proven to be a significant challenge for behavioral neurophysiology.

The problems discussed above should give the reader some impression of the complexities encountered in interpreting neurophysiological results at the single-cell level. In the following sections, I will apply those considerations to the selection of targets and movements, motor preparation, motor readiness, and the decision to act, in turn.

TARGETING DECISION AND MOVEMENT SELECTION

Accumulator Models

The choice of a target for movement appears to arise from competition among neural networks, arranged largely in parallel. According to some of the accumulator models mentioned above, each of these networks represents a potential action, e.g., a saccadic eye movement. They are arranged in a winner-take-all manner, which reflects the fact that, for example, it is not possible to make a saccade to the left and to the right at the same time. These networks accumulate information about stimuli that serve as potential movement targets. As they do so, their activity increases and they approach a threshold for movement initiation. At the single-cell level, the accumulation of targeting information is reflected in ramp-to-threshold activity (Fig. 1A). Accumulator models have been used in analyzing both perception and the selection of movement, although the latter will be emphasized here.

Accumulator networks perform several related functions. They discriminate objects or features in the sensory environment that might serve as targets of movement (Thompson et al., 1996, 1997; Thompson and Schall, 1999; Everling and Munoz, 2000; Gold and Shadlen, 2000). They also assign a salience, or biological importance, to a representational map of potential movement targets. This *salience map* reflects what one can expect to gain, based on experience, from choosing each potential target. Motor networks select, as the target of movement, a stimulus that has a relatively high salience (Thompson et al. 1996; Bichot and Schall, 1999; Schall and Thompson, 1999; Schall, 2000). This idea has been demonstrated most elegantly in a posterior parietal field called the lateral intraparietal area (Fig. 2), known as LIP (Platt and Glimcher, 1999), and there are similar data for the prefrontal cortex (Leon and Shadlen, 1999; Hikosaka and Watanabe, 2000; Schultz et al., 2000). In LIP, Platt and Glimcher (1999) compared neuronal activity as a monkey shifted visual fixation from a light spot at point A in its visual field to one at point B, but under various expectations of receiving reward. In one block of trials, the experimenters might reward the monkey for 80% of such fixation shifts, in another block 20%, or some other percentage. They could also vary the amount of reward. Given that the stimuli and movements were identical, they could rule out the possibility that any block-to-block difference in activity resulted from either sensory or motor factors, per se. Attention was also well controlled. They found that neuronal activity reflected reward expectation, a feature of biological salience.

The concept inherent in accumulator models is that sensation precedes the targeting decision and that the targeting decision precedes action. This formulation does not imply, however, that the first process must be complete before the second one can begin. That would imply rigid, serial processing rather than the more realistic idea that several parallel, often competing, processes compute the transformation from sensation to action.

As described up to this point, accumulator models may seem to reflect only passive integration of ascending sensory information, in what has been called a *bottom-up* manner. This view would resemble traditional ideas about sensorimotor integration, which posit transmission of sensory input from receptors to a motor center, sometimes the motor cortex or the cerebellum, which extracts the pertinent information in order to send a command to the motor neurons. However, accumulator models encompass top-down sensory information processing, as well. The central nervous system uses memory, in the form of representations of event probabilities, to bias sensory analysis. Experience dictates that not all inputs are equally likely; indeed, rarely are any two inputs equally likely. In the rare case that interpretations are equally likely, one can see bistable perceptual phenomena such as the well-known Necker cube. However, in ordinary experience such perfect balance rarely occurs. Instead, memory-based interpretations of inputs involve top-down biases that represent hypotheses about the world and what is happening in it (MacKay and Gardiner, 1972). This kind of top-down influence over sensory analysis occurs in parallel with the top-down transmission of motor commands, and both processes are integral to accumulator models. In the motor domain, these biases may take the form of an altered threshold for one movement as opposed to alternatives. Accumulator models posit a steady increase in network activity as its threshold is approached. This leads to what I call ramp-to-threshold activity.

Ramp-to-threshold activity

In two independent series of experiments, from Shadlen's laboratory (Kim and Shadlen, 1999; Gold and Shadlen, 2000) and from Schall's (Schall, 1995; Schall and Thompson, 1999), ramp-to-threshold activity has been examined as sensory information is accumulated. Schall's laboratory exploited a well-known perceptual phenomenon known as *pop out*. This visual effect occurs when there are several objects of given type in the visual field (for example, red squares), along with one object of a different type (for example, a green square).

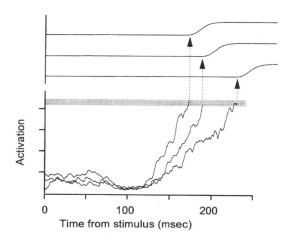

Figure 3. The ramp-to-threshold for reaction time. The activation level of a neuron the frontal eye field is shown for three conditions. In the first condition the activity rises with the largest slope and reaches the threshold for action (shaded horizontal line) fastest. This leads to the fastest reaction time (first upward-pointing arrow). From Schall and Thompson (1999), with permission, from the *Annual Review of Neuroscience*, Volume 22 © 1999 by Annual Reviews, www.AnnualReviews.org.

The green square "pops out" of the field of red squares to serve as a new target for visual fixation. Using this experimental paradigm, Schall and his colleagues have found that the representation of a movement target emerges gradually as neural signals progress through the visual system, from occipital to frontal cortex. Ultimately, the network for one movement, the one that guides an eye movement to fixate the pop-out target, ramps up to its threshold for generating a decision. That signal eventually reaches the muscles to produce the selected movement. Because of the winner-take-all architecture of the network, only one potential movement can be selected, the one that ramps to threshold first (Schall and Thompson, 1999). It appears that the network's threshold is reached when output cells in the FEF ramp up to approximately 100 impulses per second (Hanes and Schall, 1996). If a neuronal population reaches that threshold quickly, the selected movement will begin with a relatively short reaction time. If the network takes longer to ramp up to its threshold, a slower response will result (Fig. 3). This idea gains further support from the effects of cortical stimulation on reaction time in human subjects (Pascual-Leone et al., 1992a, b).

Another experimental paradigm that has been used to explore accumulator models involves the perception of motion from coherently moving light spots (Newsome et al., 1989). If hundreds of light spots move in the same direction at the same speed, people and monkeys perceive coherent motion. If only a proportion of the inputs move coherently, however, and the remainder move randomly, people and monkeys may still perceive the motion if enough light spots move together. In addition, the longer that subjects can accumulate information about the inputs, especially when the amount of coherence is low and the decision about motion is difficult, the better will be their ability to detect the direction of motion. In the coherent-motion paradigm as applied to monkeys, subjects must *report* whether the motion is to the right or to the left, and they typically must do so with an eye movement. The visual information needed to make the required decision is computed primarily by two areas: the medial temporal cortex (MT) and a nearby area called the medial superior temporal cortex (MST). The areas involved in selecting the movement dictated by the decision, i.e. the motor report, are areas such as LIP and the prefrontal cortex. The brain structures most directly involved in executing the movement are the FEF and the superior colliculus.

In work from Shadlen's laboratory (Kim and Shadlen, 1999; Gold and Shadlen, 2000), monkeys are presented with stimuli that have such low coherence that they are near the threshold for detecting the motion. This level of input is often called a perceptual threshold, which turns out to be approximately 6% to 7% coherence. The monkeys can observe the stimuli for as long as necessary to make a decision, and they initiate their report without further sensory input. Near the perceptual threshold of 6-7% coherence, the monkeys take up to a second to respond, but with as much as 35% coherence, they respond in 300 ms to 400 ms. In accord with accumulator models, strong stimulus coherence leads to a rapid increase in neuronal activity, whereas with weak coherence the ramp-to-threshold takes longer. It appears that the network's threshold for action is reached when LIP cells ramp up to approximately 65 impulses per second. For example, a neuron in LIP might show activity increases when leftward spot movement occurs. For 6% to 7% stimulus coherence, the cell's activity reaches 65 impulses per second relatively slowly, leading to reaction times on the order of one second. When stimulus coherence is 35%, the cells ramp up to this threshold within a few hundred milliseconds, corresponding to a rapid reaction time.

Similar results have been obtained by comparing neuronal activity during *antisaccades* with that during *prosaccades*. A prosaccade is an experimental term for the most typical of saccadic eye movements, those made to fixate a visible target. An antisaccade is the same eye movement, but one made directly away from a visible target rather than toward it. Ramp-to-threshold activity occurs for prosaccades and antisaccades in both the superior colliculus (Everling et al., 1999) and the FEF (Everling and Munoz, 2000). In both of these oculomotor networks, prosaccades are associated with a more rapid increase in ramp-to-threshold activity, compared to antisaccades, and this faster ramping-to-threshold correlates with faster reaction times.

An important feature of accumulator models is that there is no reason that the threshold for all potential responses must be identical. If all of the responses have roughly equal reward history, and therefore comparable salience, then equal thresholds are likely. When different potential targets of action become associated with different likelihood of reward, however, such as in the experiments described above by Platt and Glimcher (1999), the threshold for more valuable targets can be lowered, which results in a response bias based on previous success. (Indeed, the threshold can be modulated, in principle, not only selectively as just described, but generally. A generally lowered threshold would reflect an urgency to act, a possibility taken up below in the section on motor readiness.)

Suprathreshold activity is detected by the generation of a response, and all of the results summarized above have been based on such overt movements. However, the status of subthreshold decisions can also be detected experimentally, and the results are also consistent with accumulator models. For example, intracortical microstimulation in the FEF evokes eye movements that deviate systematically in the direction of the subthreshold decision (Gold and Shadlen, 2000). Further, subthreshold decisions can be countermanded, or vetoed, by subsequent decisions. Schall and his colleagues studied this phenomenon in an experiment with a "stop" signal. In their experiment, every trial

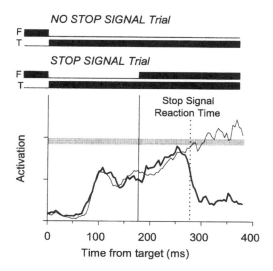

Figure 4. Veto of an instructed saccade. After the instruction stimulus appears at time 0, the activity later builds up. At the time of the solid vertical line (at 175 ms), a veto signal appears on some trials. The time that it takes for the monkey to withhold the saccade is termed the stop-signal reaction time (dotted vertical line). Note that the signals for regular trials (thin solid line) and for stop-signal trials (thick solid line) diverge shortly before the stop-signal reaction time. The gray horizontal bar shows the approximate threshold for action. Modified from Schall and Thompson (1999), with permission, from the *Annual Review of Neuroscience*, Volume 22 © 1999 by Annual Reviews, www.AnnualReviews.org.

began with the fixation point coming on, as indicated by the thick black bar extending horizontally from "F" in the top row of Figure 4. In trials of the "regular" kind, a target ("T" in Fig. 4) appeared somewhere other than the fixation point at the same time as the fixation point was extinguished. The monkey's task was to make a saccadic eye movement to fixate the new target. By contrast, the stop-signal trials had another sensory event after the new target appeared. As indicated by the third row in Figure 4, the fixation point might reappear at some variable lag time after it had been extinguished on a given trial. This event was the signal to veto the saccade. If the stop signal appeared in time, i.e., within the stop-signal reaction time, activity in the FEF decreased before the network attained threshold and the movement did not occur (Fig. 4).

As a general statement, accumulator models suggest that neural networks involved in the selection of movement test hypotheses about the state of the world. These models appear to be widely applicable, explaining findings in experiments involving pop-out targets of movement, perception of coherent motion, and the production of prosaccades versus antisaccades. Accumulator models also accord well with classical reaction-time theories (Luce, 1986), in which variation in both perceptual accuracy and reaction time depends upon stochastic variability in a neural integrator. This similarity raises two points. Because those theories apply to both oculomotor and skeletomotor action, it seems likely that the results for FEF and eye movements, outlined above, apply to premotor cortex and skeletomotor behavior, as well. In addition, because those theories involve integration in the mathematical sense, accumulation of information appears to be the critical characteristic of these models: the rate of activity growth in the networks representing the "correct" decision depends on both the strength of the signal and the time available (Gold and Shadlen, 2000). In this sense, the targeting decision reflects the accumulation of evidence bearing on an intention to make a particular movement.

MOVEMENT PREPARATION

In situations when the targeting decision has already been made, but the selected movement must be deferred until some subsequent time, neural networks maintain a state of preparation for that movement. As a consequence of that state, many neurons in the cortex and elsewhere show activity during the delay period (Fig. 1B), which is characterized variously as tonic, sustained, or persistent and named either set-related, preparatory, or delay-period activity. Readers familiar with the electroencephalographic (EEG) literature will recognize the resemblance of set-related activity to the CNV, as well as the experimental questions asked about it. A broad series of studies have shown that delay-period activity in premotor cortex reflects the preparation to make a specific movement (Wise et al., 1997). However, delay-period activity is by no means unique to premotor cortex. M1 (Evarts and Tanji, 1974; Tanji and Evarts, 1976) and the prefrontal cortex (Fuster and Alexander, 1971; Kubota and Niki, 1971; Fuster, 1973) were the first areas in which delay-period activity was described. Similar observations have been made in the SMA (Tanji et al., 1980), somatosensory cortex (Zhou and Fuster, 1996), posterior parietal cortex (Crammond and Kalaska, 1989), auditory cortex (Vaadia et al., 1982), and visual cortex (Mikami and Kubota, 1980), not to mention subcortical structures such as the striatum (Alexander, 1987) and the cerebellum (Mushiake and Strick, 1995).

So if delay-period activity is ubiquitous, or nearly so, how can one know when it reflects the preparation for movement? In the striatum and prefrontal cortex, for example, delay period activity can depend upon whether a movement will be associated with reward (Schultz et al., 2000). In the experiment that revealed this property, a monkey was shown three pictures: instruction stimulus A was associated with a hand movement that was rewarded; instruction stimulus B with a hand movement to the same target, but which was never rewarded; and instruction stimulus C with withholding of movement, which was rewarded in the same way as the affirmative motor response to stimulus A. Certain cells in the striatum show delay-period activity before rewarded movements but not before similar unrewarded movements. Not only do cells in both the striatum and the prefrontal cortex reflect the expectation of reward, but the delay-period activity reflects their relative value. That is, some cells have delay-period activity that reflects a preferred reward. They show delay-period activity when, for example, an apple is the preferred choice of two potential rewards, but not when the apple is less preferred (Tremblay and Schultz, 2000; see also Hikosaka and Watanabe, 2000). These examples show that delay-period activity does not necessarily reflect motor preparation, at least not in a relatively direct sense. So how can a role in motor preparation be demonstrated?

Premotor Cortex

Fortunately, methods have been developed to address this question (e.g., di Pellegrino and Wise, 1993). In the experiment of di Pellegrino and Wise (Fig. 5A), the monkey was required to attend to a place, which may or may not have been the target of a hand movement on any given trial. In addition, the monkey was required to continue fixating the center of a display. This imposed behavior insured that the location of the visuospatial cues would be known for all relevant coordinate frames, including retinal coordinates. The monkey had to attend to the cued place because, at some unpredictable time in the future, a cue that triggered a reaction-time response would appear at that same place. Between the initial cue and its reappearance as a trigger stimulus, an unpredictable number of intervening distractor stimuli might appear, which prevented the monkey from responding after counting some number of sensory events. The monkey was also precluded from using timing cues. The central feature of the experiment was that there were two distinct response rules, which occurred in alternating blocks of trials. According to one rule, the monkey was to reach to the place that had been cued and to which it attended (Rule 1 in Fig. 5A). According to the second rule, the monkey had to reach to a fixed, repetitive location, regardless of where it was attending (Rule 2 in Fig. 5A). These two rules were termed the compatible condition and the incompatible condition, respectively. The only feedback that the monkey received to indicate which of the two rules prevailed at any given time was a reward for correct performance. With this level of control over the sensory and attentional spatial variables, di Pellegrino and Wise could determine whether delay-period activity reflected those variables, on the one

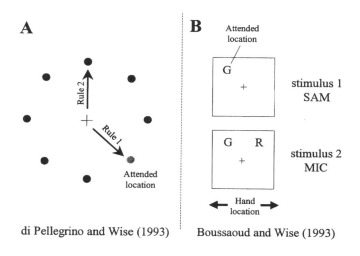

di Pellegrino and Wise (1993)   Boussaoud and Wise (1993)

Figure 5. **A.** Experiment of di Pellegrino and Wise (1993). **B.** Experiment of Boussaoud and Wise (1993a, b). These studies showed that delay-period activity in premotor cortex reflects motor preparation rather than exclusively spatial attention or sensory inputs. In both parts of the figure, the plus sign indicates the fixation point. Abbreviations: G, green stimulus; R, red stimulus.

hand, or motor preparation, on the other. If the delay-period activity was the same for the two conditions, they could conclude that either sensory factors, attentional factors, or both might account for neuronal activity. If there was a significant difference in activity between conditions, then only the motor factors could account for that difference. For most premotor cortex neurons, motor factors accounted for a significant amount of the delay-period activity. From this finding di Pellegrino and Wise concluded that those signals function in the selection and preparation of movement. Relatively few neurons in the prefrontal cortex had this characteristic, and so their activity might reflect factors common to the two rule conditions, such as spatial attention.

A different experimental design was used to examine this issue from another perspective (Boussaoud and Wise, 1993a, b). In that experiment, a spatial cue, consisting of a colored square (red or green), was presented away from the fixation point (Fig. 5B, top). Later, one colored square (red or green) would appear at the same place, and a differently colored square might (or might not) appear somewhere else in the visual field. Thus, after stimulus 1, termed the spatial/attentional mnemonic cue (SAM, Fig. 5B, top), there were either one or two cues that appeared later, as stimulus 2 (Fig. 5B, bottom). If there were two cues comprising stimulus 2, they appeared simultaneously. These later stimuli provided a motor instructional cue (MIC), which depended on the color of the stimulus. If there was only one stimulus and it was a red square (R in Fig. 5B), the monkey was to press the leftmost of two switches. If there was one green square as a MIC, the monkey was to press the rightmost switch. For trials in which two different colors, red and green, appeared simultaneously (as in Fig. 5B, bottom), the color of the stimulus that appeared at the location of the SAM cue determined the correct response. For the example illustrated in Fig. 5B, this square was green, which instructed a movement to the right switch on that trial. Note an important aspect of the experimental design: the SAM cue was identical in shape, color, and size to one of the two possible MICs, although its color was irrelevant during stimulus 1. Only the location of the SAM cue was useful to the monkey. On occasion, these two cues could appear at the same place in the visual field for a given trial. For an example, imagine the depiction of stimulus 2 in Figure 5B (bottom) but without the R. In this way, Boussaoud and Wise could test whether the delay-period activity in premotor cortex reflected spatial attention or motor preparation. For example, delay-period activity following a red SAM cue but not an identical red MIC would indicate a role in selective spatial attention. Conversely, activity following a red MIC but not a red SAM cue would show that motor preparation was the most important factor. In accord with the conclusions of di Pellegrino and Wise (1993), the activity of most cells in premotor cortex reflected motor preparation.

Posterior Parietal Cortex

The work of Andersen and his colleagues in the posterior parietal cortex demonstrates the elegance with which behavioral neurophysiology can untangle confounded spatial variables (Snyder et al., 2000). They studied the medial parietal area (MIP, see Fig. 2), also known as the parietal reach region (PRR), as well as LIP (Snyder et al., 1997, 1998). One approach was to address the problem by studying sequences of movements (Bracewell et al., 1996). They instructed a monkey to make two movements in a sequence, but delivered both targets before movement could be initiated. If the delay-period activity reflected the sensory inputs or spatial attention, all instruction cues would be encoded. Alternatively, if delay-period activity reflected the next planned target of movement, only the first of the targets would be reflected in the neuronal activity. The latter result was obtained, which suggests that delay-period activity in the parts of the posterior parietal cortex examined primarily reflects motor preparation. A similar result was obtained from premotor cortex (Wise and Mauritz, 1985).

In a separate experiment from the same laboratory, Snyder et al. contrasted attentional versus motor signals by using a similar spatial cue to guide a reaching movement or an eye movement. The monkey began a trial by fixating on a spot of light. Later, if a red spot appeared somewhere else in the monkey's visual field, the monkey needed to reach to it in order to receive a reward. However, if the light spot was green, the monkey had to make a saccadic eye movement to foveate it. If delay-period activity was much the same for arm and eye movements, then the activity could be reasonably ascribed to attentional processes. This conclusion would follow because in both cases the monkey would be expected to reorient spatial attention to the light spot away from the fixation point. However, this finding was not obtained from either LIP, which has neurons primarily related to eye movements, or from MIP (the PRR), which has neurons primarily related to arm movements. Thus, the delay-period activity in these posterior parietal areas appears to reflect motor preparation, which is, again, consistent with results from the premotor cortex. Other investigators have reached a similar conclusion (Eskandar and Assad, 1999).

Although neurons in the posterior parietal cortex have similar properties to those in premotor cortex, neurons in one part, area 5, appear to be less selective, showing delay-period activity whenever a *potential* movement target is presented (Kalaska and Crammond, 1995). Kalaska and Crammond studied a Go/No-go task in which the color of a visuospatial cue indicated whether a hand movement was to be made to that location on a given trial or, alternatively, withheld. If the instruction stimulus was green, the monkey was to make a movement to it; if red, the monkey was to withhold movement. In premotor cortex, the cells showed delay-period activity only when the monkey was instructed to make a movement toward the target. This property is exactly what would be expected of a neural signal that reflects motor preparation. In area 5, however, the cells showed delay-period activity for a location that might be a target on a subsequent trial, but not on the present one. Thus, a proportion of cells in area 5 participate in networks that identify potential targets of action, perhaps involving the salience maps mentioned above.

A separate question about delay-period activity concerns the spatial coordinate frame in which the preparation to act is encoded. Andersen and his colleagues have found that cells in both LIP and MIP encode both saccades and reaching movements in eye-centered coordinates, even for acoustic targets. They have also built neural-network models that demonstrate the feasibility of this coding scheme. The advantage of eye-centered coordinates is not entirely clear for hand movements, but three general ideas have been advanced. Having eye, head and hand movements computed in a common coordinate scheme has the advantage of parsimony. Because the visual sense dominates in primates, it is the most accurate spatial sense, and may predominate for that reason. Finally, the system may adapt to errors most efficiently by registering them in a common reference frame. There is supporting evidence from neuroimaging studies: when vision is distorted by prisms, posterior parietal cortex becomes more "activated" during motor adaptation (Clower et al., 1996). Eye-centered coordinates are not universal, however. There is evidence that in area 5 reaching movements are encoded in shoulder-centered coordinates (Burnod et al., 1999) and in a part of premotor cortex, PMv, hand-centered coordinates appear to be important (Graziano and Gross, 1994).

MOTOR READINESS

Motor readiness is defined as the increased tendency to make a movement, independent of the parameters of that movement. It appears that anticipatory activity (see Fig. 1B) plays an important role in motor readiness. After its initial observation by Niki and Watanabe (1979), anticipatory activity has been observed in prefrontal cortex (Joseph and Barone, 1987; MacKay and Crammond, 1987; Boch and Goldberg, 1989), FEF (Bruce and Goldberg, 1985), premotor cortex (Mauritz and Wise, 1986; Vaadia et al., 1988; di Pellegrino and Wise, 1993; Shen and Alexander, 1997), LIP (Colby et al., 1996), area 5 (MacKay and Crammond, 1987) and the striatum (Hikosaka et al., 1989; Apicella et al., 1992).

Moody and Wise (2000) examined the role of anticipatory activity by building an artificial neural network model and comparing its activity with that in the prefrontal cortex of a monkey. The task that both the model and the monkey needed to perform is known as a matching-to-sample task. In that task, sensory inputs follow a sample sensory input and both the monkey and model network need to produce a response when subsequent inputs match the sample. Anticipatory activity occurred robustly in both the monkey and the model. In the neural-network model, the function of this enigmatic pattern of activity could be understood with the kind of causal determinism that is impossible with neurophysiological data. Moody and Wise found that some anticipatory activity enhanced the network's general readiness to respond, i.e., to produce an output. In this sense, at least a proportion of anticipatory activity appeared to function by contributing to a process resembling motor readiness (see Brown and Robbins, 1991, for the distinction between motor preparation and motor readiness).

However, another aspect of anticipatory activity more closely resembled attentional processes. Some of the anticipatory activity in the model network acted like a filter to suppress inputs that would elicit inappropriate outputs. The idea of a filter serves as a common metaphor for attention, and the role of anticipatory activity in processes akin to both motor readiness and attention serves to underline the close relationship between these processes. Returning to the concept of accumulator models, the idea that anticipatory activity might function to lower the threshold for both inputs and outputs has several attractive features. It explains why there is so much similarity between anticipatory activity prior to predictable sensory inputs and that prior to predictable motor outputs (Fig. 1B). In both instances, the role of anticipatory activity is to generally increase the excitability of neural networks.

DECISION TO ACT

The targeting decision and the maintenance of a preparatory state do not imply that a movement will be made. Decisions can be vetoed, a fact that leads us back to a consideration of the BP. Libet and his colleagues showed that a verbal veto command causes the BP to cease increasing (see Fig. 2 of Libet, 1985), much like the stop signal does for neuronal activity in FEF (Fig. 4). Although we need to be mindful of the fact that the BP is observed before self-paced movements, rather than prior to sensorially guided movements such as those studied by Schall and his colleagues, the idea that both processes involve the ramping-to-threshold of distributed neural networks has considerable attraction. In both examples, a sensory input can initiate a process that interrupts this ramp-increase in activity, and in both cases a movement that would have been made is, instead, aborted.

The distinction between the targeting decision and the decision to act is also relevant to the division of labor among motor cortical areas. It is commonly held that medial premotor areas underlie internally generated actions, whereas lateral premotor areas subserve sensorially directed ones (Passingham, 1993; Chen et al., 1995; Thaler et al., 1995). Discussions of this idea usually emphasize the targeting decision. Perhaps a similar distinction can be made for the decision to act. According to this view, the lateral premotor areas, through their extensive interactions with posterior parietal and prefrontal cortex, function to trigger a movement on the basis of external signals. By contrast, the medial premotor areas may do so based on time-dependent factors rather than proximate sensory inputs.

In neurophysiological experiments, too, the decision to act may depend either on a later sensory event (Wise and Kurata, 1989) or on the passage of time (Kurata and Wise, 1988). In the latter circumstance, a ramp-like increase in activity develops relatively late in a self-timed delay period (Kurata and Wise, 1988). This phenomenon has been observed in the premotor cortex, including the SMA, as well as in the prefrontal cortex. As noted in the previous section, this pre-movement anticipatory activity (Fig. 1B) appears to enhance motor readiness. Let us assume that some process holds the activity of motor networks below the threshold for action during an instructed delay period. If this is the case, then perhaps the pre-movement anticipatory activity pushes a previously selected movement above the threshold for action to initiate the movement. As Kornhuber (1983) has written:

> When considering . . . voluntary decisions, it has become clear that the "what", 'how", and "when" are different components which need different kinds of information . . . . In the experimental studies used to investigate the Bereitschaftspotential preceding simple movements, the questions "What to do" and "How to do it" are already answered beforehand, so that only the third question ("When to start") remains to be decided . . . .

Perhaps it is in an internal decision to act, without an external, sensory command to do so, that the BP finds its closest correlate at the single-cell level.

ACKNOWLEDGEMENT

I thank Dr. Marie-Pierre Deiber of the University of Geneva, Switzerland, for her comments on an earlier version of this chapter.

REFERENCES

Alexander, G. E. (1987) Selective neuronal discharge in monkey putamen reflects intended direction of planned limb movements. Exp. Brain Res. 67, 623-634.

Alexander, G. E., and Crutcher, M. D. (1990a) Neural representations of the target (goal) of visually guided arm movements in three motor areas of the monkey. J. Neurophysiol. 64, 164-178.

Alexander, G. E., and Crutcher, M. D. (1990b) Preparation for movement: Neural representation of intended direction in three motor areas of the monkey. J. Neurophysiol. 64, 133-150.

Apicella, P., Scarnati, E., Ljungberg, T., and Schultz, W. (1992) Neuronal activity in monkey striatum related to the expectation of predictable environmental events. J. Neurophysiol. 68, 945-960.

Bichot, N. P., and Schall, J. D. (1999) Effects of similarity and history on neural mechanisms of visual selection. Nat. Neurosci. 2, 549-554.

Boch, R. A., and Goldberg, M. E. (1989) Participation of prefrontal neurons in the preparation of visually guided eye movements in the rhesus monkey. J. Neurophysiol. 61, 1064-1084.

Boussaoud, D., and Wise, S. P. (1993a) Primate frontal cortex: neuronal activity following attentional vs. intentional cues. Exp. Brain Res. 95, 15-27.

Boussaoud, D., and Wise, S. P. (1993b) Primate frontal cortex: effects of stimulus and movement. Exp. Brain Res. 95, 28-40.

Bracewell, R. M., Mazzoni, P., Barash, S., and Andersen, R. A. (1996) Motor intention activity in the macaque's lateral intraparietal area. 2. Changes of motor plan. J. Neurophysiol. 76, 1457-1464.

Brinkman, C., and Porter, R. (1979) Supplementary motor area in the monkey: activity of neurons during performance of a learned motor task. J. Neurophysiol. 42, 681-709.

Brown, V. J., and Robbins, T. W. (1991) Simple and choice reaction time performance following unilateral striatal dopamine depletion in the rat. Impaired motor readiness but preserved response preparation. Brain 114, 513-525.

Bruce, C. J., and Goldberg, M. E. (1985) Primate frontal eye fields. I. Single neurons discharging before saccades. J. Neurophysiol. 53, 603-635.

Burnod, Y., Baraduc, P., Battaglia-Mayer, A., Guigon, E., Koechlin, E., Ferraina, S., Lacquaniti, F., and Caminiti, R. (1999) Parieto-frontal coding of reaching: an integrated framework. Exp. Brain Res. 129, 325-346.

Carpenter, A. F., Georgopoulos, A. P., and Pellizzer, G. (1999) Motor cortical encoding of serial order in a context-recall task. Science 283, 1752-1757.

Chen, D.-F., Hyland, B., Maier, V., Palmeri, A., and Wiesendanger, M. (1991) Comparison of neural activity in the supplementary motor area and in the primary motor cortex in monkeys. Somat. Mot. Res. 8, 27-44.

Chen, Y.-C., Thaler, D., Nixon, P. D., Stern, C., and Passingham, R. E. (1995) The functions of the medial premotor cortex (SMA). II. The timing and selection of learned movements. Exp. Brain Res. 102, 461-473.

Clower, D. M., Hoffman, J. M., Votaw, J. R., Faber, T. L., Woods, R. P., and Alexander, G. E. (1996) Role of posterior parietal cortex in the recalibration of visually guided reaching. Nature 383, 618-621.

Clower, W.T., and Alexander, G.E. (1998) Movement sequence-related activity reflecting numerical order of components in supplementary and presupplementary motor areas. J. Neurophysiol. 80, 1562-1566.

Colby, C. L., Duhamel, J. R., and Goldberg, M. E. (1996) Visual, presaccadic, and cognitive activation of single neurons in monkey lateral intraparietal area. J. Neurophysiol. 76, 2841-2852.

Crammond, D. J., and Kalaska, J. F. (1989) Neuronal activity in primate parietal cortex area 5 varies with intended movement direction during an instructed-delay period. Exp. Brain Res. 78, 458-462.

Crutcher, M. D., and Alexander, G. E. (1990) Movement-related neuronal activity selectively coding either direction or muscle pattern in three motor areas of the monkey. J. Neurophysiol. 64, 151-163.

Decety, J., Jeannerod, M., Durozard, D., and Baverel, G. (1993) Central activation of autonomic effectors during mental simulation of motor actions in man. J. Physiol. (Lond). 461, 549-563.

Deiber, M.-P., Passingham, R. E., Colebatch, J. G., Friston, K. J., Nixon, P. D., and Frackowiak, R. S. J. (1991) Cortical areas and the selection of movement: a study with positron emission tomography. Exp. Brain Res. 84, 393-402.

di Pellegrino, G., and Wise, S. P. (1993) Visuospatial vs. visuomotor activity in the premotor and prefrontal cortex of a primate. J. Neurosci. 13, 1227-1243.

Donoghue, J. P., and Sanes, J. N. (1994) Motor areas of the cerebral cortex. J. Clin. Neurophysiol. 11, 382-396.

Eccles, J. C. (1982a) The liaison brain for voluntary movement: the supplementary motor area. Acta Biol., 12, 157-172.

Eccles, J. C. (1982b) The initiation of voluntary movement by the supplementary motor area. Arch. Psychiat. Nervenkr. 231, 423-441.

Eskandar, E. N., and Assad, J. A. (1999) Dissociation of visual, motor and predictive signals in parietal cortex during visual guidance. Nat. Neurosci. 2, 88-93.

Evarts, E. V. (1965) Relation of discharge frequency to conduction velocity in pyramidal tract neurons. J. Neurophysiol. 28, 216-228.

Evarts, E. V., and Tanji, J. (1974) Gating of motor cortex reflexes by prior instruction. Brain Research 71, 479-494.

Everling, S. Dorris, M.C. Klein, R.M., and Munoz, D.P. (1999) Role of primate superior colliculus in preparation and execution of anti-saccades and pro-saccades. J. Neurosci. 19, 2740-2754.

Everling, S., and Munoz, D. P. (2000) Neuronal correlates for preparatory set associated with pro-saccades and anti-saccades in the primate frontal eye field. J. Neurosci. 20, 387-400.

Fagg, A. H., and Arbib, M. A. (1992) A model of primate visual-motor conditional learning. J. Adapt. Behav. 1, 3-37.

Fetz, E.E., and Shupe, L.E. (1990) Neural network models of the primate motor system. in: R. Eckmiller (Ed.), Advanced Neural Computers, Elsevier Science Publishers, Amsterdam, pp. 43-50.

Fortier, P. A., Smith, A. M., and Kalaska, J. F. (1993) Comparison of cerebellar and motor cortex activity during reaching: directional tuning and response variability. J. Neurophysiol. 69, 1136-1149.

Fujii, N., Mushiake, H., and Tanji, J. (1998) An oculomotor representation area within the ventral premotor cortex. Proc. Natl. Acad. Sci. USA. 95, 12034-12037.

Fujii, N., Mushiake, H., and Tanji, J. (2000) Rostrocaudal distinction of the dorsal premotor area based on oculomotor involvement. J. Neurophysiol. 83, 1764-1769.

Fuster, J. M. (1973) Unit activity in prefrontal cortex during delayed-response performance: neuronal correlated of transient memory. J. Neurophysiol. 36, 61-78.

Fuster, J. M., and Alexander, G. E. (1971) Neuron activity related to short-term memory. Science 173, 652-654.

Georgopoulos, A. P., Lurito, J. T., Petrides, M., Schwartz, A. B., and Massey, J. T. (1989) Mental rotation of the neuronal population vector. Science 243, 234-236.

Georgopoulos, A. P., Taira, M., and Lukashin, A. (1993) Cognitive neurophysiology of the motor cortex. Science 260, 47-52.

Geyer, S., Zilles, K., Luppino, G., and Matelli, M. (2000) Neurofilament protein distribution in the macaque monkey dorsolateral premotor cortex. Eur. J. Neurosci. 12, 1554-1566.

Gold, J. I., and Shadlen, M. N. (2000) Representation of a perceptual decision in developing oculomotor commands. Nature 404, 390-394.

Graziano, M., and Gross, C.G. (1998) Visual responses with and without fixation: neurons in premotor cortex encode spatial locations independently of eye position. Exp. Brain Res. 118, 373-380.

Graziano, M. (1999) Where is my arm? The relative role of vision and proprioception in the neuronal representation of limb position. Proc. Natl. Acad. Sci. USA. 96, 10418-10421.

Graziano, M., and Gross, C. G. (1994) Mapping space with neurons. Curr. Dir. Physiol. Sci., 2, 1-4.

Hanes, D. P., and Schall, J. D. (1996) Neural control of voluntary movement initiation. Science 274, 427-430.

Hikosaka, K., and Watanabe, M. (2000) Delay activity of orbital and lateral prefrontal neurons of the monkey varying with different rewards. Cereb. Cortex 10, 263-271.

Hikosaka, O., Sakamoto, M., and Usui, S. (1989) Functional properties of monkey caudate neurons. III. Activities related to expectation of target and reward. J. Neurophysiol. 61, 814-832.

Horwitz, G. D., and Newsome, W. T. (1999) Separate signals for target selection and movement specification in the superior colliculus. Science 284, 1158-1161.

Ikeda, A., Lüders, H. O., Burgess, R. C., and Shibasaki, H. (1992) Movement-related potentials recorded from supplementary motor area and primary motor area. Brain 115, 1017-1043.

Jeannerod, M., and Frak, V. (1999) Mental imaging of motor activity in humans. Curr. Opin. Neurobiol. 9, 735-739.

Jenkins, I. H., Jahanshahi, M., Jueptner, M., Passingham, R. E., and Brooks, D. J. (2000) Self-initiated versus externally triggered movements II. The effect of movement predictability on regional cerebral blood flow. Brain 123, 1216-1228.

Joseph, J. P., and Barone, P. (1987) Prefrontal unit activity during a delayed oculomotor task in the monkey. Exp. Brain Res. 67, 460-468.

Kalaska, J. F., and Crammond, D. J. (1992) Cerebral cortical mechanisms of reaching movements. Science 255, 1517-1523.

Kalaska, J. F., and Crammond, D. J. (1995) Deciding not to go: neuronal correlates of response selection in a go/nogo task in primate premotor and parietal cortex. Cereb. Cortex. 5, 410-428.

Kim, J. N., and Shadlen, M. N. (1999) Neural correlates of a decision in the dorsolateral prefrontal cortex of the macaque. Nat. Neurosci. 2, 176-185.

Kornhuber, H. H., and Deecke, L. (1965) Hirnpotentialänderungen bei Willkürbewegungen und passiven Bewungen des Menschen: Bereitschaftspotential und reafferente Potentiale. Pflügers Archiv. 284, 1-17.

Krams, M., Rushworth, M. F., Deiber, M. P., Frackowiak, R. S., and Passingham, R. E. (1998) The preparation, execution and suppression of copied movements in the human brain. Exp. Brain Res. 120, 386-398.

Kubota, K., and Niki, H. (1971) Prefrontal cortical unit activity and delayed alternation performance in monkeys. J. Neurophysiol. 34, 337-347.

Kurata, K., and Wise, S. P. (1988) Premotor and supplementary motor cortex in rhesus monkeys: neuronal activity during externally- and internally-instructed motor tasks. Exp. Brain Res. 72, 237-248.

Leon, M. I., and Shadlen, M. N. (1999) Effect of expected reward magnitude on the response of neurons in the dorsolateral prefrontal cortex of the macaque. Neuron 24, 415-425.

Libet, B. (1985) Unconscious cerebral initiative and the role of conscious will in voluntary action. Behav. Brain Sci. 8, 529-566.

Lotze, M., Montoya, P., Erb, M., Hulsmann, E., Flor, H., Klose, U., Birbaumer, N., and Grodd, W. (1999) Activation of cortical and cerebellar motor areas during executed and imagined hand movements: an fMRI study. J. Cogn. Neurosci. 11, 491-501.

Luce,R.D. (1986) Response Times: Their Role in Inferring Elementary Mental Organization, Oxford University Press: New York.

Lurito, J. T., Georgakopoulos, T., and Georgopoulos, A. P. (1991) Cognitive spatial-motor processes. 7. The making of movements at an angle from a stimulus direction: studies of motor cortical activity at the single cell and population levels. Exp. Brain Res. 87, 562-580.

MacKay, D. M., and Gardiner, M. F. (1972) Two strategies of information processing. Neurosci. Res. Prog. Bull. 10, 77-78.

MacKay, W. A., and Crammond, D. J. (1987) Neuronal correlates in posterior parietal lobe of the expectation of events. Behav. Brain Res. 24, 167-179.

Matelli, M., Luppino, G., and Rizzolatti, G. (1985) Patterns of cytochrome oxidase activity in the frontal agranular cortex of the macaque monkey. Behav. Brain Res. 18, 125-136.

Mauritz, K. H., and Wise, S. P. (1986) Premotor cortex of the rhesus monkey: neuronal activity in anticipation of predictable environmental events. Exp. Brain Res. 61, 229-244.

Mel, B. W. (1991) A connectionist model may shed light on neural mechanisms for visually guided reaching. J. Cog. Neurosci. 3, 273-292.

Mikami, A., and Kubota, K. (1980) Inferotemporal neuron activities and color discrimination with delay. Brain Research 182, 65-78.

Mitz, A. R., and Wise, S. P. (1987) Somatotopic organization of the supplementary motor area: intracortical microstimulation mapping. J. Neurosci. 7, 1010-1021.

Moody, S. L., and Wise, S. P. (2000) A model that accounts for activity prior to sensory inputs and responses during matching-to-sample tasks. J. Cog. Neurosci. 12, 429-448.

Moody, S. L., and Zipser, D. (1998) A model of reaching dynamics in primary motor cortex. J. Cog. Neurosci. 10, 35-45.

Mountcastle, V. B., Lynch, J. C., Georgopoulos, A., Sakata, H., and Acuna, C. (1975) Posterior parietal association cortex of the monkey: Command functions for operations within extrapersonal space. J. Neurophysiol. 38, 871-908.

Mushiake H, Strick PL. (1995) Cerebellar and pallidal activity during instructed delay periods. Soc. Neurosci. Abstr. 21, 411.

Neshige, R., Luders, H., and Shibasaki, H. (1988) Recording of movement-related potentials from scalp and cortex in man. Brain 111, 719-736.

Newsome, W. T., Britten, K. H., and Movshon, J. A. (1989) Neuronal correlates of a perceptual decision. Nature 341, 52-54.

Niki, H., and Watanabe, M. (1979) Prefrontal and cingulate unit activity during timing behavior in the monkey. Brain Research 171, 213-224.

Ogiso, T., Kobayashi, K., and Sugishita, M. (2000) The precuneus in motor imagery: a magneto-encephalographic study. NeuroReport 11, 1345-1349.

Pascual-Leone, A., Brasil-Neto, J. P., Valls-Sole, J., Cohen, L. G., and Hallett, M. (1992a) Simple reaction time to focal transcranial magnetic stimulation. Brain 115, 109-122.

Pascual-Leone, A., Valls-Sole, J., Wassermann, E. M., Brasil-Neto, J. P., Cohen, L. G., and Hallett, M. (1992b) Effects of focal transcranial magnetic stimulation on simple reaction time to acoustic, visual and somatosensory stimuli. Brain 115, 1045-1059.

Passingham,R.E. (1993) The Frontal Lobes and Voluntary Action, Oxford University Press: Oxford.

Platt, M. L., and Glimcher, P. W. (1999) Neural correlates of decision variables in parietal cortex. Nature 400, 233-238.

Poliakov, A. V., and Schieber, M. H. (1999) Limited functional grouping of neurons in the motor cortex hand area during individuated finger movements: A cluster analysis. J. Neurophysiol. 82, 3488-3505.

Porro, C. A., Cettolo, V., Francescato, M. P., and Baraldi, P. (2000) Ipsilateral involvement of primary motor cortex during motor imagery. Eur. J. Neurosci. 12, 3059-3063.

Porter, R. (1990) The Kugelberg lecture. Brain mechanisms of voluntary motor commands–a review. Electroenceph. Clin. Neurophysiol. 76, 282-293.

Preuss, T. M., Stepniewska, I., and Kaas, J. H. (1996) Movement representation in the dorsal and ventral premotor areas of owl monkeys: A microstimulation study. J. Comp. Neurol. 371, 649-675.

Roland, P. E., Larsen, B., Lassen, N. A., and Skinhøj, E. (1980) Supplementary motor area and other cortical areas in organization of voluntary movements in man. J. Physiol. (Lond). 43, 118-136.

Rossi, A. F., Rittenhouse, C. D., and Paradiso, M. A. (1996) The representation of brightness in primary visual cortex. Science 273, 1104-1107.

Salinas, E., and Romo, R. (1998) Conversion of sensory signals into motor commands in primary motor cortex. J. Neurosci. 18, 499-511.

Schall, J. D. (1995) Neural basis of saccade target selection. Rev. Neurosci. 6, 63-85.

Schall, J. D. (2000) From sensory evidence to a motor command. Curr. Biol. 10, R404-R406.

Schall, J. D., and Thompson, K. G. (1999) Neural selection and control of visually guided eye movements. Annu. Rev. Neurosci. 22, 241-259.

Schultz, W., Tremblay, L., and Hollerman, J. R. (2000) Reward processing in primate orbitofrontal cortex and basal ganglia. Cereb. Cortex. 10, 272-283.

Shen, L., and Alexander, G. E. (1997) Preferential representation of instructed target location versus limb trajectory in dorsal premotor area. J. Neurophysiol. 77, 1195-1212.

Sheth, B. R., Sharma, J., Rao, S. C., and Sur, M. (1996) Orientation maps of subjective contours in visual cortex. Science 274, 2110-2115.

Snyder, L. H., Batista, A. P., and Andersen, R. A. (1997) Coding of intention in the posterior parietal cortex. Nature 386, 167-170.

Snyder, L. H., Batista, A. P., and Andersen, R. A. (1998) Change in motor plan, without a change in the spatial locus of attention, modulates activity in posterior parietal cortex. J. Neurophysiol. 79, 2814-2819.

Snyder, L. H., Batista, A. P., and Andersen, R. A. (2000) Saccade-related activity in the parietal reach region. J. Neurophysiol. 83, 1099-1102.

Tanji, J., and Evarts, E. V. (1976) Anticipatory activity of motor cortex neurons in relation to direction of an intended movement. J. Neurophysiol. 39, 1062-1068.

Tanji, J., Taniguchi, K., and Saga, T. (1980) Supplementary motor area: neuronal response to motor instructions. J. Neurophysiol. 43, 60-68.

Tehovnik, E. J., Sommer, M. A., Chou, I. H., Slocum, W. M., and Schiller, P. H. (2000) Eye fields in the frontal lobes of primates. Brain. Res. Rev. 32, 413-448.

Thach, W. T. (1975) Timing of activity in cerebellar dentate nucleus and cerebral motor cortex during prompt volitional movement. Brain Research 88, 233-241.

Thaler, D., Chen, Y.-C., Nixon, P. D., Stern, C., and Passingham, R. E. (1995) The functions of the medial premotor cortex (SMA) I. Simple learned movements. Exp. Brain Res. 102, 445-460.

Thompson, K. G., Bichot, N. P., and Schall, J. D. (1997) Dissociation of visual discrimination from saccade programming in macaque frontal eye field. J. Neurophysiol. 77, 1046-1050.

Thompson, K. G., Hanes, D. P., Bichot, N. P., and Schall, J. D. (1996) Perceptual and motor processing stages identified in the activity of macaque frontal eye field neurons during visual search. J. Neurophysiol. 76, 4040-4055.

Thompson, K. G., and Schall, J. D. (1999) The detection of visual signals by macaque frontal eye field during masking. Nat. Neurosci. 2, 283-288.

Tremblay, L., and Schultz, W. (2000) Reward-related neuronal activity during go-nogo task performance in primate orbitofrontal cortex. J. Neurophysiol. 83, 1864-1876.

Vaadia, E., Gottlieb, Y., and Abeles, M. (1982) Single-unit activity related to sensorimotor association in auditory cortex of a monkey. J. Neurophysiol. 48, 1201-1213.

Vaadia, E., Kurata, K., and Wise, S. P. (1988) Neuronal activity preceding directional and nondirectional cues in the premotor cortex of rhesus monkeys. Somat. Mot. Res. 6 (207-230.

Walter, W. G., Cooper, R., Aldridge, V. J., McCallum, W. C., and Winter, A. L. (1964) Contingent negative variation: An electric sign of sensorimotor association and expectancy in the human brain. Nature 203, 380-384.

Weinrich, M., and Wise, S. P. (1982) The premotor cortex of the monkey. J. Neurosci. 2, 1329-1345.

Weinrich, M., Wise, S. P., and Mauritz, K.-H. (1984) A neurophysiological analysis of the premotor cortex in the rhesus monkey. Brain 107, 385-414.

Welker, W., Johnson, J. I and Noe, A. (2001) Comparative Mammalian Brain Collection, World Wide Web, http://www.brainmuseum.org

Wexler, M., Kosslyn, S. M., and Berthoz, A. (1998) Motor processes in mental rotation. Cognition 68, 77-94.

Wise, S. P., Boussaoud, D., Johnson, P. B., and Caminiti, R. (1997) The premotor and parietal cortex: Corticocortical connectivity and combinatorial computations. Annu. Rev. Neurosci. 20, 25-42.

Wise, S. P., and Kurata, K. (1989) Set-related activity in the premotor cortex of rhesus monkeys: Effect of triggering cues and relatively long delay intervals. Somat. Mot. Res. 6, 455-476.

Wise, S. P., and Mauritz, K.-H. (1985) Set-related neuronal activity in the premotor cortex of rhesus monkeys: effects of changes in motor set. Proc. R. Soc. Lond. [Biol]. 223, 331-354.

Wise, S. P., Weinrich, M., and Mauritz, K.-H. (1986) Movement-related activity in the premotor cortex of rhesus macaques. Prog. Brain Res. 64, 117-131.

Wong, Y. C., Kwan, H. C., MacKay, W. A., and Murphy, J. T. (1977) Topographic organization of afferent inputs in monkey precentral cortex. Brain Research 138, 166-168.

Wu, C. W. H., Bichot, N. P., and Kaas, J. H. (2000) Converging evidence from microstimulation, architecture, and connections for multiple motor areas in the frontal and cingulate cortex of prosimian primates. J. Comp. Neurol. 423, 140-177.

Zhang, J., Riehle, A., Requin, J., and Kornblum, S. (1997) Dynamics of single neuron activity in monkey primary motor cortex related to sensorimotor transformation. J. Neurosci. 17, 2227-2246.

Zhou, Y. D., and Fuster, J. M. (1996) Mnemonic neuronal activity in somatosensory cortex. Proc. Natl. Acad. Sci. USA. 93, 10533-10537.

Zipser, D. (1991) Recurrent network model of the neural mechanism of short-term active memory. Neural Comp. 3, 179-193.

Zipser, D., and Andersen, R. A. (1988) A back-propagation programmed network that simulates response properties of a subset of posterior parietal neurons. Nature 331, 679-684.

Zipser, K., Lamme, V. A., and Schiller, P. H. (1996) Contextual modulation in primary visual cortex. J. Neurosci. 16, 7376-7389.

MOVEMENT PREPARATION: NEUROIMAGING STUDIES

Ivan Toni[1], Richard E. Passingham[2]

[1] F.C. Donders Centre for Cognitive Neuroimaging

University of Nijmegen

NL-6500 HB Nijmegen

The Netherlands

[2] Department of Experimental Psychology

South Park Road

Oxford OX1 3UD

U.K.

INTRODUCTION

Our interactions with the external environment involve a continuous transformation of sensory stimuli into behavioural responses. Eventually, the neural codes of stimuli and responses need to be merged. Several solutions have been proposed to account for spatial visuomotor transformations (Burnod et al. 1999; Jeannerod et al. 1995; Soechting & Flanders 1992). However, these conceptual frameworks might be inadequate to explain our ability to act according to arbitrary visuomotor associations. In this particular category of sensorimotor transformations, behaviour is guided by rules rather than objects or places (Passingham & Toni 2001; White & Wise 1999). The flexibility of these stimulus-response relationships is vast, allowing us to transcend the domain of spatially or temporally congruent associations. Such flexibility suggests that arbitrary associations can be best established and accessed through manipulation of high level representations of stimuli and responses (Passingham, Toni, & Rushworth 2000; Toni & Passingham 1999), since these internal states of an agent are not tied to a particular sensory or effector system (Markman & Dietrich 2000). For instance, movement representations are considered to be independent from the performance of an actual motor act, or from the presence of a response's target and sensory instructions (Jeannerod 1997). Since we can act according to arbitrary rules, the general mechanisms underlying sensorimotor transformations do not need to be constrained in spatial or temporal frameworks.

The Bereitschaftspotential
Edited by Jahanshahi and Hallett, Kluwer Academic/Plenum Publishers, New York, 2003

Neural correlates of movement representations have been assessed in humans, by measuring electrical or metabolic activity related to the expectation of making a motor response in the near future. This approach is justified under the assumption that the neural substrates of movement representation express a sustained discharge as long as the action has not been completed (Jeannerod 1997). This operational definition of movement representations is reflecting specific differences between sensory and motor activities, time-locked to stimuli and responses on a moment-by-moment basis; and preparatory activity, dependent on more persistent neural activity (Gold & Shadlen 2000).

The application of this criterion goes back at least to the discovery of the first electrophysiological correlates of motor preparation, like the Contingent Negative Variation (CNV, (Walter et al. 1964) and the Bereitschaftspotential (BP, (Kornhuber & Deecke 1965). These electrical signatures (or at least portions of them) have proved to be robust indices of preparatory activity during externally or internally generated movements (Cui et al. 1999; Cui et al. 2000; Verleger et al. 2000). However, neither scalp recordings alone nor imaging studies targeted at limited portions of the brain (Deiber et al. 1999; Richter et al. 1997; Weilke et al. 2001) can provide unambiguous and unbiased estimates of the neural sources of a given cognitive process (Dale et al. 2000). Considering preparatory activity in the general context of sensorimotor transformations, means that the whole neural path from perception to action is involved in this phenomenon. Therefore, it is important to assess the neural correlates of motor preparation over the whole brain in order to appreciate the integrative and dynamic characteristics of this process.

MOVEMENT PREPARATION: METABOLIC IMAGING

The spatially distributed systems at the basis of visuomotor transformations have been mapped through ^{14}C-deoxyglucose mapping (DG) (Picard & Strick 1997; Savaki & Dalezios 1999), ^{15}O positron emission tomography (PET) (Colebatch et al. 1991; Grafton et al. 1992), and functional magnetic resonance imaging (fMRI) based on the blood oxygenation level dependent (BOLD) intrinsic contrast (Rao et al. 1993; Sanes et al. 1995). These techniques provide an average and indirect index of synaptic activity of a given cerebral region, at various degrees of spatial resolution. Therefore, they are best placed to indicate differential proportions of task-related neuronal populations across areas, or sub-threshold changes in cortical excitability that might not emerge in the firing rate of single-units. However, the poor temporal resolution of these techniques has imposed heavy constraints on the experimental designs applicable. DG can resolve functional-structural relationship up to the level of cortical layers (Vanduffel, Tootell, & Orban 2000), but it is limited to "static" comparisons between two experimental conditions acquired over tens of minutes. PET and fMRI have been used to study response preparation by directly comparing conditions either involving or not involving motor preparation during visually instructed tasks involving arm (Kawashima, Roland, & O'Sullivan 1994), hand (Stephan et al. 1995), or finger movements (Deiber et al. 1996; Kawashima, Roland, and O'Sullivan 1994; Krams et al. 1998; Stephan et al. 1995). Despite such considerable procedural differences, two patterns of activity have emerged from these studies: When the timing and the type of response were specified by external cues, there was consistent activation in the posterior parietal cortex and, less reliably, in the precentral gyrus. Conversely, differential mesial frontal activities were associated with the preparation of internally generated movements, supporting the view that the early BP is generated in this region (Jahanshahi et al. 1995; Jenkins et al. 2000). These findings have been confirmed by a further set of studies where motor imagery tasks were used to obtain indirect measures of motor preparatory activity (Decety et al. 1994; Deiber et al. 1998; Gerardin et al. 2000; Parsons et al. 1995; Porro et al. 1996; Roland et al. 1980; Roth et al. 1996). Both approaches have distinguished, across blocks of trials, between areas involved in response selection and/or response execution and areas involved in response preparation. These studies have provided support to the claim that motor imagery and preparatory/executive motor processes relies on generally overlapping neural substrates (Jeannerod 1994). However, a few reports have also highlighted important differences between neural activities driven by imaged and actual movements (Deiber et al. 1998; Gerardin et al. 2000). Unfortunately, these studies have assessed the imagery performance through verbal reports, although objective psychophysical methods are also available (Decety & Jeannerod 1995; Parsons 1994). Therefore, it remains to be seen whether psychophysically-matched tasks involving imaged and actual movements would display systematic differences in their neural correlates. More generally, the scope of the inferences drawn in several imaging papers concerned with preparatory activity is severely limited by their dependence on the assumption of pure insertion of cognitive processes (Friston et al. 1996; Steinberg 1969). In the context of motor preparation, this assumption implies that preparing to move does not affect the selection and execution stages of the sensorimotor transformation. This assumption has been shown to be invalid, at the level of both single unit (Crammond & Kalaska 2000; Riehle & Requin 1989) and neuronal population (Zarahn, Aguirre, & D'Esposito 1999).

MOVEMENT PREPARATION: ELECTROPHYSIOLOGICAL IMAGING

The time-varying characteristics of visuomotor transformations have been addressed by electrophysiological studies. By introducing a delay between the presentation of instruction cues and the performance of a movement, it has been possible to temporally isolate delay-related activity on a trial by trial basis (Coles et al. 1995; Crammond and Kalaska 2000; Eimer 1995; Endo et al. 1999; Hackley & Miller 1995; Papa, Artieda, & Obeso 1991; Snyder, Batista, & Andersen 1997; Wise & Mauritz 1985). Unfortunately, the questions addressed by studies on humans and macaques have been generally framed in different contexts, e.g. discrete vs continuous or automatic vs voluntary modes of information processing in reports with human subjects; sensory vs motor frames of neuronal activity in studies with monkeys. In the domain of motor preparation, delay unpredictability is more important than delay per se. When instruction and trigger cues are separated by a variable delay period, the transformation of a stimulus into a motor response can be partitioned into temporally distinct components, since the subject needs to be ready to respond at any time but the timing of the response cannot be predicted (Klemmer 1957; Moody & Wise 2000). Under these circumstances, selection of the appropriate movement is likely to occur at the presentation of the instruction cue. In contrast, the implementation of the executive motor commands can occur only after the trigger presentation. Accordingly, only the goal of the movement is held during the delay period (Bastian et al. 1998), without the confounds of sensory and motor anticipation phenomena (Moody and Wise 2000; Requin, Brener, & Ring 1991). In this context, specific preparatory activity (i.e. dissociable from transient stimulus- or movement-locked responses and robust to the assumption of pure insertion of cognitive processes) is not related to the enactment of a movement, but rather to its representation (Jeannerod 1997) and it is likely to reflect higher cognitive aspects of the motor planning process (Wise, di Pellegrino, & Boussaoud 1996).

MOVEMENT PREPARATION: EVENT-RELATED FMRI

We designed a study aimed at distinguishing, within a trial, *specific* preparatory activity from transient stimulus- and movement-related responses. In this study, we have exploited the temporal independence of preparatory activity from overt behavioural events for imaging the neural basis of action representation. Our goal was to make use of the spatial resolution of fMRI while preserving the dynamic characteristics of visuomotor transformations. We took advantage of whole-brain event-related fMRI (er-fMRI) for dissociating transient responses time-locked to sensory or motor events in the context of a visuomotor associative task (Toni et al. 1999). Er-fMRI relies on manipulations of task contingencies on a trial by trial basis, so that contextual- and stimulus-related effects can be distinguished (Friston et al. 1999; Josephs, Turner, & Friston 1997). Er-fMRI is particularly suitable for the study of visuomotor processes, since separate inferences can be drawn on sensory, motor, or cognitive processes associated with different trial epochs (D'Esposito et al. 2000; Richter et al. 2000). Albeit er-fMRI allows for fast and random presentation rates, it still provides a static picture of the brain, although ingenious attempts have been made to extract dynamic information from the timecourse of the blood oxygenation level dependent (BOLD) signal (Friston et al. 1998; Menon, Luknowsky, & Gati 1998; Ogawa et al. 2000; Weilke et al. 2001).

Despite these limitations, it has been possible to define temporally segregated task components within each trial, such that they constituted independent partitions of a statistical model (Toni et al. in press). We trained a group of subjects to perform a conditional visuomotor associative task with instructed delays. One of four shapes (instruction cue, IC) was briefly presented (300 ms). Two shapes instructed the flexion of the index finger, the other two shapes instructed the flexion of the middle finger. A variable delay (1.3 – 20.8 s in steps of 1.3 s) was followed by a tone (trigger cue, TC – 300 ms). The subjects were asked to respond as quickly as possible after the auditory trigger cue. This experimental procedure allowed us to distinguish sustained preparatory activity from transient stimulus- or movement-locked responses. Note that more efficient approaches can be applied in different contexts, for instance when the task contingencies are temporally permutable and transient components of the response are investigated (Friston et al. 1999).

We found that *preparing* to move according to arbitrary visuomotor associations relies on parieto-frontal circuitry (Figure 1A). This result confirms that in humans, as in other primates, portions of parietal and premotor areas contribute to hold the goal of the movement during the delay period (Wise et al. 1997). The functional anatomy of these regions of the macaque brain has been characterised by several electrophysiological and anatomical studies (Ashe & Georgopoulos 1994; Fogassi et al. 1999; Geyer et al. 2000; Johnson et al. 1996; Kalaska & Crammond 1992; Matelli et al. 1998; Tanne et al. 1995; Wise et al. 1997). Visually responsive neurons have been described in a rostro-lateral sector of the dorsal premotor cortex (area F2), and forelimb movements can be elicited in this sector by intracortical microstimulation (Fogassi et al. 1999). It has also been shown that this area is specifically interconnected with restricted portions of the posterior parietal cortex (Matelli et al. 1998), namely area V6A [located in the anterior bank of the occipito-parietal sulcus (Galletti et al. 1999)] and area MIP [located in the medial bank of the intraparietal sulcus (Colby et al. 1988)]. These parietal regions show neuronal activity associated with visuo-spatial tasks (Galletti et al. 1997; Johnson et al. 1996). There are intriguing anatomical similarities between these findings on the macaque and our results on humans. This correspondence adds further support to the suggestion that the basic fronto-parietal organisation between macaques and humans has not changed (Bremmer et al. 2001; Passingham 1998).

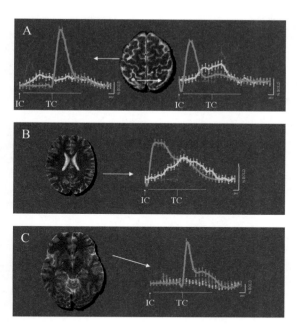

Figure 1. Evoked hæmodynamic responses. BOLD signal (mean ± se; mean adjusted group data) as a function of time and its anatomical localisation. Data from (Toni et al. in press). In red: signal evoked by the visual instruction cue; in green: signal evoked by the auditory trigger cue and the following motor response; in yellow: signal measured during the delay period. Responses have been aligned to the occurrence of the instruction cue (IC) or to the mean occurrence of the trigger cue (TC). The horizontal line above TC indicates the range of instructed delays. Responses above or below the statistical threshold (p < 0.05 corrected for multiple comparisons) are displayed with continuos or dotted lines, respectively. Signal from local maxima in (A) left intraparietal sulcus (-38, -50, 58) and left precentral gyrus (-38, -14, 64); (B) right superior temporal sulcus (62, -40, 24); (C) right inferior frontal gyrus (48, 24, -12).

However, the real advantage of whole-brain imaging over other neurobiological tools relies on its capability to assess the contribution of a whole distributed cerebral network, beyond the relatively well known involvement of the parieto-frontal visuomotor processing stream in motor preparation. We found that preparatory activity is also evoked over portions of the ventral visual stream, along the posterior part of the superior temporal sulcus (Figure 1B). A corresponding region of the macaque's brain has been shown to have mainly visual functions (Desimone & Ungerleider 1986; Yaginuma et al. 1993), and to receive convergent input from areas of both dorsal and ventral visual stream (Baizer, Ungerleider, & Desimone 1991; Distler et al. 1993; Morel & Bullier 1990). Therefore, this region might provide an anatomical bridge between object vision (i.e., infero-temporal visuo-perceptual areas) and spatial vision (i.e., fronto-parietal visuo-motor areas). These features raise the possibility that, once arbitrary visuomotor associations have been learned, the information about visual stimuli, processed in the infero-temporal cortex, might flow directly to the dorsal visual stream. This hypothesis does not contradict our previous suggestions on the involvement of the ventral prefrontal cortex in establishing the appropriate association between a particular sensory cue and an arbitrary motor response (Passingham, Toni, and Rushworth 2000; Toni et al. 2001; Toni and Passingham 1999; Toni, Rushworth, & Passingham 2001). In fact, we found instruction- and movement-related activity in inferior frontal areas across two independent studies (Figure 1C, see also Figure 8C in Toni et al. 1999). These findings suggest that the ventral prefrontal cortex could be involved in the *initial* stages of sensorimotor transformation, together with extrastriate visual areas.

However, as the visuomotor associations became automatic, the contribution of the ventral prefrontal cortex is reduced in favour of other regions or circuits (Toni et al. in press).

MOVEMENT PREPARATION: BEYOND TEMPORAL INDEPENDENCE

It might be argued that there are severe limitations in defining movement preparation simply through a temporal criterion, namely a change in neurovascular or electrical signal occurring during a delay period. A more stringent approach, in the context of neuroimaging studies of motor preparation, would be to find a predictive relationship between the intensity of the neurovascular responses, measured with fMRI, and the strength of the preparatory process, measured by reaction times. Significant covariations between neuroimaging and behavioural measurements would allow wider inferences than the simple presence of an imaging signal during the performance of a given task. This approach has already been used in the single-unit domain (Zhang, Riehle, & Requin 1997), and positive correlations have been found with PET between movement *selection* processes and cerebellar activity (Horwitz et al. 2000). In a different domain, studies on memory have exploited er-fMRI for sorting individual trials, at encoding, on the basis of retrieval performance (Brewer et al. 1998; Wagner et al. 1998).

In order to be testable, this approach would require a substantial amount of variance in task performance. This situation is not immediately compatible with the steady-state performance, free from uncontrolled learning effects or high error rates, necessary for a proper imaging experiment. However, it has been shown that different degrees of motor preparation can be elicited by manipulating the predictive value of the instruction cue (Low & Miller 1999). We have implemented such a manipulation of preparatory activity in a second imaging study (Thoenissen & Toni 2000). The rationale was to distinguish delay-related activity influenced by response probabilities from activity uncorrelated with the likelihood of providing a motor response. To this end, we trained a group of subjects to perform a conditional visuomotor associative task with instructed delays. One of four shapes (instruction cue, IC) was briefly presented (300 ms). Subjects learned before scanning that two shapes instructed the subjects to flex the index finger; the other two shapes instructed the flexion of the middle finger. A variable delay (1 - 21 s in steps of 5 s) was followed by a tone (trigger cue, TC - 300 ms). There were two different auditory cues, a high pitch tone (1700 Hz) and a low pitch tone (300 Hz). One tone instructed the subjects to press the button specified by the instruction cue (Go). The other tone instructed the subjects to cancel the response (Nogo). The colour of the instruction cue (either green or red) predicted (75% validity) the pitch of the forthcoming auditory trigger cue (i.e., its Go/Nogo value). Valid trials were labelled as Go-Go and Nogo-Nogo, whereas invalid trials were labelled as Go-Nogo and Nogo-Go. Trials for each of these four conditions occurred pseudo-randomly in equal frequency during the scanning sessions. The main comparisons in this whole-brain er-fMRI study were between activities evoked during the delay period following Go or Nogo instruction cues and preceeding the imperative trigger cues. The task was performed continuously across the scanning sessions (8.4 min) as well as during intersession intervals (~5 min). However, the predictive value of the colour of the IC was higher (85 %) during the inter-session intervals than during the scanning sessions (50 %). This allowed us to scan a similar number of trials for each condition, while preserving the overall validity of the predictive value of the colour of the IC (i.e. greater than chance level).

Formally, the task is a Go/Nogo paradigm, usually employed to study inhibitory control. However, differently from previous electrophysiological or imaging studies (Filipovic, Jahanshahi, & Rothwell 2000; Garavan, Ross, & Stein 1999; Konishi et al. 1999; Menon et al. 2001), our Go/Nogo task *with instructed delays* was only incidentally targeting the inhibition of prepotent responses. Our main goal was to vary the degree of motor preparation elicited between the instruction and the trigger cues. We achieved this by manipulating the predictive value of the instruction cue, i.e. by introducing trials where the colour of the IC did not effectively predict the Go/Nogo contingency. The behavioural measures collected during the scanning sessions (Figure 2) show that this procedure was effective.

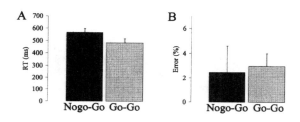

Figure 2. Behavioural data. Reaction times (RTs; median ± se) and error rate (median ± se) measured in Nogo-Go (black) and Go-Go (gray) conditions. Data from (Thoenissen and Toni 2000). RTs in the Nogo-Go condition were significantly longer (p < 0.01) than in the Go-Go condition. Error rate did not differ across conditions.

The experimental manipulation induced a significant effect on RTs (Figure 2A) independent from error rates (Figure 2B). The virtually error free performance in both Go-Go and Nogo-Go trials indicates that the information on the movement to be performed was carried over the delay independently from the likelihood of using such information. The homogeneous error rate across conditions excludes disparities in movement selection: equally correct stimulus-driven choices were made in response to both Go and Nogo instruction cues. Interestingly, when we debriefed our subjects at the end of the experiment, they invariably reported to have performed the task by ignoring the colour of the instruction cues. Despite their verbal reports, we can infer that the subjects took into account the colour of the instruction cue, preparing significantly less when the instruction cue predicted a Nogo contingency. This induced longer RTs when a Nogo instruction cue was followed by an imperative Go trigger cue. Therefore, we can interpret the difference in RTs between Go-Go and Nogo-Go trials as reflecting differences in preparatory activity expressed during the delay period, independently from subjects' strategies.

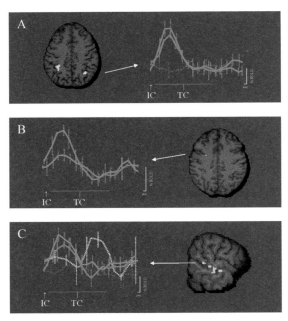

Figure 3. Evoked hæmodynamic responses. BOLD signal (mean ± se; mean adjusted group data) as a function of time and its anatomical localisation. Data from (Thoenissen and Toni 2000). In green: delay-related activity during the Go trials; in red: delay related activity during the Nogo trials; in yellow: signal evoked by the auditory trigger cue during the Go-Nogo trials, aligned to the mean occurrence of the auditory trigger cue (TC); in cyan: signal evoked by the auditory trigger cue and the subsequent movement during the Go-Go trials, aligned to TC. Other conventions as in figure 1. Signal from local maxima in (A) right intraparietal sulcus (34, -62, 46); (B) left precentral sulcus (-26, 0, 36); (C) right superior temporal sulcus (62, -46, 14).

We found that the delay-related activity of parietal clusters, along the intraparietal sulcus, is not affected by the likelihood of executing a movement (Figure 3A). This pattern of activity can be interpreted in the framework of "motor intentions" (Kalaska & Crammond 1995; Snyder, Batista, and Andersen 1997). When sensory stimuli are associated with a behavioural response, the appearance of such stimuli might automatically evoke an intention to move, independently from the actual execution of a movement. Therefore, motor preparation might be a composite phenomenon, articulated in bottom-up processes, that is a stimulus-driven intention to move, and top-down processes, that is a context-dependent command to move (or to withhold the response). The present results suggest that portions of posterior parietal cortex are interested not only on preparing actual movements but also in evaluating the potential motor significance of sensory stimuli.

There might be alternative explanations for this pattern of activity. It might be argued that these areas are involved in sensory working memory, or in movement inhibition. We have addressed and discarded the working memory confound in a separate behavioral experiment (Toni et al. 2002). We also tested for a potential involvement of this parietal region in movement inhibition. We considered the stimulus-related responses obtained during the Go-Nogo trials, aligned to the occurrence of the trigger cue. In these trials the subjects needed to withhold the response, in spite of having being pre-cued to provide it. There was no response in such trials at the time of the trigger cue presentation (Figure 3A, yellow curve), suggesting that the delay-related responses evoked in this region are genuinely associated with motor intentions.

We also confirmed the dissociation between parietal and frontal responses found in the macaque (Kalaska and Crammond 1995) during the performance of a similar task (Figure 3B). While posterior parietal cortex appears to be equally interested in both Go and Nogo trials, we found portions of premotor cortex that are sensitive to the differences between the two contingencies. This suggests that premotor cortex is closer to the executive processes than posterior parietal cortex, as would be expected on the basis of their relative cortico-spinal connectivity (Galea & Darian-Smith 1994; Murray & Coulter 1981).

Delay-related activity was also found in the posterior part of the superior temporal sulcus (Figure 3C). This region was equally active during the delay period for both Go and Nogo trials. However, there was also a response to the Nogo trigger cue in GoNogo trials. It might be argued that this response is related to the sensory stimulation provided by the trigger cue, rather than to its inhibitory meaning. We were able to discard this possibility, since this area did not respond during GoGo trials, where an auditory trigger cue was provided, but not an inhibitory instruction (Figure 3C). Therefore, it emerges that posterior temporal areas appear to be involved in implementing the associative value of sensory stimuli in the context of a visuomotor task.

FUTURE DEVELOPMENTS

Action representations are related to goals, rather than specific movements (Jeannerod 1997). Therefore, it is important to assess the neural correlates of preparatory activity across different effectors (e.g. hand, eye, foot). Recent imaging studies have reported effector-independent activity during preparation (van der Lubbe et al. 2000) and execution (Connolly et al. 2000; Jahanshahi et al. 1995; Jahanshahi et al. 2000; Jenkins et al. 2000) of eye or hand movements. It is important to further extend this line of research to assess the neural basis of effector-independent and effector-dependent activities, as well as their relative dynamics, during motor preparation.

Another topic left unaddressed by imaging studies of motor preparation is the frame(s) of reference used by the brain to represent movements. Single-unit studies have provided several insights into this basic question (Andersen et al. 1997; Batista et al. 1999; Boussaoud & Bremmer 1999; Caminiti, Johnson, & Urbano 1990; Kalaska et al. 1997; Snyder et al. 1998). These electrophysiological studies have exploited their spatio-temporal resolution, together with complex manual task (i.e. performed in 2D or 3D space), in order to distinguish between frames of reference employed by frontal and parietal areas. This approach cannot be transferred as such to the imaging domain. Unfortunately, neurons with different receptive, motor or memory fields are not spatially segregated and therefore do not provide a coherent imaging signal. Nevertheless, the introduction of articulated responses might provide sufficient task-related variance for modelling an optimal frame of reference explaining not only movement performance (McIntyre et al. 2000) but also regional brain activity. Imaging has helped to define the spatial segregation of cerebral functions and representations; the approach delineated above might prove useful for addressing the content of such representations. Finally, the combination of these techniques with interference and stimulation methods (Strafella & Paus 2001) appears essential to test whether the brain is actually using a given representation in a given frame of reference.

Although we might like to describe a given behaviour as based on a set of rules, this does not mean that the brain is using these rules. Studies have shown that rule-based and list-based behaviour might be context-dependent (Dassonville et al. 1999) and that the encoding of visuomotor associations can shift, within neurons and trials, between spatial and non-spatial modes (Johnson et al. 1999). These phenomena call for sophisticated tasks that allow monitoring behaviour across several response levels, so that fewer approximation are necessary in coding a continuous variable (neural activity) into a categorical domain (statistical inference).

ACKNOWLEDGEMENTS

I.T. was supported by the F.C. Donders Centre and by the Hermann von Helmholtz-Gemeinschaft. R.E.P. was supported by the Wellcome Trust.

REFERENCES

Andersen, R. A., Snyder, L. H., Bradley, D. C., and Xing, J. (1997) Multimodal representation of space in the posterior parietal cortex and its use in planning movements. Annu.Rev.Neurosci. 20, 303-330.

Ashe, J. and Georgopoulos, A. P. (1994) Movement parameters and neural activity in motor cortex and area 5. Cereb.Cortex 4, 590-600.

Baizer, J. S., Ungerleider, L. G., and Desimone, R. (1991) Organization of visual inputs to the inferior temporal and posterior parietal cortex in macaques. J Neurosci. 11, 168-190.

Bastian, A., Riehle, A., Erlhagen, W., and Schoner, G. (1998) Prior information preshapes the population representation of movement direction in motor cortex. Neuroreport 9, 315-319.

Batista, A. P., Buneo, C. A., Snyder, L. H., and Andersen, R. A. (1999) Reach plans in eye-centered coordinates. Science 285, 257-260.

Boussaoud, D. and Bremmer, F. (1999) Gaze effects in the cerebral cortex: reference frames for space coding and action. Exp Brain Res. 128, 170-180.

Bremmer, F., Schlack, A., Shah, N. J., Zafiris, O., Kubischik, M., Hoffmann, K.-P., Zilles, K., and Fink, G. R. (2001) Polymodal motion processing in posterior parietal and premotor cortex: a human fMRI study strongly implies equivalencies between humans and monkeys. Neuron 29, 287-296.

Brewer, J. B., Zhao, Z., Desmond, J. E., Glover, G. H., and Gabrieli, J. D. (1998) Making memories: brain activity that predicts how well visual experience will be remembered. Science 281, 1185-1187.

Burnod, Y., Baraduc, P., Battaglia-Mayer, A., Guigon, E., Koechlin, E., Ferraina, S., Lacquaniti, F., and Caminiti, R. (1999) Parieto-frontal coding of reaching: an integrated framework. Exp Brain Res. 129, 325-346.

Caminiti, R., Johnson, P. B., and Urbano, A. (1990) Making arm movements within different parts of space: dynamic aspects in the primate motor cortex. J Neurosci. 10, 2039-2058.

Colby, C. L., Gattass, R., Olson, C. R., and Gross, C. G. (1988) Topographical organization of cortical afferents to extrastriate visual area PO in the macaque: a dual tracer study. J Comp Neurol. 269, 392-413.

Colebatch, J. G., Deiber, M. P., Passingham, R. E., Friston, K. J., and Frackowiak, R. S. (1991) Regional cerebral blood flow during voluntary arm and hand movements in human subjects. J Neurophysiol. 65, 1392-1401.

Coles, M. G., Smid, H. G. O., Scheffers, M. K., and Otten, L. J. (1995) Mental chronometry and the study of human information processing. In: Rugg, M. D. and Coles, M. G. (Eds.) Electrophysiology of mind.Oxford Psychology series, 86-125. Oxford: Oxford University Press

Connolly, J. D., Goodale, M. A., Desouza, J. F., Menon, R. S., and Vilis, T. (2000) A comparison of frontoparietal fMRI activation during anti-saccades and anti-pointing. J Neurophysiol. 84, 1645-1655.

Crammond, D. J. and Kalaska, J. F. (2000) Prior information in motor and premotor cortex: activity during the delay period and effect on pre-movement activity. J.Neurophysiol. 84 , 986-1005.

Cui, R. Q., Egkher, A., Huter, D., Lang, W., Lindinger, G., and Deecke, L. (2000) High resolution spatiotemporal analysis of the contingent negative variation in simple or complex motor tasks and a non-motor task. Clin.Neurophysiol. 111, 1847-1859.

Cui, R. Q., Huter, D., Lang, W., and Deecke, L. (1999) Neuroimage of voluntary movement: topography of the Bereitschaftspotential, a 64-channel DC current source density study. Neuroimage 9, 124-134.

D'Esposito, M., Ballard, D., Zarahn, E., and Aguirre, G. K. (2000) The role of prefrontal cortex in sensory memory and motor preparation: an event-related fMRI study. Neuroimage. 11, 400-408.

Dale, A. M., Liu, A. K., Fischl, B. R., Buckner, R. L., Belliveau, J. W., Lewine, J. D., and Halgren, E. (2000) Dynamic statistical parametric mapping: combining fMRI and MEG for high- resolution imaging of cortical activity. Neuron 26, 55-67.

Dassonville, P., Lewis, S. M., Foster, H. E., and Ashe, J. (1999) Choice and stimulus-response compatibility affect duration of response selection. Brain Res.Cogn Brain Res. 7, 235-240.

Decety, J. and Jeannerod, M. (1995) Mentally simulated movements in virtual reality: does Fitts's law hold in motor imagery? Behav.Brain Res. 72, 127-134.

Decety, J., Perani, D., Jeannerod, M., Bettinardi, V., Tadary, B., Woods, R., Mazziotta, J. C., and Fazio, F. (1994) Mapping motor representations with positron emission tomography. Nature 371, 600-602.

Deiber, M. P., Honda, M., Ibanez, V., Sadato, N., and Hallett, M. (1999) Mesial motor areas in self-initiated versus externally triggered movements examined with fMRI: effect of movement type and rate. J.Neurophysiol. 81, 3065-3077.

Deiber, M. P., Ibanez, V., Honda, M., Sadato, N., Raman, R., and Hallett, M. (1998) Cerebral processes related to visuomotor imagery and generation of simple finger movements studied with positron emission tomography. Neuroimage. 7, 73-85.

Deiber, M. P., Ibanez, V., Sadato, N., and Hallett, M. (1996) Cerebral structures participating in motor preparation in humans: a positron emission tomography study. J.Neurophysiol. 75, 233-247.

Desimone, R. and Ungerleider, L. G. (1986) Multiple visual areas in the caudal superior temporal sulcus of the macaque. J Comp Neurol. 248, 164-189.

Distler, C., Boussaoud, D., Desimone, R., and Ungerleider, L. G. (1993) Cortical connections of inferior temporal area TEO in macaque monkeys. J.Comp Neurol. 334, 125-150.

Eimer, M. (1995) Stimulus-response compatibility and automatic response activation: evidence from psychophysiological studies. J.Exp.Psychol.Hum.Percept.Perform. 21, 837-854.

Endo, H., Kizuka, T., Masuda, T., and Takeda, T. (1999) Automatic activation in the human primary motor cortex synchronized with movement preparation. Brain Res.Cogn Brain Res. 8 , 229-239.

Filipovic, S. R., Jahanshahi, M., and Rothwell, J. C. (2000) Cortical potentials related to the nogo decision. Exp Brain Res. 132, 411-415.

Fogassi, L., Raos, V., Franchi, G., Gallese, V., Luppino, G., and Matelli, M. (1999) Visual responses in the dorsal premotor area F2 of the macaque monkey. Exp Brain Res. 128, 194-199.

Friston, K. J., Fletcher, P., Josephs, O., Holmes, A., Rugg, M. D., and Turner, R. (1998) Event-related fMRI: characterizing differential responses. Neuroimage. 7, 30-40.

Friston, K. J., Price, C. J., Fletcher, P., Moore, C., Frackowiak, R. S., and Dolan, R. J. (1996) The trouble with cognitive subtraction. Neuroimage. 4, 97-104.

Friston, K. J., Zarahn, E., Josephs, O., Henson, R. N., and Dale, A. M. (1999) Stochastic designs in event-related fMRI. Neuroimage. 10, 607-619.

Galea, M. P. and Darian-Smith, I. (1994) Multiple corticospinal neuron populations in the macaque monkey are specified by their unique cortical origins, spinal terminations, and connections. Cereb.Cortex 4, 166-194.

Galletti, C., Fattori, P., Kutz, D. F., and Battaglini, P. P. (1997) Arm movement-related neurons in the visual area V6A of the macaque superior parietal lobule. Eur.J Neurosci. 9, 410-413.

Galletti, C., Fattori, P., Kutz, D. F., and Gamberini, M. (1999) Brain location and visual topography of cortical area V6A in the macaque monkey. Eur.J Neurosci. 11, 575-582.

Garavan, H., Ross, T. J., and Stein, E. A. (1999) Right hemispheric dominance of inhibitory control: an event-related functional MRI study. Proc.Natl.Acad.Sci.U.S.A 96, 8301-8306.

Gerardin, E., Sirigu, A., Lehericy, S., Poline, J. B., Gaymard, B., Marsault, C., Agid, Y., and Le Bihan, D. (2000) Partially overlapping neural networks for real and imagined hand movements. Cereb.Cortex 10, 1093-1104.

Geyer, S., Zilles, K., Luppino, G., and Matelli, M. (2000) Neurofilament protein distribution in the macaque monkey dorsolateral premotor cortex. Eur.J Neurosci. 12, 1554-1566.

Gold, J. I. and Shadlen, M. N. (2000) Representation of a perceptual decision in developing oculomotor commands. Nature 404, 390-394.

Grafton, S. T., Mazziotta, J. C., Woods, R. P., and Phelps, M. E. (1992) Human functional anatomy of visually guided finger movements. Brain 115 (Pt 2), 565-587.

Hackley, S. A. and Miller, J. (1995) Response complexity and precue interval effects on the lateralized readiness potential. Psychophysiology 32, 230-241.

Horwitz, B., Deiber, M., Ibanez, V., Sadato, N., and Hallett, M. (2000) Correlations between reaction time and cerebral blood flow during motor preparation. Neuroimage. 12, 434-441.

Jahanshahi, M., Dimberger, G., Filipovic, S. R., Jones, C., Fuller, R., Frith, C. D., and Barnes, G. (2000) Willed vs externally-triggered saccades. Soc Neurosc Abstracts.

Jahanshahi, M., Jenkins, I. H., Brown, R. G., Marsden, C. D., Passingham, R. E., and Brooks, D. J. (1995) Self-initiated versus externally triggered movements. I. An investigation using measurement of regional cerebral blood flow with PET and movement-related potentials in normal and Parkinson's disease subjects. Brain 118 (Pt 4), 913-933.

Jeannerod, M. (1994) The representing brain: neural correlates of motor intention and imagery. Behav.Brain.Sci. 17, 187-245.

Jeannerod, M., Arbib, M. A., Rizzolatti, G., and Sakata, H. (1995) Grasping objects: the cortical mechanisms of visuomotor transformation. Trends Neurosci. 18, 314-320.

Jeannerod, M. 1997, The cognitive neuroscience of action Blackwell, Oxford.

Jenkins, I. H., Jahanshahi, M., Jueptner, M., Passingham, R. E., and Brooks, D. J. (2000) Self-initiated versus externally triggered movements. II. The effect of movement predictability on regional cerebral blood flow. Brain 123 (Pt 6), 1216-1228.

Johnson, M. T., Coltz, J. D., Hagen, M. C., and Ebner, T. J. (1999) Visuomotor processing as reflected in the directional discharge of premotor and primary motor cortex neurons. J Neurophysiol. 81, 875-894.

Johnson, P. B., Ferraina, S., Bianchi, L., and Caminiti, R. (1996) Cortical networks for visual reaching: physiological and anatomical organization of frontal and parietal lobe arm regions. Cereb.Cortex 6, 102-119.

Josephs, O., Turner, R., and Friston, K. J. (1997) Event-related fMRI. Hum.Brain Mapp. 5, 243-248.

Kalaska, J. F. and Crammond, D. J. (1992) Cerebral cortical mechanisms of reaching movements. Science 255, 1517-1523.

Kalaska, J. F. and Crammond, D. J. (1995) Deciding not to GO: neuronal correlates of response selection in a GO/NOGO task in primate premotor and parietal cortex. Cereb.Cortex 5, 410-428.

Kalaska, J. F., Scott, S. H., Cisek, P., and Sergio, L. E. (1997) Cortical control of reaching movements. Curr.Opin.Neurobiol. 7, 849-859.

Kawashima, R., Roland, P. E., and O'Sullivan, B. T. (1994) Fields in human motor areas involved in preparation for reaching, actual reaching, and visuomotor learning: a positron emission tomography study. J.Neurosci. 14, 3462-3474.

Klemmer, E. T. (1957) Simple reaction time as a function of time uncertainty. J Exp Psychol 54, 195-200.

Konishi, S., Nakajima, K., Uchida, I., Kikyo, H., Kameyama, M., and Miyashita, Y. (1999) Common inhibitory mechanism in human inferior prefrontal cortex revealed by event-related functional MRI. Brain 122 (Pt 5), 981-991.

Kornhuber, H. H. and Deecke, L. (1965) hirnpotentialänderungen bei Willkürbewegungen und passiven Bewegungen des Menschen: Bereitschschaftspotential und reafferente Potentiale. Pflügers Arch 284, 1-17.

Krams, M., Rushworth, M. F., Deiber, M. P., Frackowiak, R. S., and Passingham, R. E. (1998) The preparation, execution and suppression of copied movements in the human brain. Exp.Brain Res. 120, 386-398.

Low, K. A. and Miller, J. (1999) The usefulness of partial information: effects of go probability in the choice/Nogo task. Psychophysiology 36, 288-297.

Markman, A. B. and Dietrich, E. (2000) In defense of representation. Cognit.Psychol 40, 138-171.

Matelli, M., Govoni, P., Galletti, C., Kutz, D. F., and Luppino, G. (1998) Superior area 6 afferents from the superior parietal lobule in the macaque monkey. J Comp Neurol. 402, 327-352.

McIntyre, J., Stratta, F., Droulez, J., and Lacquaniti, F. (2000) Analysis of pointing errors reveals properties of data representations and coordinate transformations within the central nervous system. Neural Comput. 12, 2823-2855.

Menon, R. S., Luknowsky, D. C., and Gati, J. S. (1998) Mental chronometry using latency-resolved functional MRI. Proc.Natl.Acad.Sci.U.S.A 95, 10902-10907.

Menon, V., Adleman, N. E., White, C. D., Glover, G. H., and Reiss, A. L. (2001) Error-related brain activation during a Go/NoGo response inhibition task. Hum.Brain Mapp. 12, 131-143.

Moody, S. L. and Wise, S. P. (2000) A model that accounts for activity prior to sensory inputs and responses during matching-to-sample tasks. J.Cogn Neurosci. 12, 429-448.

Morel, A. and Bullier, J. (1990) Anatomical segregation of two cortical visual pathways in the macaque monkey. Vis.Neurosci. 4, 555-578.

Murray, E. A. and Coulter, J. D. (1981) Organization of corticospinal neurons in the monkey. J Comp Neurol. 195, 339-365.

Ogawa, S., Lee, T. M., Stepnoski, R., Chen, W., Zhu, X. H., and Ugurbil, K. (2000) An approach to probe some neural systems interaction by functional MRI at neural time scale down to milliseconds. Proc.Natl.Acad.Sci.U.S.A 97, 11026-11031.

Papa, S. M., Artieda, J., and Obeso, J. A. (1991) Cortical activity preceding self-initiated and externally triggered voluntary movement. Mov Disord. 6, 217-224.

Parsons, L. M. (1994) Temporal and kinematic properties of motor behavior reflected in mentally simulated action. J Exp.Psychol.Hum.Percept.Perform. 20, 709-730.

Parsons, L. M., Fox, P. T., Downs, J. H., Glass, T., Hirsch, T. B., Martin, C. C., Jerabek, P. A., and Lancaster, J. L. (1995) Use of implicit motor imagery for visual shape discrimination as revealed by PET. Nature 375, 54-58.

Passingham, R. E. (1998) The specializations of the human neocortex. In: Milner, A. D. (Ed.) Comparative Neuropsychology., 271-298. Oxford: Oxford University Press

Passingham, R. E. and Toni, I. (2001) Contrasting the Dorsal and Ventral Visual Systems: Guidance of Movement versus Decision Making. Neuroimage 14, S125-S131.

Passingham, R. E., Toni, I., and Rushworth, M. F. (2000) Specialisation within the prefrontal cortex: the ventral prefrontal cortex and associative learning. Exp Brain Res. 133, 103-113.

Picard, N. and Strick, P. L. (1997) Activation on the medial wall during remembered sequences of reaching movements in monkeys. J Neurophysiol. 77, 2197-2201.

Porro, C. A., Francescato, M. P., Cettolo, V., Diamond, M. E., Baraldi, P., Zuiani, C., Bazzocchi, M., and di Prampero, P. E. (1996) Primary motor and sensory cortex activation during motor performance and motor imagery: a functional magnetic resonance imaging study. J Neurosci. 16, 7688-7698.

Rao, S. M., Binder, J. R., Bandettini, P. A., Hammeke, T. A., Yetkin, F. Z., Jesmanowicz, A., Lisk, L. M., Morris, G. L., Mueller, W. M., and Estkowski, L. D. (1993) Functional magnetic resonance imaging of complex human movements. Neurology 43, 2311-2318.

Requin, J., Brener, J., and Ring, C. (1991) Preparation for Action. In: Jennings, J. R. and Coles, M. G. (Eds.) Handbook of Cognitive Psychophysiology., 357-448. New York: John Wiley & Sons

Richter, W., Andersen, P. M., Georgopoulos, A. P., and Kim, S. G. (1997) Sequential activity in human motor areas during a delayed cued finger movement task studied by time-resolved fMRI. Neuroreport 8, 1257-1261.

Richter, W., Somorjai, R., Summers, R., Jarmasz, M., Menon, R. S., Gati, J. S., Georgopoulos, A. P., Tegeler, C., Ugurbil, K., and Kim, S. G. (2000) Motor area activity during mental rotation studied by time-resolved single-trial fMRI. J Cogn Neurosci. 12, 310-320.

Riehle, A. and Requin, J. (1989) Monkey primary motor and premotor cortex: single-cell activity related to prior information about direction and extent of an intended movement. J.Neurophysiol. 61, 534-549.

Roland, P. E., Skinhoj, E., Lassen, N. A., and Larsen, B. (1980) Different cortical areas in man in organization of voluntary movements in extrapersonal space. J Neurophysiol. 43 , 137-150.

Roth, M., Decety, J., Raybaudi, M., Massarelli, R., Delon-Martin, C., Segebarth, C., Morand, S., Gemignani, A., Decorps, M., and Jeannerod, M. (1996) Possible involvement of primary motor cortex in mentally simulated movement: a functional magnetic resonance imaging study. Neuroreport 7, 1280-1284.

Sanes, J. N., Donoghue, J. P., Thangaraj, V., Edelman, R. R., and Warach, S. (1995) Shared neural substrates controlling hand movements in human motor cortex. Science 268, 1775-1777.

Savaki, H. E. and Dalezios, Y. (1999) 14C-deoxyglucose mapping of the monkey brain during reaching to visual targets. Prog.Neurobiol. 58, 473-540.

Snyder, L. H., Batista, A. P., and Andersen, R. A. (1997) Coding of intention in the posterior parietal cortex. Nature 386, 167-170.

Snyder, L. H., Grieve, K. L., Brotchie, P., and Andersen, R. A. (1998) Separate body- and world-referenced representations of visual space in parietal cortex. Nature 394, 887-891.

Soechting, J. F. and Flanders, M. (1992) Moving in three-dimensional space: frames of reference, vectors, and coordinate systems. Annu.Rev.Neurosci. 15, 167-191.

Steinberg, S. (1969) The discovery of processing stages: extensions of Donder's method. Acta Psychologica 30, 276-315.

Stephan, K. M., Fink, G. R., Passingham, R. E., Silbersweig, D., Ceballos-Baumann, A. O., Frith, C. D., and Frackowiak, R. S. (1995) Functional anatomy of the mental representation of upper extremity movements in healthy subjects. J.Neurophysiol. 73, 373-386.

Strafella, A. P. and Paus, T. (2001) Cerebral blood-flow changes induced by paired-pulse transcranial magnetic stimulation of the primary motor cortex. J Neurophysiol. 85, 2624-2629.

Tanne, J., Boussaoud, D., Boyer-Zeller, N., and Rouiller, E. M. (1995) Direct visual pathways for reaching movements in the macaque monkey. Neuroreport 7, 267-272.

Thoenissen, D. and Toni, I. (2000) Movement preparation and motor intention: an event-related fMRI study. Soc Neurosc 178.

Toni, I. and Passingham, R. E. (1999) Prefrontal-basal ganglia pathways are involved in the learning of arbitrary visuomotor associations: a PET study. Exp.Brain Res. 127, 19-32.

Toni, I., Ramnani, N., Josephs, O., Ashburner, J., and Passingham, R. E. (2001) Learning arbitrary visuomotor associations: temporal dynamic of brain activity. Neuroimage. 14, 1048-1057.

Toni, I., Rushworth, M. F., and Passingham, R. E. (2001) Neural correlates of visuomotor associations: Spatial rules compared with arbitrary rules. Exp.Brain Res. 141, 359-369.

Toni, I., Schluter, N. D., Josephs, O., Friston, K., and Passingham, R. E. (1999) Signal-, set- and movement-related activity in the human brain: an event-related fMRI study [published erratum appears in Cereb Cortex 1999 Mar;9(2):196]. Cereb.Cortex 9, 35-49.

Toni, I., Shah, N. J., Fink, G. R., Thoenissen, D., Passingham, R. E., and Zilles, K. (in press) Multiple movement representations in the human brain: an event-related fMRI study. J Cogn Neurosci.

Toni, I., Thoenissen, D., Zilles, K., and Niedeggen, M. (2002) Movement preparation and working memory: a behavioural dissociation. Exp.Brain Res. 142, 158-162.

van der Lubbe, R. H., Wauschkuhn, B., Wascher, E., Niehoff, T., Kompf, D., and Verleger, R. (2000) Lateralized EEG components with direction information for the preparation of saccades versus finger movements. Exp Brain Res. 132, 163-178.

Vanduffel, W., Tootell, R. B., and Orban, G. A. (2000) Attention-dependent suppression of metabolic activity in the early stages of the macaque visual system. Cereb.Cortex 10, 109-126.

Verleger, R., Wauschkuhn, B., van der Lubbe, R. H., Jaskowski, P., and Trillenberg, P. (2000) Posterior and anterior contribution of hand-movement preparation to late CNV. J Psychophysiology 14, 69-86.

Wagner, A. D., Schacter, D. L., Rotte, M., Koutstaal, W., Maril, A., Dale, A. M., Rosen, B. R., and Buckner, R. L. (1998) Building memories: remembering and forgetting of verbal experiences as predicted by brain activity. Science 281, 1188-1191.

Walter, W. G., Cooper, R., Aldridge, V. J., McCallum, W. C., and Winter, A. L. (1964) Contingent negative variation: an electric sign of sensorimotor association and expectancy in the human brain. Nature 203, 380-384.

Weilke, F., Spiegel, S., Boecker, H., von Einsiedel, H. G., Conrad, B., Schwaiger, M., and Erhard, P. (2001) Time-Resolved fMRI of Activation Patterns in M1 and SMA During Complex Voluntary Movement. J Neurophysiol. 85, 1858-1863.

White, I. M. and Wise, S. P. (1999) Rule-dependent neuronal activity in the prefrontal cortex. Exp.Brain Res. 126, 315-335.

Wise, S. P., Boussaoud, D., Johnson, P. B., and Caminiti, R. (1997) Premotor and parietal cortex: corticocortical connectivity and combinatorial computations. Annu.Rev.Neurosci. 20, 25-42.

Wise, S. P., di Pellegrino, G., and Boussaoud, D. (1996) The premotor cortex and nonstandard sensorimotor mapping. Can.J.Physiol Pharmacol. 74, 469-482.

Wise, S. P. and Mauritz, K. H. (1985) Set-related neuronal activity in the premotor cortex of rhesus monkeys: effects of changes in motor set. Proc.R.Soc.Lond B Biol.Sci. 223, 331-354.

Yaginuma, S., Osawa, Y., Yamaguchi, K., and Iwai, E. (1993) Differential functions of central and peripheral visual field representations in monkey prestriate cortex. In: Ono, T. et al. (Eds.) Brain mechanisms of perception and memory., 3-15. Oxford: Oxford University Press

Zarahn, E., Aguirre, G. K., and D'Esposito, M. (1999) Temporal isolation of the neural correlates of spatial mnemonic processing with fMRI. Brain Res.Cogn Brain Res. 7, 255-268.

Zhang, J., Riehle, A., and Requin, J. (1997) Analyzing neuronal processing locus in stimulus-response association tasks. J Mathem Psychol 41, 2119-2236.

HUMAN FREEDOM, REASONED WILL, AND THE BRAIN: THE BEREITSCHAFTSPOTENTIAL STORY

Lüder Deecke,[1] and Hans Helmut Kornhuber[2]

[1] Department of Clinical Neurology, University of Vienna, and Ludwig Boltzmann Institute for Functional Brain Topography, Vienna, Austria

[2] Professor Emeritus, formerly Head, Department of Neurology University of Ulm, Germany

THE DISCOVERY OF THE BEREITSCHAFTSPOTENTIAL

The Bereitschaftspotential story began when on a sunny Saturday in the spring of 1964 Hans Helmut Kornhuber (then *Dozent* and chief physician at the Department of Neurology, head, Professor Richard Jung, University Hospital, Freiburg/Breisgau) and Lüder Deecke (his doctoral student) went for lunch to the 'Gasthaus zum Schwanen' at the foot of the Schloßberg hill. Sitting alone in the beautiful garden they discussed their frustration with research on the brain as a passive system prevailing worldwide and their desire to investigate self-initiated action (Kornhuber & Deecke, 1990).

The possibility to do research on electrical brain potentials preceding voluntary movements came with the advent of the 'computer of average transients,' the first, primitive averager available at that time in the Freiburg laboratory. In the electroencephalogram little is to be seen preceding actions, except an inconstant diminution of the α- (or μ-) rhythm. The young researchers stored the electroencephalogram and electromyogram of self-initiated movements (fast finger flexions) on tape and analyzed the cerebral potentials preceding movements time-reversed with the start of the movement as the trigger, literally turning the tape over for analysis since they had no reversal playback or programmable computer. A potential preceding human voluntary movement was discovered and published in the same year (Kornhuber & Deecke, 1964). After detailed investigation and control experiments such as passive finger movements the citation 'Classic' with the term Bereitschaftspotential (BP) was published (Kornhuber & Deecke, 1965).

The Bereitschaftspotential (Figure 1) is ten to hundred times smaller than the α-rhythm of the EEG and it becomes apparent only by averaging, relating the electrical potentials to the onset of the movement. The pre-motion positivity is even smaller, and the motor-potential which starts about fifty to sixty milliseconds before the onset of movement and has its maximum over the contralateral precentral hand area is still smaller. Thus, great care is required to record these potentials: exact triggering by the real onset of movement is important, which is especially difficult preceding speech movements (Grözinger et al., 1975; 1980). Furthermore artefacts due to head-, eye-, lid-, mouth-movements and respiration have to be eliminated before averaging because such artefacts may be of a magnitude which makes it difficult to render them negligible even after hundreds of sweeps. In the case of eye movements, eye muscle potentials have to be distinguished from cerebral potentials (Becker et al., 1971; Kornhuber, 1973b). In some cases animal experiments were necessary to clarify the origin of potentials such as the respiratory wave ('R-wave', Diekmann et al., 1980; Kriebel et al., 1990). Therefore, it took many years until some of the other laboratories were able to confirm the details of our results.

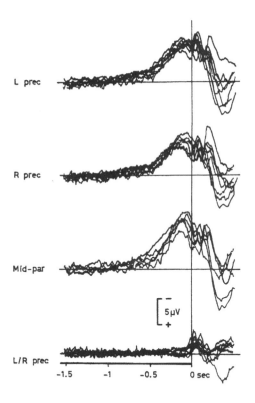

Figure 1. Slow shifts of the cortical DC potential (Bereitschaftspotential, BP) preceding volitional, rapid flexions of the right index finger (t= 0 s, vertical line). Recording positions are left precentral (L prec, C3), right precentral (R prec, C4), mid-parietal (Pz). Unipolar recordings with linked ears as reference. The difference between the BP in C3 and in C4 is displayed in the lowest graph (L/R prec). Superimposed are the results of eight experiments as obtained in the same subject (B.L.) on different days. Note that the BP has two components, the early one (BP1) lasting from app. -1.2 to -0.5; the late component (BP2) from -0.5 to shortly before 0 s [Data obtained by Deecke, Grözinger, Kornhuber 1976a *Biol Cybern* Springer Berlin, etc., with permission]

The BP was received with great interest by the scientific community as reflected by Sir John Eccles' comment: 'There is a delightful parallel between these impressively simple experiments and the experiments of Galileo Galilei who investigated the laws of motion of the universe with metal balls on an inclined plane' (Eccles & Zeier, 1980). The interest was even greater in psychology and philosophy because volition is traditionally associated with the belief in human freedom. The spirit of the time, however, was hostile to freedom since it was believed that freedom is an illusion. The tradition of behaviourism and Freudism was deterministic. While will and volition were frequently leading concepts in psychological research papers before and after the first world war and even during the second war, after the end of the second world war this focus declined, and by the mid-sixties these key words completely disappeared and were abolished in the thesaurus of the American Psychological Association (Heckhausen, 1987).

THE CONCEPT OF FREEDOM

It is not surprising therefore, that the real beginning of the story was much earlier. One of us who as a youth was interested in astronomy and mainly in chemistry became a prisoner of war early in 1945 and was kept in a Soviet camp for slave work until late in 1949. The longer this time of imprisonment, the less hope for return home: a situation suited to develop a motivation independent of external reward, a will in agreement with the self and in doing things right for the fellow prisoners. As a schoolboy, he had already participated in scientific research at the Chemical Institute of Königsberg University, and it had seemed clear to him that he would be a researcher in chemistry. But after return from the prison camp he decided to become a physician to help people. While studying medicine, he read Herakleitos and Kungtse at night time, Plato and Kant, Shakespeare and Montesquieu, Goethe and Hölderlin, Kierkegaard and Thoreau, Nietzsche and Albert Schweitzer. He became a student of the philosopher Karl Jaspers in Basel, and he did his thesis work under Kurt Schneider in Heidelberg on the release of endogenous depressions by shocking experience (Kornhuber, 1955). When already a neurologist in Freiburg he wrote a handbook article on the psychology of the prisoner of war (Kornhuber, 1961). Since 1962, he gave lectures on freedom for students of all faculties at Freiburg University. LD was one of these students, this way also becoming interested in freedom which is one of the old great questions preoccupying mankind.

HHK's concept of freedom was not at all contrary to nature (Kornhuber, 1978e; 1984b; 1987; 1988c; 1992; 1993). What he had in mind was 'positive' freedom, freedom to deeds (creativity, reasoned will, self-control, authenticity). He believed that most of the freedom from hunger, fear, disease, ignorance, dictatorship and so on ('negative' freedom in Kant's terms) was the result of positive freedom. Freedom is relative: adult man is usually more free than a toddler, man more free than a chimpanzee, and a chimpanzee more free than a worm. Man has an intellectual and moral development with influence of the individual person on its own education. In the brain there is not only causality from bottom up (for instance from circadian regulation on thinking), but also from top (for instance conscience) to bottom. Signs of diminished freedom also point towards freedom which was lost. Walter Rudolf Hess (Hess, 1954; 1956) for instance stimulated a center of hunger in the diencephalon of a cat. The animal became so hungry that it tried to eat a wooden stick or a metal cable - it showed compulsive behaviour. A balanced system of drives belongs to the bases of freedom, and obviously we can do something to keep it near equilibrium. In many ways one can do something for freedom, for instance by supplementing a newborn child with thyroxin in case of dysfunction of the thyroid gland, in order to prevent cretinism. A healthy lifestyle is a condition for freedom in the long run. And truth is a basis for freedom. Freedom is a task for human life and civilization to be solved stepwise by research, arts, justice and practice. Obviously the belief in complete determinism is contradictory in itself because without freedom for research, truth can not be distinguished from error. On the other hand, a freedom against nature (as sometimes understood with the term free will) seems meaningless: on what information should this free will be based, and to what end outside the natural world? Although Immanuel Kant had the right sense of freedom, Kant's freedom belonged to the mundus intelligibilis, a world different from the mundus sensibilis, the natural world in which Kant (with the physicists of his time) believed in complete determinism. The two worlds, however, are in reality one: the world of intelligent living beings. The phenonema of freedom occur in our everyday life: civilization as a result of human creativity, the development of freedom in children, the destruction of freedom by diseases of the brain, the circadian changes of our freedom depending on wakefulness and fatigue and so on. We are even obliged to distinguish states of freedom, for instance regarding driving under the influence of drugs or alcohol. In the ancient world, positive freedom was called virtue. In ancient times, the great story of becoming free (told by Xenophon) was the myth of Herakles at the crossroads. The importance of reasoned will and choice was self-evident in the ancient world; the key word was sophrosyne (one of Plato's 'basic virtues'). The lack of the concept of will in Plato's writings is only seemingly so, because for Socrates it was self-evident that man has will: Socratic strength of will was proverbial at that time. Aristotle (Ethika Nikomacheia) pointed towards exercise as a source of freedom. He wrote: Ethical virtue comes from habit-formation. Virtues arise in us neither by nature alone nor contrary to nature; we have natural dispositions for them, but we perfect them by practice. We become just by doing what is just, brave by doing brave deeds. Sophocles in his famous chorus of 'Antigone' on the greatness of man pointed to freedom in the positive sense as a consequence of human creativity. But while he pointed to civilization he also reminded us that the creativity of man may be dangerous. Positive freedom has a Janus-face: it is the ability for good and evil. It is in line with this that both creativity and conscience (or self-criticism) are juxtaposed in the human frontal lobe. When after the medieval age the Hellenic spirit was revived in the renaissance, Pico della Mirandola explained human freedom again by creativity (De hominis dignitate, 1496)

Freedom is neither outside nature nor a chance event nor hedonistic behaviour. Positive freedom is related to honesty and authenticity. It is not indifferent how we think about freedom. The overemphasis on 'negative' freedom such as the protest against the Vietnam war and of an easy life without duties towards children (the 'pill' came in 1965) led 1968 to a big experiment: the hedonistic society. The number of drug-addicts rose in Germany sixty-fold from 1968 to 1971, and there were many more negative consequences of hedonism such as a more than tenfold increase of alcoholism in women: a loss of freedom.

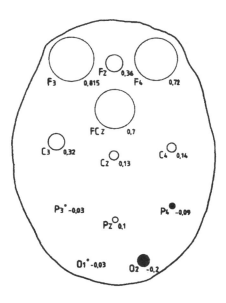

Figure 2. The correlations between learning (error reduction in motor performance) and enhancement of negative cortical potential in a visuomotor learning task shows high and significant correlations only at frontolateral and frontomedial areas. Thus, the effort of the will is accompanied by extra excitation of the frontal lobe cortex only. [From Lang et al 1983 *Pflügers Arch Eur J Physiol,* Springer Berlin etc., with permission]

EXPERIMENTS ON LEARNING WITH EFFORT AND A THEORY OF MEMORY FORMATION

The BP pointed to the supplementary motor area (SMA) and the anterior cingulate gyrus. The following experiments on motor learning requiring the effort of will revealed the entire convexity of the frontal lobe to be involved (Lang et al., 1983) (Figure 2). At the time when we started the experiments which led to the BP, the centers of motivation were thought to be localized in subcortical structures such as the diencephalon and hypothalamus (Hess, 1954; 1956). This corresponded to Freudian theory. The neurophysiological data came from Walter Rudolf Hess's stimulation experiments in cats showing signs of hunger, fear, aggression and so on. A prominent author in neuropsychology at that time, Hans Lukas Teuber, considered frontal lobe function to be a riddle. He speculated about the frontal lobe as the source of a corollary discharge - an information for the sensory systems. At that time, these systems dominated the interests of neuropsychologists. The prefrontal cortex, however, is the most humane part of human anatomy. Learning, self-criticism and reasoned will act throughtout life and shape our personality. It is a widespread mistake that the development of the character is determined solely by the genes and the milieu. Man himself is a third factor: he also shapes his fate (see also Asendorpf, 1999; Ernst & von Luckner, 1985; Göppinger, 1987). Today it seems obvious to us that the frontal lobe (which contains most of the association cortex of the human brain) is the leading part of man, the highest center of motivation. But because of a 'fear' of freedom and reasoned will and the difficulty of experimentally studying these, many neuropsychologists still speak instead of working memory as the function of the frontal lobes.

Because of our interest in the influence of motivation on the selection and storage of information in man, we also developed a theory of information processing and memory. It is clear that the abundant sensory signals received by our receptors must be highly selected. Some of this occurs already at the level of the spinal cord, but most in the forebrain. Attention is one of the functions in which convergent sensory signals meet with motivational impulses from the limbic system and the frontal lobe for the selection of important information and for long term storage (Kornhuber, 1973a; 1978d).

THE VERSATILITY OF THE BRAIN, THE COMPONENTS OF VOLITION, AND THE FUNCTIONS OF THE FRONTAL LOBE

The methods showing cerebral activity with good localization using the measurement of regional cerebral blood flow with positron emission tomography, single photon emission computerized-tomography and functional magnetic resonance imaging have usually a temporal resolution in the order of seconds or more so that they cannot distinguish between processes before or after an action. The analysis of electrical potentials or magnetic fields, on the other hand, has a temporal resolution in the order of one millisecond. Therefore we tried to use electrical potentials to investigate the versatility of cerebral functions under different conditions of volition, attention, perception and action. In a hand tracking experiment, for instance, in which the time course of stimuli was known to the subject whereas the direction of the moving stimulus (which changes suddenly at a certain time) is unpredictable, the SMA showed anticipatory behaviour: the BP declined half a second before the expected change, whereas the directed attention potential (which had its maximum over the parietal area) continued to remain high until two hundred milliseconds after the stimulus reversal when the sensory processing was completed (Figure 3). Thus, the frontal lobe, after deciding what to do, delegated further action to posterior cortical areas which are competent to utilize visual stimuli and to decide in detail how to perform the task. From these and similar experiments and from previous results on lesions (Kleist, 1934; Shallice, 1982) we developed a theory on the components of volition and their funcional localization in the frontal lobes (Kornhuber, 1984a; Lang et al., 1984; Deecke et al., 1985). One of these stages is planning, and for planning, working memory is required.

The brain is a cooperative system, but with strategic organization. There is little doubt that in man the prefrontal-orbital cortex is the highest level of planning and decision-making. The frontal lobe is the most humane part of man. Language was an earlier step in hominid evolution; *homo erectus* and probably *homo habilis* had language, but only the creative abilities of the large frontal lobe brought *homo sapiens* to arts and civilisation. The Neanderthals although they had language and even hundred grammes more brain than recent man made no progress in tools through 200 000 years and did not develop arts as far as we know; they had a smaller frontal lobe (Kornhuber, 1993; 1995).

A THEORY OF THE MOTOR SYSTEM

Being neurologists and interested in rehabilitation after cerebral lesions we were not happy with the dominance of the precentral motor cortex in the textbooks; in this strip of cortex all motor abilities where believed to be concentrated; which had paralyzing effects on the intention to restore function by re-learning and exercise. Fortunately one of us had expertise also in the oculomotor system (Kornhuber, 1974b; 1978a) and in the somatosensory system (Kornhuber, 1962; 1972); this helped in drawing a more realistic picture of the motor system (Kornhuber, 1974a). The frontal eye-field, also included in the precentral motor cortex at that time, is different from the motor cortex; according to its thalamic nucleus and its visual cortical afferents it belongs to the prefrontal cortex. The motor cortex was developed for fine tactile control of those movements that need this information; therefore the large representations of the lips and the tongue in rodents and the large representations of the fingers and toes in primates. Somatosensory afferents are the most prominent of neurons of the precentral motor cortex. By itself the motor cortex is unable to do anything; it needs afferents from the basal ganglia and the cerebellum. On the other hand, the motor cortex also receives information from motivational processes in the limbic system, basal ganglia and prefrontal cortex which in turn get highly processed sensory information from the parietal, temporal and occipital sensory association areas. Thus, the motor cortex is not the master of the basal ganglia and cerebellum as was thought at that time, but on the contrary it needs the help of both of them; among other things they seem to serve as function generators. The akinesia of the Parkinsonian syndrome shows the helplesness of the motor cortex when deprived of the input from the basal ganglia. The BP is reduced over the motor cortex in this case (Deecke et al., 1976b; 1977). Similar to the cerebellum, it receives afferents from large parts of the cerebral cortex via the pontine nuclei, and in man a great deal of the cerebellar output goes via the thalamus to the motor cortex. These views which seem almost self-evident today were so new at that time that two prominent neurophysiologists and textbook authors at that time, Richard Jung and Robert F. Schmidt rejected the new picture. After lesions, however, the distributed organization of the motor system is one of the bases for the restoration of function under active learning and exercise (the other basis is the plasticity of the cortex).

FREEDOM AS A PRACTICAL TASK

Our belief in human freedom and in the duty of man to promote it (civilization, after all, lives from cooperation of free humans) was not only theoretical. We engaged, for instance, in the prevention of cigarette smoking (Kornhuber, 1982) and of alcohol consumption (Kornhuber, 1984c; 1990; 2001) and against the abuse of benzodiazepines (Binder et al., 1984, Kornhuber, 1988a). For the last 25 years (Kornhuber, 1977) we propose a health levy on alcohol and cigarettes (revenues to the health insurance system); in Germany now the

VISUAL TRACKING

Supplementary motor area, midline

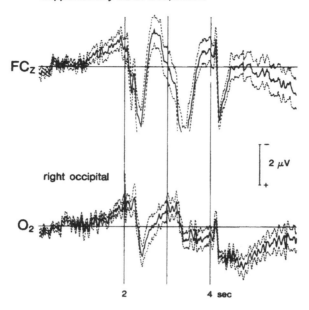

Figure 3. Visuomotor tracking experiment as in Figure 2. The comparison of the time course of potentials at FCz (overlying the supplementary motor area) and O2 (right occipital lead) shows the anticipatory behaviour of the negative potential at FCz which ends before the stimulus reversal (s 3), about 500 ms prior to the end of the directed attention potential in the occipital lead. Grand average from the 16 subjects, with standard error (dotted lines); note the small intersubject variability of the potentials. [From Deecke et al. (1984) *Ann NY Acad Sci* Vol 425: 450-464, with permission]

political parties and the labour unions agree that such a levy is desirable: A law seems within reach, despite the counter-propaganda of the lobby. We used epidemiological methods in order to clarify the causes of arterial hypertension which is the most important cause of stroke. At that time it was believed that the main causes of hypertension were salt consumption and psychological stress. We did not confirm either of these hypotheses, instead we found alcohol consumption and obesity to be the main causes of hypertension, with abdominal obesity (and the metabolic syndrome) in men largely being the consequence of daily alcohol (Kornhuber, 1984c; Kornhuber, Backhaus et al., 1989; Kornhuber, 2001). We introduced the measurement of blood pressure into the teaching of primary schools and high schools for earlier treatment of hypertension (Kornhuber & Kornhuber, 1980), and we tried to improve prevention by developing a preventive branch in every part of clinical medicine, coining the term preventive neurology (Kornhuber, 1983). We developed methods for decentralized care and rehabilitation at home through advice and practical courses for the families of the chronically ill (Geiselmann & Kornhuber, 1981, Kornhuber & Riebler, 1984) and constructed, for example, a portable sonograph for the ambulatory measurement of residual urine in cases of neurogenic bladder disorder (Kornhuber, Widder, Christ, 1980, Widder et al., 1983) and a reliable baby-protector against sudden infant death (Renner & Kornhuber, 1987, Kornhuber, 1989). Our advice (Kornhuber, 1986b), communicated directly to the Federal Minister of Health, led to the law which improved the nursing care for the chronically ill. We developed effective methods for the treatment of multiple sclerosis (Kornhuber & Mauch, 1995). We designed methods to find drugs for rapid treatment of acute stroke and cerebral haematomas, and found Flunarizine to be effective (Kleiser et al., 1990). To prevent Alzheimer's disease we pointed to the microangiopathy underlying the disease in most cases (Kornhuber, 1997a; 1998). In the course of search for a better treatment of schizophrenia the role of glutamate was elucidated (Kim et al., 1980, 1981). This led to the use of glutamate antagonists in therapy, for example, of dementia. To promote the treatment of neuromuscular diseases a multidisciplinary center was established with success, and for the therapy of epilepsy, an epilepsy center which was among the first worldwide that was able to remove selectively the most epileptogenic structures of the brain, amygdala and hippocampus, without side-effects.

To improve education, we established a general studies program with lectures of eminent scientists, philosophers, poets, economists and politicians and with courses in humanities and arts to widen the mind of the students in the medical, science and technical schools at Ulm University (Kornhuber, 1978c; Waldvogel, 1982) and we founded a school for language- and speech-therapists, the first in Germany which preceded all the following ones by its curriculum including therapy for aphasia (Schwäb. Zeitung, 1978). We campaigned for better legislation for families with children because we consider the family to be the primary source of education and morals as well as care for the ill and the elderly (Kornhuber, 1978b, 1985, 1986a). We proposed a better promotion of research by more objective, retrospective evaluation instead of preceding reviews of applications by peers who actually are competitors (except, of course, for very large projects) (Kornhuber, 1988b, 1994, 1997b). We founded a journal which avoids plagiarism and nepotism (Huber & Kornhuber, 1993). In Vienna, a stroke registry was established, and a municipal stroke emergency ambulance system was organized (Lang & Lalouschek, 2001). A 'help-MS-Fonds' was founded to help MS patients in their domestic environment and improve their activities of daily living (Deecke & Gebhard, 2001).

THE BP AND MAGNETOENCEPHALOGRAPHY

In 1982, HHK sent LD on leave to Vancouver in order to work with Hal Weinberg on one of the first magnetoencephalography (MEG) systems, which at that time was a 1-channel system. This meant repositioning the dewar and repeating the same experiment over and over until the mapping over the scalp was completed. Such topography isofield lines was to show two extrema, one where the field lines went into the skull, and one extremum where they came out of the skull (for review of the method cf. Beisteiner et al., 1998). Figure 4 shows the first recordings of the MEG equivalent of the BP, which we termed 'Bereitschaftsmagnetfeld' or, shorter, Bereitschaftsfeld (Deecke et al., 1982). At present two terms are prefered Bereitschaftsfeld (BF) and Readiness field (RF).

Experiments in movement using the MEG were continued by studying the BF with movements of the lower extremity (foot and toe, Deecke et al., 1983). The advantage of MEG over EEG is better localization. The BF late component (BF2) has been localized in MI (Brodmann´s area 4, Deecke et al., 1983). It reveals the typical somatotopic organization known from Foerster (1936; cf. Figure 11) and Penfield and Rasmussen (1950), that is for movements of different parts of the body, the equivalent current dipoles corresponding to BF2 were distributed in a homuncular pattern (Cheyne et al., 1991, from where Figure 5 is taken).

So far so good for the primary motor area (MI) generator (BF2). MEG localization of the BF1 generator was less readily achieved. We were wondering from the beginning why BF onset times were shorter (0.8 to 0.5 s before movement) than the usual BP onset times (1 to 1.5 s before movement). The reason is that with all the MEG recordings of our own and other groups, the early BF or BF1 was not recorded. In the healthy subject, both SMA generators are active also in the case of unilateral movement. Due to the anatomical localization of the SMA on the mesial surface of the two hemispheres, the two SMAs face eachother and their fields partially cancel one another. How is it then with lesions of the SMA?

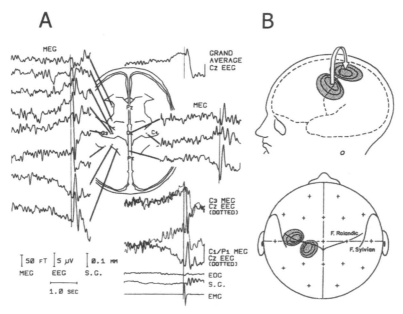

Figure 4. First magnetoencephalographic recordings of the BP equivalent, Bereitschaftsfield (BF) or readiness field. A: In prerolandic scalp positions, upward deflection of the BF is observed, which means that field lines come out of the head. In retrorolandic positions downward BF deflection is recorded which indicates that field lines go into the head. Between the two extrema, standing 90° on the connecting line, the equivalent current dipole (ECD) is to be assumed, which fits very well into the rolandic region. B: Diagrams illustrating the two extrema, field lines coming out of the head and those going into the head. FT = femtoTesla, 10^{-15} Tesla, S.G. = Strain Gauge device for measuring head displacement that can cause artifacts. [Modified from Deecke, Weinberg, Brickett 1982 *Exp Brain Res* Springer Berlin, etc., with permission].

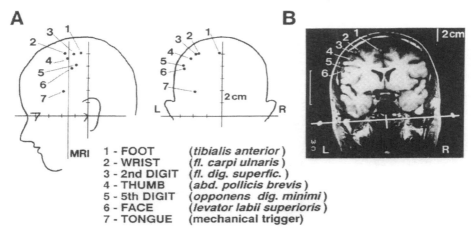

1 - FOOT	(*tibialis anterior*)
2 - WRIST	(*fl. carpi ulnaris*)
3 - 2nd DIGIT	(*fl. dig. superfic.*)
4 - THUMB	(*abd. pollicis brevis*)
5 - 5th DIGIT	(*opponens dig. minimi*)
6 - FACE	(*levator labii superioris*)
7 - TONGUE	(mechanical trigger)

Figure 5. 'Homuncular' distribution of the MEG Bereitschaftsfield.
A: Three-dimensional dipole source locations for 7 different movement conditions as indicated. Sources of neural activity - identified using the non-invasive measurements of cerebral magnetic fields - were found to confirm the somatotopic organization of primary motor cortex for movements of different parts of the body in normal human subjects. The somatotopic map produced with this technique revealed slight differences to the classic homunculus obtained from studies using invasive cortical stimulation in epileptic patients (Penfield and Rasmussen, 1950).
B: These source locations were projected onto a coronal magnetic resonance image (MRI) taken from the same subject. The typical homuncular distribution of foot-, wrist-, digit-, face- and tongue movements indicated that the generators of BF2/RF2 lie in Brodmann's area 4. An exception is the dipole for voluntary tongue movements. It showed more variability due to mechanical triggering and other reasons. [From Cheyne, Kristeva & Deecke, 1991 *Neurosci Lett* Elsevier Amsterdam, with permission]

THE BP AND SMA LESIONS

We have seen that activation of the SMA – and as already mentioned also the cingulate motor area (CMA) – leads the activation of MI in time, that is it is upstream in the final volitional cascade when it comes to the transition of intention into actual execution of movement. HHK and LD had drawn attention to the important observation when studying Parkinsonian patients that the SMA participates in the preparation and organization of voluntary movement (Deecke & Kornhuber, 1978). In this paper the *starting function* of voluntary action had been attributed to the SMA.

It was, therefore, of great interest to know how conditions are when one of the two SMAs was 'missing' or in other words, what happens with *dysfunction of the SMA*? In a first series of experiments (Deecke et al., 1987), about 20 000 computerized tomograms of the CT center of the Neurological Hospital of Ulm University at Dietenbronn were evaluated to detect cases where chronic unilateral lesions of one SMA were the only lesion. The patients were contacted, and a number of them volunteered to be readmitted to the Neurological Hospital in order to participate in a simple voluntary movement paradigm (BP paradigm, comparing the crossed hemisphere hand system having the SMA lesion with the healthy, not affected system). In the end, 12 successful recordings were available. The results are shown in Figure 6 – 6A showing the lesions, and 6B the BP. What was perhaps 'disappointing" for the investigators, was a good message for the patients in the chronic state: There were no major differences in the BP amplitudes comparing the affected system with the non-affected one (cf. Figure 6B). This is typical for the frontal lobe, that unilateral lesions can be compensated rather well from the other side.

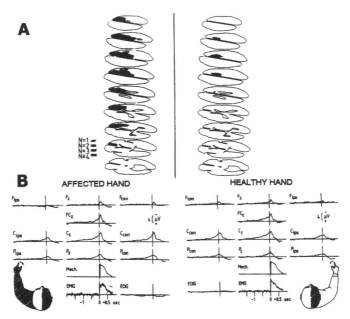

Figure 6. Bereitschaftspotential and SMA lesions 1.
A: First series of patients having a lesion of one SMA only (app. 20 000 computerized tomograms of the CT center of the Neurological Hospital of Ulm University at Dietenbronn were evaluated). From the mailing, a group of patients was willing to be readmitted to the Neurological Hospital for the experiment. From 12 patients usable data were obtained.
B: Data from the 12 experiments using the voluntary movement BP paradigm. Compared is the crossed hemisphere hand system carrying the SMA lesion ('Affected Hand') with the healthy not lesioned system ('Healthy Hand'). Ipsilateral leads (F_{ips}, C_{ips}, P_{ips}) have to be compared with ipsilateral ones, and contralateral leads (F_{con}, C_{con}, P_{con}) with contralateral ones, midline (Fz, FCz, Cz, Pz) can be compared with midline. Note that for these patients in the chronic state with simple index finger flexions, there was no difference in the BP comparing the affected system with the healthy one. [Modified from Deecke et al. (1987) *J Neurol Neurosurg Psychiat* Brit Med Ass London, with permission]

However, could it be that we did not ask the specific 'SMA- question' with these simple movements? Simple movements do not particularly 'challenge' the SMA; for simple movements the SMA can 'delegate' the job to the primary motor area (MI), and a person having a chronic lesion of only one SMA can do simple movements steered by the remaining other SMA alone. Consequently we started investigating unilateral SMA lesions with a more complex task involving different kinds of bimanual motor sequences (Asenbaum, Oldenkott, Lang, Lindinger & Deecke, 1990). This was done on the basis of scanning through about 30 000 CTs of the CT center of the Neurological University Hospital Vienna. Clinically, these patients presented with bradykinesia contralateral to the lesion, deceleration of initial movement, switching from sequential to simultaneous movement and frequent failure to initiate or inhibit a movement on one side or the other. This is shown in Table 1: The clinical syndrome of SMA lesion is termed *motor dysrhythmia* or *dyschronokinesia*. The BP (prior to movement) and the negativity of performance (N-P, post movement onset) were larger over the lesioned hemisphere (Figure 7). This seeming paradox is due to the location of the remaining SMA´s dipole on the mesial surface of the hemisphere. The bradykinesia of these frontal patients resembled that seen in Parkinson's disease.

Table I: Motor dysrhythmia, dyschronokinesia. Disturbance of movement performance with simultaneous vs. sequential movements in unilateral SMA lesions:

I. Disturbance of movement initiation
(1) Slowing of the first segment of movement .... N=7
(2) Delay (left).. N=9
(3) Omission of one or more segments of movement
(a) unilateral (left).. N=8
(b) bilateral.. N=5
(c) break off.. N=4
II. Disturbance of movement rhythmicity
(*motor dysrhythmia, dyschronokinesia*)
(1) Transition into simultaneous action......................... N=5
(2) Disturbed right/left alternation............................... N=4
III. Disturbance of movement performance
Contralateral bradykinesia.. N=11

A third experiment with SMA lesions was carried out. The Lang et al., (1991) study was an MEG experiment on a patient with a right SMA lesion was asked to participate in an MEG experiment. This was one possibility to come around the cancellation problem as mentioned above (the CMA activity could be a pure radial dipole that escapes MEG detection). The results are shown in Figure 8. The patient having only one remaining SMA (the left), performed voluntary flexions with his right index finger. The results were quite convincing. In the early phase of the RF (1200 to 800 ms prior to the onset of movement, corresponding to BP1, Figure 8A) field lines were going out of the head at the vertex (Cz) and were going into the head at a frontal position between F3 and Fz. They thus enveloped an electric dipole on the mesial surface of the left hemisphere in the left (intact) SMA. A similar dipole was found during the period of 800 to 600 ms prior to the onset of movement (Figure 8B), still corresponding to the BP1- SMA system. However, in the late period of the RF (during BP2), in this case measured between 200 and 0 ms prior to movement onset (Figure 8C), activity had shifted.

A

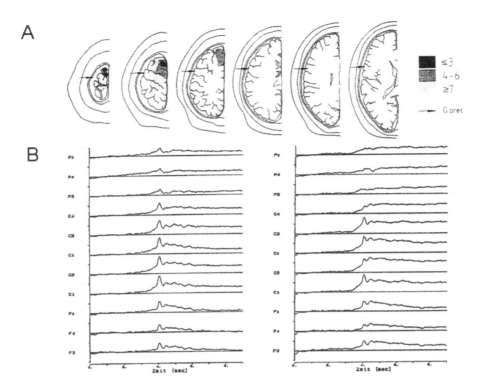

B

Figure 7. The BP and SMA lesions 2.
A: A second series of patients was recruited (app. 30 000 computerized tomograms of the CT center of the Neurological University Hospital of Vienna were scanned). Shown are only the patients with SMA lesions in the left hemisphere (N=7). Peripherally, these patients 'at first sight' had a normal neurological status, no reduction of force, no overt clumsiness in the contralateral hand, etc. With the sequential movements under study, however, typical patterns of dysfunction were seen as summarized in Table 1.
B: Grand average curves from the 7 recordings using the voluntary movement (BP) paradigm. Left, simultaneous task, right sequential task. Note after movement onset, sustained negativity termed N-P (negativity of performance) especially for the sequential task (right) particularly over Cz (vertex) and the para-vertex leads, while the N-P for simultaneous movement (left) tends to return to baseline. Compare C4 with C3, particularly with the sequential movements (right): The C3 lead on the side of the lesion shows the larger potential compared to C4. This 'paradox' is explained by the location of the dipole on the mesial surface of the hemisphere. The only remaining (right) SMA is active. Its dipole points with its negative pole towards the lesioned hemisphere, where it influences the surface topography accordingly. Similar findings can be obtained with foot and toe movements, with evoked potentials of the lower extremity (e.g. tibial nerve EPs) or with visual hemifield stimulation. Ordinate: Leads from Pz through F3, amplitudes in µV, negative up. Abscissa, time ('Zeit') in sec. [From Asenbaum et al. (1990) Tilburg University Press, with permission]

Magnetic field lines now left the head at FC3 and entered the head at FCz. This indicated an electrical dipole in the left area 4 hand representation (MI). This again supports our hypothesis that the SMA leads the MI motor cortex activation prior to human voluntary movement. We believe that any self-initiated action is preceded by SMA activity (Deecke et al., 1976a, and – largely founded on our BP results – Eccles, 1982). The SMA/CMA system is obviously also needed for simple movements in order to prepare for the voluntary, endogenous movement (it is necessary for self-initiation), however more so for complicated movements and actions.

The final proof that SMA activity can be seen in the MEG came with the experiments of Erdler et al. (2000). There we have employed another strategy to try to find the SMA activity prior to movement in the MEG, that is in the early BP component, BP1 or its MEG equivalent BF1. It should be possible to detect it in the intact subject as well. Although both SMAs are normally always active even preceding unilateral movement, the *contralateral*

SMA activity should be a bit stronger than the ipsilateral one for unilateral movement. A self-initiated finger tapping task of the right fingers was employed in 8 normal volunteers. The BF was recorded by using our 143-channel whole scalp MEG system (CTF Inc. Port Coquitlam, Canada) accommodated in a magnetically shielded room (Vacuumschmelze, Hanau, Germany) in the early morning hours, when the strong dipoles of the streetcar overhead contact line had been switched off. Under these quiet conditions, the pre-movement SMA activity of BF1 from about 1.5 to 0.5 s prior to movement was recorded. As evident from Figure 9, the left SMA dipole was stronger than the right one (dipole moment contralateral 2.4 nAm (nano Ampère meters) as compared to 1.6 nAm ipsilateral (Erdler et al., 2000).

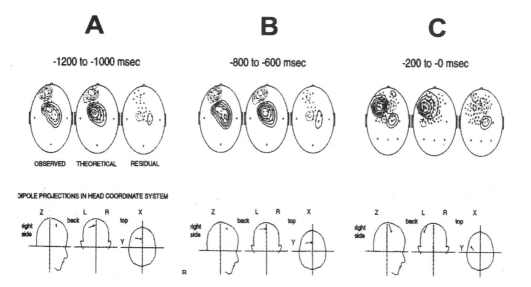

Figure 8. The BP in SMA lesions 3. MEG recordings of the Bereitschaftsfield (BF). Three different 200 ms time windows of the foreperiod in a patient having an SMA infarction (right anterior cerebral artery). The patient performed brisk flexions of his right index finger in a volitional self-initiated manner (voluntary movement BP paradigm).

A: Early phase of the readiness field (BP1 or BF1) 1200 to 1000 ms prior to the first EMG activity related to the voluntary contraction. Field lines come out of the head (solid iso-field lines) at the vertex Cz, they go into the head (stippled) at the frontal region corresponding to a position F1 (between F3 and Fz). Magnetic field lines envelop an electric dipole situated on the mesial surface of the left hemisphere (see projections of the dipole into the head coordinate system at the bottom). In the center (coronal) head sketch the dipole points to the left, that is negative *intra*cellular currents point to the left, which means that negative *extra*cellular currents point to the right. This is consistent with negative activity in superficial cortical layers of the left SMA. Note dipole in head coordinate system at the bottom.

B: Same situation a little later in the foreperiod -800 to -600 ms prior to the onset of movement, still BP1 period (or BF1 period, respectively). SMA dipole still active and well-pronounced (very little residual variance). Note dipole remaining in same place as in A (SMA) in head coordinate system at the bottom.

C: In the late phase of the readiness field BP2 or BF2, respectively (between -200 and 0 ms prior to the onset of movement), field lines come now out of the head (solid field lines) at FC3 and go into the head at FCz (stippled), corresponding to an electric dipole in the left precentral area (MI hand area). Note completely different dipole location as in A and B in head coordinate system at the bottom. This late dipole is less unequivocal, which is probably due to the fact that the early SMA dipole is not completely silent but still shows some activity. [Unpublished data from the experiment by Lang, Cheyne, Kristeva, Beisteiner, Lindinger, Deecke, 1991 *Exp Brain Res* Springer Berlin etc.,]

THE BP AND ELECTROENCEPHALOGRAPHY

The association of the early BP component with the pre-movement activation of the SMA and the late BP component with the – subsequent – activation of MI was proposed by Deecke and Kornhuber (1978). The importance of the BP lies in the fact that it is the only clinical-neurophysiological correlate of brain functions such as motivation, preparation, intention and initiation of voluntary movements in man.

This is why the BP has now become a well-established tool in the motor physiological laboratory, in clinical neurophysiology and in research (Marsden, Deecke, Freund et al., 1996). We were the first who used real DC recording conditions, and certainly, this should be current practice in all BP laboratories, that is recording the BP using an infinite time constant ($\tau = \infty$), which we call high resolution DC-EEG [64-channel computer-assisted DC-EEG amplifier (Lindinger et al., 1991)]. While, as mentioned above, the expectancy wave (CNV, Walter et al., 1964) occurs in conjunction with a reaction time experiment to exogenous stimuli, volitional self-initiated acts are endogenous and are preceded by the BP. Thus, the CNV paradigm is associated with *re-actions*, while the BP paradigm is associated with *actions*. Like the CNV, the BP has both an early and a late component (cf. Figs. 1 and 10). The early component is termed BP1 and lasts from the very beginning of the BP (1-2 s or more prior to movement onset depending on the complexity of the task) to about 0.5 s prior to movement onset. The late component is termed BP2 (see Figs. 1 and 10) and lasts from 0.5 s before to shortly before movement (s 0 in all Figures is the onset of movement indicated by the EMG). BP1 is symmetrical even for unilateral movement, while BP2 shows a characteristic contralateral dominance.

Figure 10 gives an example of the BP preceding voluntary right index finger flexions averaging a large number (N=1100) of self-initiated movements. We see the slow negativity of the BP starting already 1.2 s prior to the onset of movement (0, vertical line). Similar to Figure 1, the two BP components are clearly seen. At the mid-parietal recording in Figure 10, we have labelled the early, more shallow negative component, BP1, and the late, steeper negative component, BP2. It is our hypothesis supported by many pieces of experimental evidence that the principal generator of BP1 is the mesial prefrontal cortex including the SMA and also the CMA, while the principal generator of BP2 is the primary motor cortex (MI, area 4, precentral gyrus, 'motor strip').

The BP as such has an early start of 1 to 2 s prior to movement onset. For complex movements, onset time is even 3 s (e.g. writing or drawing, Schreiber et al., 1983). For simple movements as in Figure 10, the BP is positive in frontal leads, negative in central and parietal leads as well as at the mid-parietal recording. According to the theory of the EEG, negativity can be related to activity of the cortical areas under study, while positivity is related to inactivity (Creutzfeldt, 1983). The BP paradigm (voluntary movement paradigm) investigates internally produced movement, that is willed actions (Deecke et al., 1969; 1976a; Deecke, 1987; 1990; Deecke & Lang, 1996). For these volitional movements one can envisage an 'act of will' necessary to self-initiate the movement (Kornhuber & Deecke et al., 1989). This certainly is a 'frontal act', and the decision of 'when to move' is thought to be elaborated by the mesial fronto-central cortex.

It is the frontal lobe, where the SMA is located, and there is evidence that not only the SMA but also adjacent parts of the cingulate gyrus are activated prior to movement onset ('cingulate motor area', CMA, (Shima et al., 1991). Figure 11 gives Figures 3 a and b from Foerster (1936), where we have indicated by handwriting the present nomenclature: 'SMA proper' corresponds to Foerster's area 6aα, 'pre-SMA' to 6aβ. It is the pre-SMA where we think BP1 is generated and also from the CMA located below 6aβ (Figure 11).

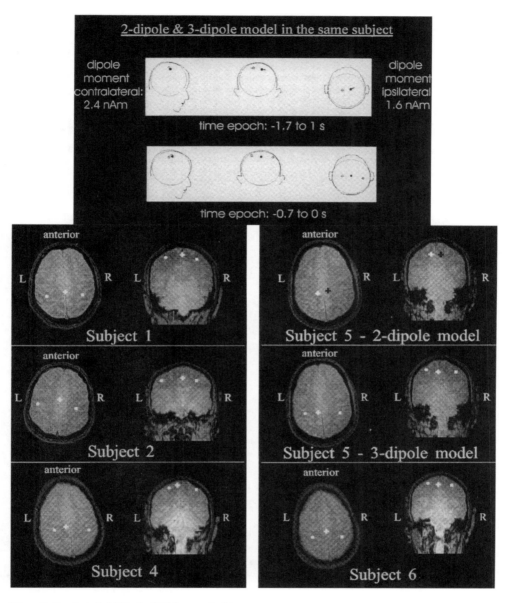

Figure 9. Bereitschaftsfield in the SMA in normal subjects.

Top part: Localization of the dipoles in the head coordinate system for subject 2 showing separation of the left and the right SMA dipoles for finger tapping of the right hand, corresponding to bottom part right side top row. The left (contralateral) SMA showed a dipole moment of 2.4 nAm, the ipsilateral a dipole strength of 1,6 nAm [unpublished data from Erdler et al., 2000]

Bottom part: Left side: Overlays of 3-dipole models on axial and coronal anatomical MRIs for subjects 1, 2, and 4. The cross represents the SMA dipole. The circle the contralateral MI dipole, and the square the ipsilateral MI dipole. Dipoles are visualized through different MRI slices. L, left; R, right.

Right side: For subject 2 two different dipole models are overlaid on an axial and coronal MRI. The upper row shows the overlay of the 2-dipole model, the white cross representing the contralateral SMA, and the black cross the ipsilateral SMA. The middle row shows the 3-dipole model for subject 5, and the bottom row the 3 dipole model for subject 6, meaning of cross, circle, and square as above. [Modified from Erdler, Beisteiner, Mayer, Kaindl, Edward, Windischberger, Lindinger & Deecke (2000) *NeuroImage* Academic Press London, with permission]

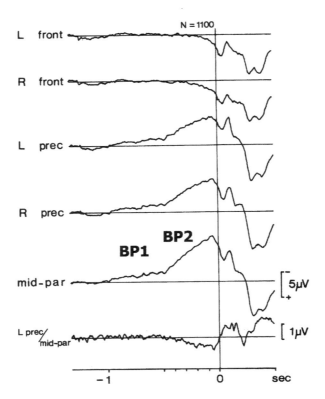

Figure 10. The early and late BP. Movement-related potentials in different recording locations of a subject performing flexions of the right index finger (Average of 1100 self-initiated movements at irregular intervals). The BP is precentro-parietally negative, and frontally positive. BP1 (early component of the BP) from about 1.2 to 0.5 sec prior to movement onset. BP2 (late component of the BP) from 0.5 s to shortly before movement onset (0s , vertical line). Pre-motion positivity (PMP, from culmination point 90 to 80 ms prior to movement onset in all centro-parietal leads, type C-subject). Motor potential (MP) in bipolar recording L/R precentral 60 to 50 ms prior to movement onset. Reafferent potentials (RAP) after movement onset. [Modified from Deecke 1974 *Habilitationsschrift*]

The justification for our use of the term SMA/CMA is the following: Marsden, Deecke et al., (1996; p 477) agreed that the distinction between the anterior SMA or pre-SMA and the posterior SMA or SMA-proper should be emphasized. In the human, the posterior SMA or SMA proper probably corresponds to the medial Brodmann area 6aα, while the anterior SMA or pre-SMA corresponds to medial area 6aβ as was drawn on Foerster's stimulation map (Figure 11). In addition, a large pyramidal cell cluster in the upper bank of the cingulate sulcus corresponds to the CMA (Tanji, 1994). The anterior SMA and posterior SMA in humans are roughly divided by the VAC line, drawn vertically through the anterior comissure at 90° to the AC-PC line connecting the anterior and posterior comissures. In short, when we say 'SMA' we mean the pre-SMA, the medial Brodmann area 6aβ and when we say 'CMA' we mean the precentral part of the cingulate gyrus (the upper bank of the cingulate sulcus). For the sake of simplicity, the term SMA/CMA is used for the two zones of activity in the early pre-movement period prior to self-initiated movement or action. This early BP component is called BP1. There has been a debate in the literature about a hierarchy and some groups have said that they cannot envisage the SMA as a so-called 'supramotor area.'

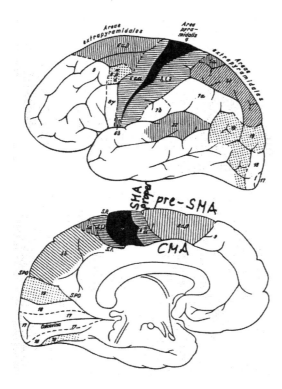

Figure 11. Foerster´s (1936) stimulation mapping illustration with our present, pencil-marked, nomenclature: Foerster´s 6aα is what we now call SMA proper. However, this is not the area we think the early BP comes from. This is rather area 6aβ, which is presently termed 'pre-SMA.' Below the pre-SMA is the cingulate motor area (CMA). Translation of Foerster´s German legend: Motor cortical areas of the human. Area 4 or area gigantopyramidalis, in short area pyramidalis is drawn in black. The areas extrapyramidales 6aα, 6aβ, 6b, 3, 1, 2, 5a, 5b, 22 are hatched. The eye movement areas 8 α β γ, and 19 are stippled [Modified from Foerster (1936) Springer Berlin, with permission]

We do not interpret the SMA as a supramotor area. According to our hypothesis, the SMA is a premotor area, both premotor in time and premotor in cortical topography. Surely it is upstream of the MI, that is, it starts activation earlier in the premovement period of self-initiated actions than do the MIs.

This is the situation with unilateral voluntary movements, but how are bilateral movements organized? Experiments investigating unilateral versus bilateral movements can tell us something about the *temporal order of the time instants* at which activation starts in the two principal generators that produce the BP, SMA/CMA as the principal generator for BP1 and MI as the principal generator for BP2. When does activation start? There are three possibilities: Do both generators start activation at the same time, as some groups believe (Ikeda et al., 1993), or is MI earlier than SMA, which nobody believes, or is SMA activation earlier than MI activation, which we believe to be the case (cf. Deecke, 1987)? Our hypothesis that SMA is upstream in the final motor cascade when it comes to channeling motivation, intention or the act of will into motor execution is demonstrated in Figure 12.

In Figure 12A, the BP preceding simple monophasic *bilateral* index finger flexions is shown. At the fronto-central midline (vertex, Cz), negativity starts already at -1.1 s, while at the primary motor cortex (MI, C'3, C'4), the BP starts significantly later, at -0.7 s. In Figure 12B a recent confirmation of our early observation can be seen: Experiments involving *bi-*

lateral simultaneous voluntary self-initiated index finger extensions against different loads were carried out by Cui and Deecke (1999). We see an early BP onset over the SMA over the fronto-central midline, FCz and Cz at -1.8 s, and a significantly later BP onset over primary motor cortex (MI, C3 and C4) at -0.8 s. The generally earlier BP onset times are due to the fact that the loaded movements of Figure 12B are complex movements as opposed to the simple ones of Figure 12A. We have shown that BP onset times increase with the level of complexity of the movement, the longest onset times having been observed with speaking (Deecke et al., 1986) or writing and drawing (Schreiber et al., 1983).

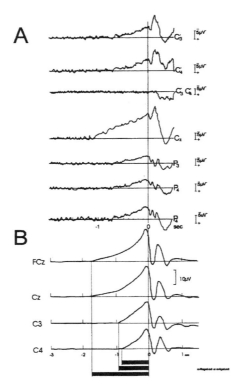

Figure 12. Bilateral movement.
A: Cerebral potentials preceding *bilateral* simultaneous simple monophasic index finger flexions. Monophasic implies as in all our experiments that the subject performed self-initiated brisk voluntary flexions of the two index fingers that remained in flexed position returning to a resting   position only during the ITI (inter trial interval). Bereitschaftspotential earliest and largest at vertex. Note that at the fronto-central midline (vertex, Cz), negativity starts already at -1.1 s, while at the primary motor cortex (MI, C'3, C'4), BP starts significantly later: at -0.7 s (movement onset at 0 s, vertical line). 128 artifact-free trials. Monopolar recordings against linked mastoids. C'3 or C'4 1cm anterior of C3 or C4. [Part of Fig. 1 from Kristeva & Deecke (1980) Elsevier Amsterdam, with permission].
B: Cerebral potentials preceding *bilateral* simultaneous voluntary self-initiated index finger extensions against different loads. Note similar finding as in A: Early BP onset over the supplementary motor area (SMA, fronto-central midline, FCz and Cz) at -1.8 s; significantly later BP onset over primary motor cortex (MI, C3 and C4) at -0.8 s. The generally earlier BP onset times are due to the fact that the loaded movements are complex movements as opposed to the simple movements of A. [Modified from Cui, Deecke (1999) *Brain Topography* Human Sciences Press, with permission]

THE BP AND CURRENT SOURCE DENSITY

Perhaps one can say that as a result of being challenged by the new technique of MEG, the electroencephalographers tried harder. Using special mathematics first applied by Donald Mackay (1984), they have improved their EEG considerably, where spatial resolution, localisation power and reliability are concerned. Current source density analysis is a method that allows evaluation of the topography of current sources and sinks on the scalp. Current source density is proportional to the sum of partial second derivatives of the potential field (Mackay, 1984). Its values are proportional to the current entering and exiting the scalp just like magnetic field lines in the MEG. The method is independent of the location of the reference electrode used for recording. Therefore, current source density analysis arrives at an absolute and quantitative description of the field distribution, showing a more precise localization of the electrical activities than the raw potential distribution (Nagamine et al., 1992).

Also for cortical DC potentials current source density (Laplacian transformation and spline interpolation) is useful (Lindinger et al., 1994). Thus, the cortical DC potential was topographically displayed using current source density mapping algorithms that enable surface topographies which are quite advanced relative to conventional mapping of 'raw' potentials. Figure 13 was derived in such a way. It is composed from data of Cui et al. (2000). In this study, we investigated the BP at 56 scalp positions when 23 healthy subjects performed finger tapping against the thumb of both hands comparing two tasks, a simple sequence 2-2-2 and a complex one 2-5-2. The current source density Laplacians revealed that a distinct current sink (= surface negativity) of the BP appeared on the scalp as early as 2.3 s before the onset of movement in the EMG (see Figure 13). Note that this SMA/CMA activation starts earlier for the complex than for the simple sequence. Thus, the sink at the midline is earlier and stronger for the complex finger sequence (right in Figure 13). 2.3 s as the onset time of the BP is early for simple movements (usually about 1.5 s) but not for complex sequential movements as in Figure 13.

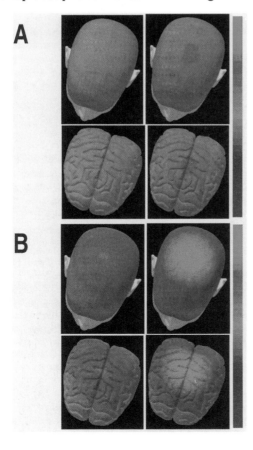

Figure 13. Current source density (CSD) maps of the BP from the grand averages of two experimental tasks. Left images, simple finger sequence 2-2-2; right images, complex sequence 2-5-2. The negative scale's maximum (highest current *sink* density) is represented by red. The positive scale's maximum is represented by blue (highest current *source* density). The two upper images in each section A and B show the current source density using the "head model," the two lower ones employing the "cortex model" of our laboratory (cf. Lindinger et al., 1991; 1994).

A: Early BP phase showing the BP at its dawn. Grand average of 16 subjects analyzed at 2.3 s prior to movement onset in the EMG. Note that with the complex sequential movement (right images), the BP at the fronto-central midline (SMA/CMA) is just starting, whereas for the simple sequence not yet (left images).

B: Late BP phase. BP2 starts at about -0.5 s on the average. Consequently CSD sink (negativity) starts shifting towards the contralateral MI, with the present bilateral movements to both MIs. The lower section, BP2, is "shot" at 0.45 s prior to EMG onset [Modified from Cui et al., 2000 Exp Brain Res, Springer Berlin etc., with permission].

We also studied different inertial loads on a finger movement (100 g as opposed to 200 g, Cui et al., 1996). The aim of the investigation was to clarify, whether the pre-movement neuronal activation in the form of the BP was different for example in topography for the different inertial loads. Furthermore, we wanted to know, whether the activation of the SMA/CMA does occur earlier than that of the MI. The BP was recorded on the scalp using a 64 channel DC amplifier system. The results showed that the amplitudes of the BP were significantly different between the vertex (SMA/CMA) and the contralateral MI (measured at C5/P3 [midway between 10-20 electrode positions C5 and P3]). BP amplitudes were higher in Cz than they were in C5/P3 for both tasks, especially the early component, BP1. The comparison of the 3 motor areas (SMA/CMA, contralateral and ipsilateral MI) showed that the BP was largest over the SMA/CMA and smallest over the ipsilateral MI in both tasks.

The late component, BP2, starting about 0.5 s prior to movement onset, which for unilateral movement is lateralized towards the contralateral hemisphere, accordingly showed a lateralized current sink. We have called this the contralateral preponderance of negativity, CPN (Deecke et al., 1976a; for review cf. Deecke, 1987, 1990). The CPN holds for hand movements, while foot movements are preceded by an ipsilateral preponderance of negativity (IPN [Boschert et al., 1983; Boschert & Deecke, 1986]). The location of the motor potential's (MP) current sink was similar to that of BP2, but the areas and the densities were higher. It could be concluded that the more force required for finger extension against a load, the larger the BP in motor areas. This is probably due to the greater *effort* necessary for preparing the voluntary movement against force. BP1 clearly started in the SMA (see Figure 13), while BP2 and the MP were lateralized in the form of the CPN. The activation of the SMA occurred considerably earlier than that of MI. We had stressed early on the importance of the SMA in voluntary movement (Kornhuber, 1974a; Deecke & Kornhuber, 1978; Deecke, 1987; Kornhuber, Deecke et al., 1989) long before it was recognized as a key structure in the cortico-basal ganglia-thalamo-cortical motor loop (DeLong, 1990).

In recent experiments (Cui et al., 1999), the spatial pattern of the BP has been analyzed in short time intervals (35 and/or 70 ms) starting 2.51 s before movement onset. For each time segment a spherical model of the BP was calculated by using spline interpolation. Subsequently, the spatial distribution of the electric potential at the scalp surface was transformed into a spatial distribution of current source densities (CSD map). In 13 of 17 subjects, there was a significant current sink in the scalp area located over the fronto-central mesial cortex (SMA/CMA) in the absence of a significant current sink over the lateral motor cortex. In three subjects significant current sinks were present at both sites and in another three subjects a current sink was only observed over the lateral motor cortex. We concluded that there is a large group of subjects (13/17) in whom the early component of the BP is associated with a significant current sink over SMA and CMA. At a later time interval (0.6 to 0.5 s before movement onset), current sinks in mesial and lateral positions were found in most subjects (10 of 17). We consider that these data corroborate the hypothesis that in the majority of subjects the midline motor areas SMA/CMA are activated earlier than the primary motor cortex when preparing for a voluntary movement. It is clear then that our postulate that the SMA/CMA activity is upstream of MI activity holds for the vast majority of subjects but not for all, which might explain the different views of other groups (e.g. Ikeda et al., 1993). In new experiments we are now attempting to clarify whether this difference is an intrinsic inter-subject characteristic or rather due to different strategies: purposeful conscious actions on the one hand or more automatic routine performance on the other.

THE BP AND IMAGERY — MENTAL REPRESENTATION OF MOTOR ACTS

These studies deal with the investigation of motor imagery, that is our ability to see a movement or action 'in our mind's eye.' For example athletes make use of motor imagery ('mental rehearsal') before they start a competition in order to improve performance, such as diving. Their 'mental representation of motor acts' is so good that they know beforehand whether a stroke for instance in golf will be a hit or a miss. We conducted a topographical study on the mental imagery of motor and other tasks (Goldenberg et al., 1989, Uhl et al., 1990, also cf. Deecke, 1996).

The imagery paradigm: Experiments were carried out in which human mental activity was recorded and mapped through scalp electrodes using DC amplifiers (Uhl et al., 1990).

The problem: Mental activity usually occurs spontaneously and unpredictably. The recording of cortical DC potentials requires averaging. How can the averaging process be triggered by an unpredictable event?

The solution: The subject starts the trial by a self-initiated voluntary movement (preceded by a BP), that is the subject presses the buttons of a slide projector presenting himself with a slide. This initiating movement provides the trigger for time-locked potential averaging (at 0 s in Figure 14). Instantaneously, a slide appears on the screen in front of the subject for 200 ms (ending at 0.2 s). With the contents of the slide, the subject was instructed which of three categories – colours, faces or a spatial map – would follow. After 3 s, a recorded voice announced the item that was to be imagined by mental imagery for a duration of 0.2 s (from 3.2 to 4.4 s in Figure 14). Subjects were instructed to create a visual image of the item, that is to see it in their mind's eye. They were asked to generate this image as soon as possible after they had heard the word and to hold it until the experimenter announced the end of the trial. This was done at irregular intervals in order to prevent the occurence of an expectancy wave (CNV, Walter et al., 1964). When taking the measurement of the cortical DC level between 7 and 9 s, one has a high probability that the cooperative subject actually performs the mental task of imagery, which can then be recorded and mapped over the scalp. The results showed that the bilateral movement of pressing buttons, occuring at 0 s, was preceded by a BP. This showed the usual central maximum and was bilaterally symmetrical. However, the BP was also present in occipital leads, where it is usually absent preceding simple finger movement tasks. Thus, already in the distribution of the BP preceding it, the visual task displayed itself in a modality-specific manner. After the onset of movement, a visual evoked potential in response to the self-presentation of the slide occured. Thereafter, we see a typical expectancy wave (CNV) in anticipation of the auditory stimulus announcing the item to be imagined, followed by the auditory (verbal) evoked potential in response to the item presentation. While the early potential complex seemed to be rather normal, the late components appeared to be quite remarkable: The peak that appeared at 4.4 s (1.2 s after the onset of the loudspeaker signal) showed reversal; it was negative over frontal areas (F3 to C4) and positive over posterior areas (P3) at 2 s. Thus, it pointed towards a generator in deeper structures that is perhaps related to word comprehension and memory storage.

Slow potentials *during* imagery: Since subjects were instructed to immediately start imagining the item announced and maintain the image, the subsequent DC potential shifts are undoubtedly related to creating and maintaining mental imagery. At the beginning (around 4 s), the frontal areas (and the central areas) show strong initial DC negativity. The

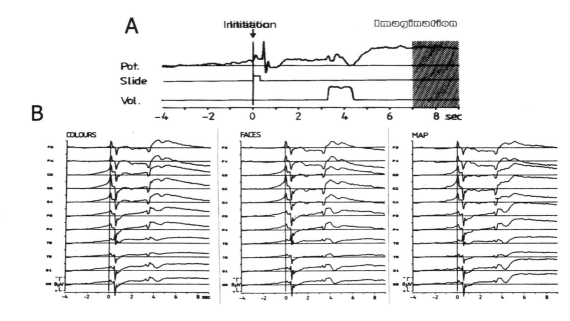

Figure 14. Experiments in movement imagery.
A: *Imagery paradigm.* Negative DC shifts (upper trace) were recorded from 4 s before to 9 s after the volitional initiation of the trial (subjects presented themselves with a slide by self-initiated button presses with both index fingers simultanously). Center trace, time marker indicating presentation of the slide revealing the condition (COLOUR, FACE or MAP). The lower trace depicts the moment of the auditory item presentation (particular national politician, or particular colour or route from 1 to 2 on the spatial map the subjects had been familiarized with). The hatched area marks the time range from 7 to 9 s of the epoch for calculating the mean amplitude of DC negativity during mental imagery.
B: DC potential waveforms averaged across all 28 subjects for imagining COLOURS (left), FACES (centre) and the spatial MAP (right). Negativity up. The vertical line represents the moment when subjects initiated the slide projector by pressing buttons with both index fingers simultaneously. Note that COLOURS created higher amplitudes (generators on the convexity) than did FACES (generators at the brain base, infra-temporal and infra-occipital locations). Highest amplitudes with MAP due to occipito-parietal generators (cf. separate visual pathways for object vision and spatial vision according to Mishkin). [Modified from Uhl et al., (1990) *Neuropsychologia* Pergamon Press, with permission].

reason is that these areas, in particular frontal ones, fulfill an important role in imagery. If we consider that imagery is also an act of volition, it is quite conceivable that these frontal areas are needed for the act of bringing about the imagery. Imagery, thus, is a willful act. This experiment is so interesting, because it shows that our 'motivational brain' is involved - not only when it 'exerts itself' in the form of movement or action - but also when it comes to 'endogenous acts' (pure mental acts) such as the formation of imagery. We all know from introspection that the generation of an image in our 'mind's eye' may need considerable *effort.*

Thus, while frontal areas, are responsible for the *generation* of imagery, the imagery as such and the *maintainance* of it seems to occur in retrorolandic brain areas. This is expressed in the DC recordings of Figure 14B by the fact that the pre-rolandic leads already show a decline of the negativity, while there was sustained DC negatvity over the retrorolandic brain. Only over areas showing such sustained negativity we assumed imagery to actually occur, measurements being taken between 7 and 9 s (hatched area in Figure 14A). And here we found interesting topographical differences between the 3 categories of imagined items. The level of negativity was almost lowest for faces, medium for colours highest for the map. It is surprising that imagination of images as complex as faces generated less negativity than did imagining colours (plain uni-coloured A4 sheets). The explanation might be that imagining faces engages more infero-temporobasal areas than do colours and - in particular - the map. Negative DC shifts in basal cortical areas were not seen by our electrodes arranged over the convexity of the skull, and, furthermore, these areas can even inject a positive bias on the DC shift recorded via the linked ears reference electrodes. The fact that the map caused the maximum overall DC negativity also fits this assumption because now imagery particularly involved areas of the convexity, namely the parietal lobes.

We also analyzed the cortical DC potential (prevailing between 7 and 9 s of the epoch - imagery period) using current source density mapping algorithms (Laplacian transformation and spline interpolation; Lindinger et al., 1994). With this display (colour figures in Uhl et al., 1990, not shown here) the 2 experimental conditions involving *visual imagery* (faces and colours) differed from the *spatial imagery* condition (map). For faces and colours, regional DC negativities were distributed over the occipital cortex and over temporal areas, showing a preponderance over the left hemisphere. For the spatial map the topography was different, regional DC negativity was distributed occipitally like the other ones, but somewhat more anterior and predominantly *parietally* (not temporally). A slight left preponderance was also observed here. From these observations we conclude that visual imagery and spatial imagery differ in their regional DC negativity pattern, the first having an occipital and temporal distribution the latter having an occipital and parietal topography. Since the same difference was found for visual and spatial perception, we considered our data as providing further evidence for the proposal that imagery takes place in the same areas that elaborate the *percepts*. These data fit with the concept that object vision and spatial vision are processed in the brain via two different pathways: Object vision (and object imagery) goes from the occipital cortex towards the temporal lobe, whereas spatial vision (and spatial imagery) goes from the occipital cortex towards the parietal lobe (Mishkin et al., 1983).

The same cohort of subjects participated in an *HMPAO-SPECT and mental imagery* experiment (30 subjects -28 of whom were identical with those of the DC potential study). The results showed that - this time as expected - the imagination of *faces* caused distinctly more regional activation than with any of the other two imagery conditions, as compared to the resting state. Statistically, imagining faces was also the only condition that led to an increase of activity in - particularly the left - inferior occipital region which has been suggested by previous studies as being a crucial area for visual imagery (Goldenberg, 1987; Goldenberg et al., 1989; Uhl et al., 1990). Also for colours, basal temporal and occipital regions are activated but less so than for faces. In contrast, the imagery of routes on a map showed a tendency towards more activation in visual association areas *on the convexity* of the brain, particularly in parietal areas. This is another piece of evidence for a 'visual/spatial dichotomy' as suggested by Mishkin et al. (1983). We now know that the visual modality has many submodalities such as form, disparity, colour, motion perception and others that are envisaged as being represented in different visual subsystems, some of which occupy distinct areas of the visual association cortex V1 to V7. Mishkin's distinction between *object vision* and *spatial vision* being represented in two separate visual pathways is a particularly intriguing concept which is partially corroborated by the two imagery studies reported here.

Furthermore, the finding in our SPECT study that imagining faces involves basal cortical areas to a greater extent than does colour and map imagery - the latter rather activating the parietal convexity - can explain the above 'paradoxical' result of our DC potential study that the imagination of faces yielded less overall negativity than did colours. Since the information content of faces is by orders of magnitude higher than the one of plain uni-coloured A4 sheets, this paradox can be explained by the greater degree of 'basality' (inferotemporal, infero-occipital) of faces as compared to colours. It is important to consider that there are large cortical regions which the electroencephalographer does not see. This again calls for the conduction of joint studies. From the SPECT study on imagery we can conclude as well that *imagery* occurs in those brain areas where *perception* takes place.

THE BEREITSCHAFTSFIELD AND FUNCTIONAL MAGNETIC RESONANCE IMAGING

The technique of functional magnetic resonance imaging (fMRI) has turned out to be superior to SPECT for the measurement of activation-related cerebral blood flow changes occurring with specific tasks. However, the spatial relationship between neuronal activity and functional cerebral blood flow changes is not yet known, and the temporal resolution is poor. A study was carried out in order to compare the center of *neuronal* activation (measured by MEG) with that of the *blood flow* responses, measured by fMRI (Beisteiner et al., 1995a,b). A group of subjects (N=8) performed a typical BP-

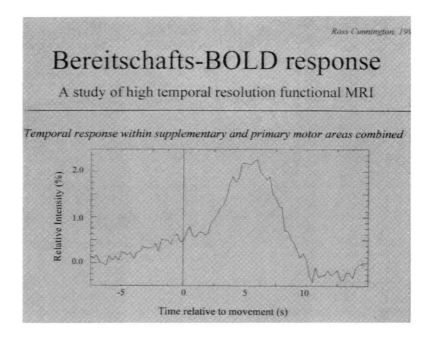

Figure 15. Bereitschafts-BOLD response 1. The subject performed a single self-initiated finger to thumb opposition sequence, touching each finger to the thumb once. Echo-planar functional MR images were acquired every 250 ms, from about 10 s before the onset of movement to 20 s after movement onset (3 Tesla Bruker fMRI). Fifteen identical trials were averaged. The graph gives the temporal response (relative intensity in %, abscissa; time relative to movement (s abscissa) pooled over supplementary and primary motor areas [From Cunnington et al (1999) *Biomed Technik* Springer Berlin etc., with permission]

paradigm of tapping with their right index finger, and the same subjects participated in both an MEG and an fMRI experiment. The common region of *neuronal* activation (which was taken as the mean of the motor and sensory cortex MEG regions) and the region of the blood flow response were both in the contralateral rolandic hand area but about 1½ cm apart from each other, that is the 'neuronal signal' (MEG) and the 'cerebrovascular signal' (fMRI) were not at exactly the same location, which demanded methodological improvements. We assume that the error is on the side of the fMRI method, and one source of localization error may be pixels with large signal amplitudes, since these pixels could stem from larger vessels that may be even remote from the region of neuronal activation (large vessel effect). Thus such dual approach, using MEG – 'neuronal signal' – in addition to the 'cerebrovascular signal' (fMRI) in a simple finger tapping task, should improve fMRI brain mapping results (Beisteiner et al., 1997).

In line with the large vessel hypothesis, we found a deterioration of fMRI localization quality with increasing signal amplitudes. Thus, for fMRI evaluation it is recommended that pixels with large signal amplitudes be excluded in order to reduce the large vessel effect and to get closer to the neuronal signal which is the important parameter in functional topography of the brain.

Recently a Multicenter Study was carried out with experiments conducted in Heidelberg, Copenhagen, and Vienna; the latter providing also the data processing. The results showed that in addition to the rolandic ('homuncular') motor area MI, fMRI revealed the participation of the SMA in the voluntary movement task. Thus, we have another technique that provided evidence that the SMA is activated in conjunction with a voluntary movement.

Last year, Ross Cunnington in our laboratory in collaboration with the Department of Diagnostic Radiology and the Department of Medical Physics of the University of Vienna examined the issue of the BP paradigm using what is now called 'event-related fMRI' (Cunnington et al., 1999). Figure 15 shows that it is feasable to 'record' a BP equivalent in the haemodynamic response of the fMRI. This was realized by using a 3 Tesla Medspec S300 wide-bore scanner equipped with a whole body gradient system and birdcage head coil (Bruker Medical Ettlingen, Germany), using a phase-corrected single-shot gradient-echo EPI sequence. By the use of single event fMRI and fuzzy clustering analysis, as is seen in Figure 15, the 'Bereitschafts-BOLD response' in the form of the haemodynamic response time course can be analyzed, and it even resembles the BP or BF.

Also the spatial distribution of the preparatory and movement-related blood flow changes was elaborated (Figure 16). As is seen, activity starts in the midline (pre-SMA and CMA), and only later comes the lateral motor hand area (MI) into play, a nice confirmation – using another method - of our hypothesis about the sequence of occurence of BP1 and BP2 (Deecke & Kornhuber, 1978).

Subsequently, Cunnington put the BP paradigm versus the CNV paradigm to direct comparison (Cunnington, Windischberger, Deecke & Moser, 2002, Figure 21) examining activity associated with a brief finger sequence movement in twelve right-handed healthy subjects. The task consisted of three rapid alternating button-presses (made with the index-middle-index finger) performed either in response to an unpredictably timed auditory cue (externally-triggered movement [a 'truncated CNV paradigm' which lacked the warning stimulus]) or with self-initiated intervals (BP paradigm). When lumped together over the whole epoch – as usually done in these kinds of 'activation studies' – both movement conditions involved strong activation of medial motor areas including the pre-SMA, SMA-proper and rostral cingulate cortex which did not differ significantly between conditions. Most significantly, activation of similar magnitude for both movement conditions was found in the frontocentral midline including the pre-SMA, SMA proper and the posterior end of the rostral cingulate motor cortex. This perhaps reflects a significant degree of internal representation and control which may have allowed optimal performance given the goal of the task (to perform the complete motor sequence as quickly as possible with no additional external cues to guide submovements) both when the timing of initial onset of the movement sequence was externally-triggered or self-determined. There was, however, a significant difference in the *timing* of activation of the pre-SMA between movement conditions. Pre-SMA activity began significantly earlier for self-initiated movements, with a mean peak latency of 1.48 s prior to that for externally-triggered movements. This reflects involvement of the pre-SMA in early stage movement preparation processes which precede the onset of voluntary movements.

Another important finding of this study was that activity is also recorded from the basal ganglia (cf. Figure 21). Rektor recently reported from his intracranial recordings that a BP can be recorded from the basal ganglia (Rektor, this book, Rektor et al., 2001). This finding is very important and quite conceivable in view of the cortico-basal ganglia-thalamocortical loop (cf. Kornhuber, 1974a; DeLong, 1990; Alexander, 1994). The activity travelling over this loop comes from the SMA/CMA and goes to MI, on its way to informing the basal ganglia that a movement is about to come. In this context, the 'chunking hypothesis" of the Hallett group is very attractive (Gerloff et al., 1997; Gerloff this book). These nice experiments are exactly in line with our SMA hypothesis, where we envisage the SMA as a job distributor and supervisor. The SMA organizes sequential work in such a way that it organizes the sequences into handy pieces and reserves the appropriate time slots for their launch. This is what we understand by spatial and temporal coordination. The interesting finding of Cunnington et al. (2002, Figure 21) was, however, that activity in the lentiform nucleus (along the junction between the putamen and external segment of the globus pallidus) was found *for self-initiated movements only.* For externally-triggered movements, there was no evidence of increased activation within the basal ganglia.

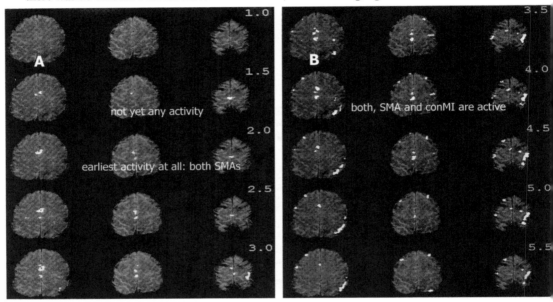

Figure 16. Bereitschafts-BOLD response 2. The spatial distribution of activation patterns over time is shown in a correlation time-series analysis overlayed on high resolution anatomic images. Intensity changes in these images (white) represent changes in deoxyhaemoglobin concentration (BOLD response), which reflect changes in cortical activation [Modified from Cunnington et al (1999) *Biomed Technik* Springer Berlin etc., with permission]

THE BP IN PARKINSON AKINESIA

Our early study suggested that in Parkinsonian akinesia, the BPs are normal over the SMA but reduced and delayed over MI (Deecke et al., 1977). Also in Dick et al.'s (1989) recordings, BP amplitudes at the vertex are larger than those over MI. On closer examination, the early BP component, BP1, is reduced, while a normal or even larger late BP component, BP2, helps the patients to catch up, and at movement onset reach the same amplitudes as the healthy controls – even overshooting them after movement onset (Dick et al., 1989; Harasko-van der Meer et al., 1996; Lang & Deecke, 1998). This is shown in Figure 17.

Eight Parkinson patients and eight age-matched healthy controls participated in a typical voluntary movement BP paradigm. They performed brisk monophasic extensions or flexions of the right index finger against slight resistence, kept the target position over at least 5 s, and during the inter-trial interval (ITI) moved the finger back to its initial position. Brain potentials are shown in Figure 17 at the vertex (Cz, over the SMA) and at C3 (over the left MI hand area). As is seen, in both extension and flexion, the BP over SMA (Cz) in

Parkinson patients (thick line, later onset) starts significantly later and is significantly smaller than it is in the controls (thin line, earlier onset). After the onset of movement, the situation is reversed. Amplitudes for Parkinson patients are higher than they are for controls. Over the contralateral motor cortex (C3), the BP in Parkinson's disease also starts later but then gains even higher amplitudes than the controls. As can also be seen in Figure 17, extension creates higher amplitudes than does flexion in both Parkinson patients and normals. We have previously reported on this concept using normal subjects (Deecke et al., 1980). As seen from the movement trajectories at the bottom, movements of the Parkinson patients are slower (upper graphs) as compared to the healthy controls (lower graphs) for both extension (left) and flexion (right). The findings are 1.) direct evidence that in patients with Parkinson's disease there is a reduced cortical excitability in the organization of voluntary movement. 2.) they revealed that in Parkinson's disease postsynaptic changes occur as well as presynaptic ones (dopamine depletion).

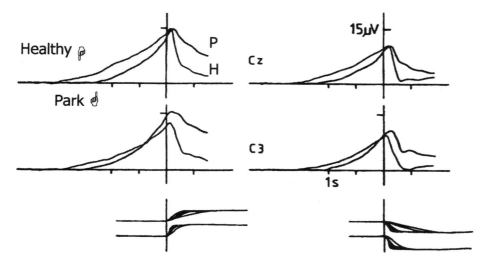

Figure 17. The BP in *Parkinson* patients and age-matched healthy controls. Subjects (N=8) and patients (N=8) performed in a typical monophasic voluntary movement BP paradigm (they performed brisk extensions [left-sided graphs] or flexions [right-sided graphs] of their right index finger against slight resistence, kept the target position over at least 5 s, and then, in the inter trial interval (ITI) moved the finger back to initial position. Movement onset at vertical line (t = 0 s). Brain potentials were recorded at the vertex (Cz, over the SMA, upper graphs) and at C3 (over the left MI hand area, lower graphs). Negativity is up. Note that in both conditions, extension (left) and flexion (right), the BP over SMA (Cz) in Parkinson patients (thick line) starts significantly later and is significantly smaller than in the controls (thin line). After the onset of movement, the situation is reversed: Amplitudes for Parkinson patients are higher than they are for controls. Over the contralateral motor cortex (C3) BP in Parkinson's disease also starts later but then becomes higher in amplitude than in the controls. Also note that extension creates higher amplitudes than does flexion in both Parkinson patients and normals. For the latter this has previously been reported (Deecke et al 1980). At the bottom, the trajectories of the movement are given: Movements of the Parkinson patients are slower (upper graphs) as compared to the healthy controls (lower graphs) for both extension (left) and flexion (right). The findings of Fig. 17 are the first direct evidence that in patients with Parkinson's disease there exists reduced cortical excitability in the organization of voluntary movement. [Modified from Lang & Deecke (1998) Enzyklopädie der Psychologie, Hogrefe Göttingen, with permission (data from Harasko-van der Meer et al., 1996)]

This implies dysfunction of *cortical* structures as a 'hodological' consequence of basal ganglia dopaminergic deficiency. That is why Parkinson patients have trouble in self-initiation (action) while still being able to respond to external stimuli (re-action, Deecke et al., 1977; Deecke & Kornhuber, 1978; Dick et al., 1989; Harasko-van der Meer et al. 1996, Jahanshahi et al., 1995; Lang & Deecke, 1998; Simpson, Khuraibet, 1986; 1987). These postsynaptic dysfunctions seem to be partially reversible with L-Dopa substitution (Dick et al., 1989; Feve et al., 1992). Analyses of BP topography revealed that the contribution of the SMA to the BP was reduced in Parkinson's disease (Dick et al., 1989; Harasko-van der

Meer et al., 1996; Lang & Deecke, 1998; Jahanshahi et al., 1995; Cunnington et al., 1995; Marsden & Obeso, 1994). PET studies confirmed a diminution of movement-related SMA activation (Playford et al., 1992; Jenkins et al., 1992; Rascol et al., 1992; Jahanshahi et al., 1995).

In primate models of Parkinson's disease, it is now clear that the feedback circuits between cortex, basal ganglia, and thalamus (where the SMA plays a key role) are disturbed in this disease (Alexander et al., 1986; Alexander, 1994; DeLong, 1990; Juncos & DeLong, 1997). To be precise it is the lack of inhibition of the inhibitory action of the Globus pallidum internum (GPi) onto the excitatory (glutamatergic) thalamo-cortical pathway. Inhibition of inhibition results in excitation, and it is the intrinsic pathophysiology of Parkinsonian akinesia that the inhibition of inhibition, which is dopaminergic, is deficient: Parkinson patients are akinetic because they are caught in the state of this GPi-'hyperinhibition', which in the untreated severe cases down-regulates their motor capacity to almost zero.

Pallidotomy can markedly improve akinesia (DeLong, 1990). We can now understand why it is effective: It removes this detrimental hyperinhibition that rests at the thalamocortical output circuit. The SMA is one of the major projection areas of this circuit receiving even stronger input than MI. It has been shown that post pallidotomy, movement-related SMA activity as assessed by PET is restored in Parkinson patients (Ceballos-Baumann et al., 1994; Grafton et al., 1995). It is likely that the Parkinson-typical changes of the BP will also return to normal post pallidotomy, but recent evidence (Limousin et al., 1999) suggested that this is not completely so, since only the amplitude of the late BP (BP2) is increased after surgery. However, it is clear that Parkinsonian akinesia is the result of the dysfunction of a complicated feedback circuit system, in which not only the basal ganglia but also the SMA play an important role. The role of the CMA in this circuit has still to be defined. Recently, modification of the indirect pathway by subthalamic nucleus (STN) lesion has been shown equal to or even better than pallidotomy. Also a trend towards prefering deep brain stimulation (DBS) over lesioning is observed. DBS of both STNs is performed at the Neurosurgery Department, University of Vienna in collaboration with us with excellent results (Alesch & Deecke, 1995; Gerschlager et al., 1999; 2001).

CONTINGENT NEGATIVE VARIATION IN PARKINSON PATIENTS ON DEEP BRAIN STIMULATION

Recently we recorded the CNV in Parkinsonian patients who had undergone stimulator implantation by F. Alesch, Department of Neurosurgery, in both subthalamic nuclei (STN) versus a normal control group. PD patients were recorded both when STN stimulation was on and when off, enabling immediate comparison in the same patient and session. This is seen in Figure 18, where the stimulator 'on' condition generated considerably higher CNV amplitudes – mostly over frontal areas – than with the stimulator 'off' condition (Gerschlager et al., 1999; 2001). It is known that Parkinson's disease involves impaired activation of frontal cortical areas, including the SMA and prefrontal cortex (Playford et al., 1992; Jahanshahi et al., 1995), resulting from impaired thalamocortical output from the basal ganglia. Electrophysiologically, such impaired cortical activation may be seen as a reduced amplitude of the CNV. Surgical interventions aimed at increasing basal ganglia-thalamic outflow to the cortex, such as electrical stimulation of the STN with chronically implanted electrodes, have been shown to be effective at improving clinical symptoms of Parkinson's disease. Our group was able to show changes in cortical activity, as reflected in the CNV, associated with bilateral STN stimulation in Parkinson's disease. The CNV was recorded from 10 patients with bilateral STN stinulation 'on' and 'off', and compared them with 10 healthy control subjects. Without STN stimulation, the patients showed reduced CNV amplitudes over frontal and fronto-central regions compared with control subjects. With bilateral STN stimulation, however, CNV amplitudes over frontal and frontocentral regions were significantly increased (see Figure 18). These results suggest that impaired cortical functioning in Parkinson's disease, particularly within frontal and premotor areas including SMA, is improved with STN stimulation in parallel with the clinical improvement due to this therapy.

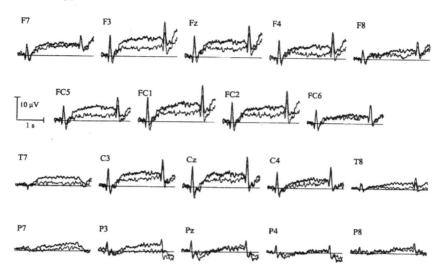

Figure 18. Contingent negative Variation (CNV) recordings in Parkinson patients with deep brain stimulation. During stimulation 'on' (thick lines) significantly higher CNV amplitudes are seen than during stimulation 'off', particularly over frontal areas. [From Gerschlager et al., 1999 *Brain* Oxford University Press, with permission]

One may wonder why we have not recorded the BP with stimulation on and off in the STN stimulation patients. The reason is that Brown et al. (1999) reported there is no significant difference (also see Praamstra, Jahanshahi, Rothwell chapter in this book). However, we did pilot experiments (Cunnington, Erdler, Mayer, Asenbaum & Deecke, 2000) investigating a few hemiparkinsonian patients in the MEG.

Left hand movement....SMA (right hand affected)

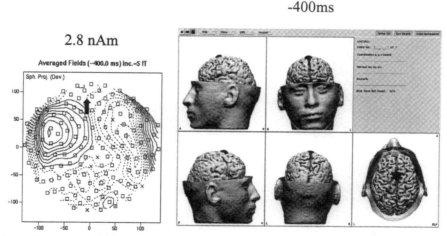

Figure 19. Bereitschaftsfield (BF) in a hemi-Parkinson patient (right side of the body) using the non-affected (left) hand. Note a moderate strength SMA dipole of 2.8 nAm. [Unpublished data of Cunnington et al., 2000 adapted to the Curry software by Dr. Peter Walla, Ludwig Boltzmann Institute of Functional Brain Topography Vienna]

Right hand movement....SMA (right hand affected)

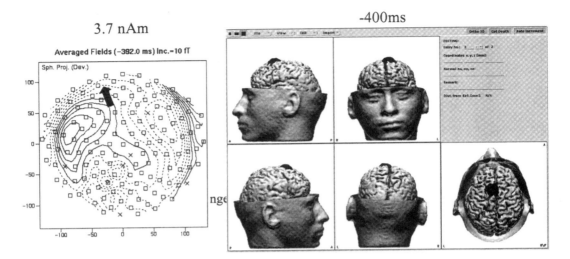

Figure 20. Bereitschaftsfield (BF) in a hemi-Parkinson patient (right side of the body) using the affected (right) hand. Note a stronger SMA dipole of 3.7 nAm, due to more *effort* needed when moving the impaired system as compared to the healthier one. [Source same as Figure 19]

An interesting trend was observed, namely that with movements of the affected side of the body ('most affected side') more activity of the SMAs was found than with movement of the 'less affected side.' This implies that the SMA dipole was not strongest with the healthy system, as one would perhaps expect. On the contrary, the SMA exhibited more activity when the patient moved with his more impaired system (compare Figures 19 and 20).

This is another piece of evidence in favour of the 'effort' hypothesis. In simple terms: the side with the impairment has to try harder in fulfilling the experimental task than has the 'healthier' side. The normal controls showed only little RF at the SMAs, since for them the task was easy. The explanation in terms of 'effort' was the most plausible one also in earlier experiments together with Rumyana Kristeva (Kristeva et al., 1979; Deecke et al., 1978; Kristeva & Deecke, 1980; Kristeva chapter in this book), in which we investigated handedness. We had started testing the hypothesis, that with movement of the 'dominant' crossed hand hemisphere system, a larger BP should occur than with movement of the non-dominant system (also cf. Eccles, 1990). However, the contrary turned out to be the case: For movement of the non-dominant system, BP amplitude was higher! We explained this at that time as we do explain it presently that the less skilled system has to use more effort (recruit more 'energy' [perhaps more neuronal populations]) in order to cope with the more skilled system. This holds as well for hemiParkinsonism (and hemi-paresis?), conditions that are currently under investigation.

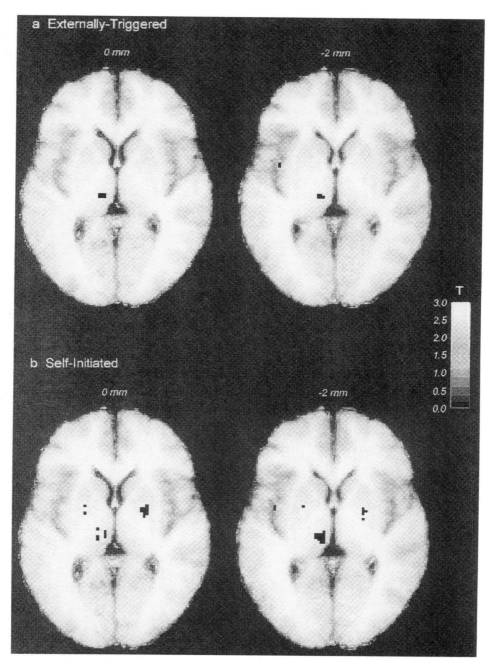

Figure 21. Movement-related activation Bereitschafts-BOLD activity in the basal ganglia Areas of subcortical activation (black squares, $P_{uncorrected} < 0.0001$) located within the left midbrain (0 to 2 mm below the AC-PC plane) for both externally triggered (top) and self-initiated movements (bottom) and bilaterally within the lentiform nucleus of the basal ganglia (junction putamen-external pallidum) for self-initiated movements only. [From Cunnington et al. (2002) *Neurol-mage*, Academic Press, London, with permission]

BP AND CONSCIOUS EXPERIENCE OF THE WILL TO MOVE

To combine objective neurobiological measurements with subjective psychological methods is often enlightening. We also used it for clarification of sensory processes (Talbot et al., 1968; Kornhuber, 1972; Bechinger et al., 1972; Deecke et al., 1979; Deecke 1996). The importance of consciousness has been underestimated by the behaviourists. It is not an epiphenomenon. If after brain injury consciousness is regained, the lesion can (partially) be compensated, however never without consciousness. Conscious awareness is a shining light of freedom, although it should not be overestimated either. For most of the vital functions consciousness is not necessary, and it is not the only sign of freedom. The experiment of Libet et al. (1983), which showed that the BP is not, right from the beginning, accompanied by a consciousness about the intention of movement, is taken at present as the main argument for advocating a total determinism, a complete lack of freedom of man (Roth, 2001); this position is not tenable. What would be necessary to study, is the original planning and decision; this, however, has been completed already before the beginning of the experiment. Repetitions of stereotyped simple movements are not suitable for such an investigation. Thoughts of planning and motivation - if not obscured by fear or wrath - are as we all know performed in the light of consciousness. The conscious awareness of wanting to make a movement does occur, in investigations of the BP, about 200 ms prior to the muscle contraction (Libet, 1985). This is roughly the same time span needed for a motor reaction to an expected auditory stimulus. Although the decision to act has been made already earlier, consciousness is switched on in order to be able to make changes of the movement if necessary (for example not executing the action) and to be able to learn from the success of the movement. In both cases the following brain areas are activated: the SMA, the pre-SMA, the anterior portion of the cingulate cortex and a part of the motor cortex (Cunnington et al., 2002). However, with the self-initiated movement, additionally the basal ganglia are activated before the movement (cf. Figure 21, Cunnington et al., 2002,). This preparatory process for the spontaneous movement, through which the readiness for movement in the SMA builds up, remains unconscious for the first 400 ms (Libet, 1985).

That in the brain something happens unconsciously, is not unusual, as seen in sensory systems. In the motor system, processes that are initially conscious can become unconscious through automatisation; this relieves consciousness. Otherwise the brain would be extremely overcharged with the immense flow of information. The unconscious processes do not lessen freedom; on the contrary, they form its primary basis.

EFFECTS ON PSYCHOLOGY AND PHILOSOPHY

As already mentioned, the interest in freedom and will had declined in psychology in the years after the second world war. In social research the interest in freedom was also dead at this time (Noelle-Neumann, 1978) although freedom (in the sense of democracy, freedom of the press, etc.) appeared frequently of course in political adresses. But after the concepts of will and 'positive' freedom were restored to life in neurophysiology, a new interest in these concepts appeared in psychology, now under the terms volition, self-control and intention. Volition as a control term (key word) in the psychological literature appeared four times in 1968, six times in 1974, eight times in 1979, fourteen times in 1987 and twenty-one times in 1999. The term self-control was completely absent until 1971; it appeared twice in 1972, six times in 1973, nine times in 1974, twelve times in 1975, nineteen times in 1976, twenty-five times in 1978. The concept of intention was absent in titles of psychological papers until 1967. It appeared in a title once in 1968, twice in 1971, four times in 1974, six times in 1981, eight times in 1983, ten times in 1985, fifteen times in 1991, nineteen times in 1993, twenty-four times in 1999. Correspondingly, freedom was rediscovered in empirical social research in 1978 (Noelle-Neumann). By means of factor analysis, it was shown that the freedom to make decisions in working life was significantly associated with happiness whereas the amount of free time and of freedom in the sense of *'libertinage'* did not correlate with happiness and this was true in both blue and white collar workers. A similar development occurred in philosophy. While some scientists tried to make complete determinism even more waterproof (Roth, 2001), more and more philosophers turned toward a concept of 'positive' freedom which is compatible with nature (Bieri, 2001).

Important discussions on will and freedom, triggered by the discovery of the BP, were the books edited by Lindauer and Schöpf, 1984; Heckhausen et al., 1987; Hershberger, 1989; Libet et al., 1999, and the numerous papers by Libet and others in 'The Behavioural and Brain Sciences,' 1985.

FUTURE OUTLOOK

The method of investigation that led to the discovery of the BP aimed at exploring the intentionality of man. In this direction it probably will continue to be useful, since bioelectric and biomagnetic techniques are powerful in time resolution. However, for the study of cognitive processes it must be further refined by single trial methodology and combination with tools that are more powerful in localisation (for example fMRI). Pathological processes as well as animal experiments and pharmacological influences will continue to be investigated.

If we are asked how research will go on, we presume that the study of the human brain in the future will not disregard activity and creativity, and the internal control of these innovative and, therefore, also potentially 'dangerous' abilities. We expect that neurobiological and other research will continue to contribute to the self-recognition of the most humane in us, including the aim to steer against dysfunctional developments and to help treat pathological processes. One will see again that man is more than a complex of genetic and environmental influences and that our reasoned will has goals beyond ourselves.

REFERENCES

Alesch F, Deecke L (1995) Chirurgische Behandlungsmethoden bei Morbus Parkinson. *Neuropsychiatrie* 9 Suppl 1: S42-S44

Alexander GE, DeLong MR, Strick PL (1986) Parallel organization of functionally aggregated circuits linking basal ganglia and cortex. *Ann Rev Neurosci* 9: 357-381

Alexander G (1994) Basal ganglia - thalamocortical cortical circuits: Their role in control of movements. *J Clin Neurophysiol* 11: 420-431

Asenbaum S, Oldenkott B, Lang W, Lindinger G, Deecke L (1990) Motor sequences in patients with unilateral SMA lesions: Movement-related potentials and motor disorders. In: Brunia CHM, Gaillard AWK, Kok A, eds.,: *Psychophysiological Brain Research*. Vol. I, Tilburg University Press, pp 119-123

Asendorpf JB (1999) *Psychologie der Persönlichkeit*. Springer, Berlin, Heidelberg.

Bechinger D, Kongehl G, Kornhuber HH (1972) Eine Hypothese für die physiologische Grundlage des Größensehens: Quantitative Untersuchungen der Informationsübertragung für Längen und Richtungen mit Punkten und Linien. Arch Psychiat Nervenkr 215: 181-189

Becker W, Hoehne O, Iwase K, Kornhuber HH (1971) Bereitschaftspotential, prämotorische Positivierung und andere Hirnpotentiale bei sakkadischen Augenbewegungen. *Vision Res* 12: 421 - 436.

Beisteiner R, Gomiscek G, Erdler M, Teichtmeister C, Moser E, Deecke L (1995a) Funktionelles Magnet Resonanz Imaging (fMRI) - Ergebnisse und Kombinationsmöglichkeiten mit der Magneto- und Elektroenzaphalographie. In: W Lang, L Deecke, HC Hopf, eds.,: *Topographische Diagnostik des Gehirns*. Springer Wien Vh Dt Ges Neurol 9: 189-192

Beisteiner R, Gomiscek G, Erdler M, Teichtmeister C, Moser E, Deecke L (1995b) Comparing localization of conventional functional magnetic resonance imaging and Magnetoencephalography. *Eur J Neurosci* 7: 1221-1224

Beisteiner R, Erdler M, Teichtmeister C, Diemling M, Moser E, Edward V, Deecke L (1997) Magnetoencephalography may help to improve functional MRI brain mapping. *Eur J Neurosci* 9: 1072-1077

Beisteiner R, Vrba J, Deecke L (1998) Methods and applications of magnetoencephalography. Chapter 10 in: GK von Schulthess, J Hennig, eds., *Functional imaging*. Lippincott-Raven pp 409-431

Bieri P (2001) *Das Handwerk der Freiheit. Über die Entdeckung des eigenen Willens*. Hanser, München.

Binder W, Kornhuber HH, Waiblinger G (1984) Benzodiazepinsucht, unsere iatrogene Seuche - 175 Fälle von Benzodiazepinabhängigikeit. *Öff Gesundh Wes* 46: 80 - 86.

Boschert J, Hink RF, Deecke L (1983) Finger movement versus toe movement-related potentials: further evidence for supplementary motor area (SMA) participation prior to voluntary action. *Exp Brain Res* 52(1): 73-80

Boschert J, Deecke L (1986) Cerebral potentials preceding voluntary toe, knee and hip movements and their vectors in human precentral gyrus. *Brain Res* 376(1): 175-9

Brown RG Limousin-Dowsey P Brown P Jahanshahi M Pollak P Benabid AL Rodriguez-Oroz MC Obseo J Rothwell JC (1999) Impact of deep brain stimulation on upper limb akinesia in Parkinson's disease. *Ann Neurol* 45, 473-488.

Ceballos-Baumann AO, Obeso JA, Vitek JL, DeLong MR, Bakay R, Linazasoro G, Brooks DJ (1994) Restoration of thalamocortical activity after postero-ventral pallidotomy in Parkinson´s disease. *Lancet* 344: 814

Cheyne D, Kristeva R, Deecke L (1991) Homuncular organization of human motor cortex as indicated by neuromagnetic recordings. *Neurosci Lett* 122: 17-20

Christ KJ, Kornhuber HH (1980) Treatment of neurogenic bladder dysfunction in multiple sclerosis by ultrasound-controlled bladder training. *Arch Psychiat Nervenkr* 228: 191 – 195.

Creutzfeldt OD (1983) *Cortex cerebri*. Springer Berlin, Heidelberg, New York, Tokyo pp 145-149

Cui RQ, Huter D, Lang W, Lindinger G, Beisteiner R, Deecke L (1996) Multichannel DC current source density mapping of the Bereitschaftspotential in the supplementary and primary motor area preceding differently loaded movements. *Brain Topography* 9(2): 83-94

Cui RQ, Deecke L (1999) High resolution DC-EEG analysis of the Bereitschaftspotential and post movement onset potentials accompanying uni- or bilateral voluntary finger movements. *Brain Topography* 11: 233-248

Cui RQ, Huter D, Lang W, Deecke L (1999) Neuroimage of voluntary movement: Topography of the Bereitschaftspotential, a 64-channel DC current source density study. *NeuroImage* 9: 124-134

Cui RQ, Huter D, Egkher A, Lang W, Lindinger G, Deecke L (2000) High resolution DC EEG mapping of the Bereitschaftspotential preceding simple or complex bimanual sequential finger movement. *Exp Brain Res* 134: 49-57

Cunnington R, Iansek R, Bradshaw JL, Phillips JG (1995) Movement-related potentials in Parkinson´s disease: Presence and predictability of temporal and spatial cues. *Brain* 118: 935-950

Cunnington R, Windischberger C, Deecke L, Moser E (1999) The use of single event fMRI and fuzzy clustering analysis to examine haemodynamic response time courses in supplementary motor and primary motor cortical areas. *Biomed Technik* 44 (Suppl 2): 116-119

Cunnington R, Erdler M, Mayer D, Asenbaum S, Deecke L (2000) Premovement cortical activity in hemi-Parkinson´s disease: a study of whole-scalp magnetoencephalography. *Movement Disorders* 15 (Suppl 3): 82

Cunnington R, Windischberger C, Deecke L, Moser E (2002) The preparation and execution of self-initiated and externally-triggered movement: a study of event-related fMRI. *NeuroImage* 15: 373-385.

Deecke L (1974) *Die corticalen Potentiale des Menschen vor raschen willkürlichen Fingerbewegungen*. Habilitationsschrift. University of Ulm Library

Deecke L (1987) Bereitschaftspotential as an indicator of movement preparation in supplementary motor area and motor cortex. In: R. Porter (Ed), *Motor Areas of the Cerebral Cortex*. Wiley. Chichester, Ciba Found Symp 132: pp 231-250

Deecke L (1990) Electrophysiological correlates of movement initiation. *Rev Neurol* 146/10: 612-619

Deecke L (1996) Planning, preparation, execution, and imagery of volitional action. (Introduction/Editorial) in: Deecke L, Lang W, Berthoz A, eds., *Mental representations of motor acts. Cogn Brain Res* 3 (2, Special Issue): 59-64

Deecke L, Scheid P, Kornhuber HH (1969) Distribution of readiness potential, pre-motion positivity and motor potential of the human cerebral cortex preceding voluntary finger movements. *Exp Brain Res* 7: 158 - 168.

Deecke L, Grözinger B, Kornhuber HH (1976a) Voluntary finger movement in man: Cerebral potentials and theory. *Biol Cybern* 23: 99-119

Deecke L, Schmitt G, Kornhuber HH (1976b) Bereitschaftspotential in parkinsonian patients. pp 169 - 171 in McCallum WC, Knott JR, eds., *The responsive brain*. Bristol, John Wright.

Deecke L, Englitz H-G, Kornhuber HH, Schmitt G (1977) Cerebral potentials preceding voluntary movement in patients with bilateral or unilateral Parkinson akinesia pp 151-163 in Desmedt JE, ed., *Attention, voluntary contraction, and event-related cerebral potentials. Prog Clin Neurophysiol* 1, Karger, Basel

Deecke L, Kornhuber HH (1978) An electrical sign of participation of the mesial 'supplementary' motor cortex in human voluntary finger movement. *Brain Res* 159: 473-476

Deecke L, Grözinger B, Kristeva R (1978) Hemispherical differences of cerebral potentials preceding right and left unilateral and bilateral finger movements in right-handed subjects. *Pflügers Arch* 373 (Suppl): R74

Deecke L, Becker W, Jürgens R, Mergner T (1979) Interaction of vestibular and somatosensory afferents for perception and postural control. Editorial. *Agressologie* 20, C: 179-184

Deecke L, Eisinger H, Kornhuber HH (1980) Comparison of Bereitschaftspotential, pre-motion positivity and motor potential preceding voluntary flexion and extension movements in man. In: Kornhuber HH, Deecke L, eds., *Motivation, motor and sensory processes of the brain: Electrical potentials, behaviour and clinical use*. Amsterdam, Elsevier, *Prog Brain Res* Vol 54: pp 171-176

Deecke L, Weinberg H, Brickett P (1982) Magnetic fields of the human brain accompanying voluntary movement. Bereitschaftsmagnetfeld. *Exp Brain Res* 48: 144-148

Deecke L, Boschert J, Weinberg W, Brickett P (1983) Magnetic fields of the human brain (Bereitschaftsfeld) preceding voluntary foot and toe movements. *Exp Brain Res* 52: 81-86

Deecke L, Heise B, Kornhuber HH, Lang M, Lang W (1984) Brain potentials associated with voluntary manual tracking: Bereitschaftspotential, conditioned pre-motion positivity, directed attention potential, and relaxation potential. Anticipatory activity of the limbic and frontal cortex. In: Karrer R, Cohen J, Tueting P, eds., *Ann NY Acad Sci* Vol 425: 450-464

Deecke L, Kornhuber HH, Lang W, Lang M, Schreiber H (1985) Timing functions of the frontal cortex in sequential motor- and learning tasks. *Hum Neurobiol* 4: 143 – 154.

Deecke L, Engel M, Lang W, Kornhuber HH (1986) Bereitschaftspotential preceding speech after holding breath. *Exp Brain Res* 65: 219-223

Deecke L, Lang W, Heller HJ, Hufnagl M, Kornhuber HH (1987) Bereitschaftspotential in patients with unilateral lesions of the supplementary motor area. *J Neurol Neurosurg Psychiat* 50: 1430-1434

Deecke L, Lang W (1996) Generation of movement-related potentials and fields in the supplementary sensorimotor area and the primary motor area. In: Lüders HO, ed., *Supplementary Sensorimotor Area - Advances in Neurology*. Vol 70 Lippincott/Raven Publishers pp 127-146

Deecke L, Gebhard J, (2001) Multiple Sklerose: Ein Leiden mit 'tausend Gesichtern.' *KURIER* 29. 1. p 31

DeLong MR (1990) Primate models of movement disorders of basal ganglia origin. *Trends Neurosci* 13: 281-285

Dick JPR, Rothwell JC, Day BL, Cantello R, Buruma O, Gioux M, Benecke R, Berardelli A, Thompson PD, Marsden CD (1989) The Bereitschaftspotential is abnormal in Parkinson´s disease. *Brain* 112: 233-244

Diekmann V, Grözinger B, Kornhuber HH, Kriebel J, Bock H (1980) Evidence for the cerebral origin of the R-wave in the pig. pp 103 - 108 in Kornhuber HH, Deecke L, eds., *Motivation, motor and sensory processes of the brain: Electrical potentials, behaviour and clinical use. Progr Brain Res* 54, Amsterdam, Elsevier.

Eccles JC, Zeier H (1980) *Gehirn und Geist*. Zürich, Kindler.

Eccles JC (1982) The initiation of voluntary movements by the supplementary motor area. *Arch Psychiat Nervenkr* 231: 423-441

Eccles JC (1990) The evolution of cerebral asymmetry. In: L Deecke, JC Eccles, VBMountcastle, eds., *From Neuron to Action*. Springer Berlin etc.,, pp. 315-328

Erdler M, Beisteiner R, Mayer D, Kaindl T, Edward V, Windischberger C, Lindinger G, Deecke L (2000) Supplementary motor area activation preceding voluntary movement is detectable with a whole scalp magnetoencephalography system. *NeuroImage* 11: 697-707

Ernst C, von Luckner N (1985) *Stellt die Frühkindheit die Weichen?* Enke, Stuttgart.

Feve AP, Bathien N, Rondot P (1992) Chronic administration of L-Dopa affects the movement-related cortical potentials of patients with Parkinson´s disease. *Clin Neuropharmacol* 15: 100-108

Foerster O (1936) Motorische Felder und Bahnen. In: O Bumke, O Foerster, eds., *Handbuch der Neurologie Vol IV* p 4

Geiselmann B, Kornhuber HH (1981) Dezentrale Therapie der Multiplen Sklerose durch Hausärzte, Gemeindeschwestern und Angehörige. *Dtsch Med Wschr* 106: 15 - 18.

Gerloff C, Corwell B, Chen R, Hallett M, Cohen LG (1997) Stimulation over the human supplementary motor area interferes with the organization of future elements in complex motor sequences. *Brain* 120: 1587-1602

Gerschlager W, Alesch F, Cunnington R, Deecke L, Dirnberger G, Endl W, Lindinger G, Lang W (1999) Bilateral subthalamic nucleus stimulation improves frontal cortex function in Parkinson´s disease. An electrophysiological study of the contingent negative variation. *Brain* 122: 2365-2373

Gerschlager W, Bloem BR, Alesch F, Lang W, Deecke L, Cunnington R (2001) Bilateral subthalamic nucleus stimulation does not improve prolonged P300 latencies in Parkinson´s disease. *J Neurol* 248: 285-289

Göppinger H (1987) *Lifestyle and criminality*. Springer Berlin, etc.

Goldenberg G (1987) *Neurologische Grundlagen bildlicher Vorstellungen*. Springer Wien, New York

Goldenberg G, Podreka I, Uhl F, Steiner M, Willmes K, Deecke L (1989) Cerebral correlates of imagining colours, faces and a map - I. SPECT of regional cerebral blood flow. *Neuropsychologia* 27: 1315-1328

Grafton ST, Waters C, Sutton J, Lew MF, Coudwell W (1995) Pallidotomy increases activity of motor association cortex in Parkinson´s disease: A positron emission tomographic study. *Ann Neurol* 37: 776-783

Grözinger B, Kornhuber HH, Kriebel J (1975) Methodological problems in the investigation of cerebral potentials preceding speech: Determining the onset and suppressing artefacts caused by speech. *Neurpsychologia* 13: 263 - 270.

Grözinger B, Kornhuber HH, Kriebel J, Szirtes J, Westphal KP (1980) The Bereitschaftspotential preceding the act of speaking. Also an analysis of artefacts. pp 798 - 804 in Kornhuber HH, Deecke L, eds., *Motivation, motor and sensory processes of the brain: electrical potentials, behaviour and clinical use. Progr Brain Res* 54, Amsterdam, Elsevier.

Harasko-van der Meer C, Gerschlager W, Lalouschek W, Lindinger G, Deecke L, Lang W (1996) Bereitschaftspotential preceding onset and termination of a movement is abnormal in Parkinson´s disease. *Movement Disorders* 11 Suppl 1: 84

Heckhausen H (1987) Perspektiven einer Psychologie des Wollens. pp 121 – 142 in Heckhausen H, Gollwitzer PM, Weinert FE, eds., *Jenseits des Rubikon: Der Wille in den Humanwissenschaften*. Springer-Verlag Berlin, Heidelberg, New York, London, Paris, Tokyo

Hershberger WA, ed (1989) *Volitional action: Conation and control*. North Holland, Amsterdam, New York, Oxford, Tokyo

Hess WR (1954) *Das Zwischenhirn*. Basel, Benno Schwabe.

Hess WR (1956) *Hypothalmus und Thalamus*. Stuttgart, Thieme.

Huber G, Kornhuber HH, eds., (since 1993) *NEUROLOGY, PSYCHIATRY AND BRAIN RESEARCH*, Springer, Berlin, Heidelberg, New York, Tokyo, now Ulm University Press.

Ikeda A, Lüders HO, Burgess RC, Shibasaki H (1993) Movement-related potentials associated with single and repetitive movements recorded from human supplementary motor area. *Electroenceph Clin Neurophysiol* 89: 269-277

Jahanshahi M, Jenkins HI, Brown RG, Marsden CD, Passingham RE, Brooks CD (1995) Self-initiated versus externally-triggered movements: I. An investigation using regional cerebral blood flow and movement-related potentials in normals and in patients with Parkinson's disease. *Brain*, 118, 913-933.

Jenkins IH, Fernandez W, Playford ED, Lees AJ, Frackowiak REJ, Passingham RE, Brooks DJ (1992) Impaired activation of the supplementary motor area in Parkinson´s disease is reversed when akinesia is treated with apomorphine. *Ann Neurol* 32: 749-767

Juncos J, DeLong MR (1997) Parkinson´s disease and basal ganglia movement disorders. Chapter XV in 11 Neurology in: DC Dale, DD Federman, eds.,: *Scientific American Medicine*. New York, Scientific American

Kim JS, Kornhuber HH, Schmid-Burgk W et al. (1980) Low cerebrospinal fluid glutamate in schizophrenic patients and a new hypothesis on schizophrenia. *Neurosci Lett* 20: 379-382

Kim JS, Kornhuber HH, Brand U, Menge HG (1981) Effects of chronic amphetamine treatment on the glutamate concentration in the cerebrospinal fluid and brain: implications for a theory of schizophrenia. *Neurosci Lett* 24: 93-96

Kleiser B, van Reemts J, Horn E, Kornhuber HH (1990) Experimental intracerebral haematoma and treatment with Flunarizine in rats. pp 527 - 529 in Deecke L, Eccles JC, Mountcastle VB, eds., *From Neuron to Action*. Springer Berlin, etc.

Kleist K (1934) *Gehirnpathologie*. Barth, Leipzig
Kornhuber A, Kornhuber HH (1980) Blutdruckmessung im Schulunterricht und Blutdruckscreening durch Schulkinder. *Therapiewoche* 30/47: 7874.
Kornhuber HH (1955) Über Auslösung cyklothymer Depressionen durch seelische Erschütterungen. *Arch Psychiat Z Neurol* 193: 391 – 405.
Kornhuber HH (1961) Psychologie und Psychiatrie der Kriegsgefangenschaft. pp 631 – 742 in Gruhle HW, Jung R, Mayer-Gross W, Müller M, eds., *Psychiatrie der Gegenwart*, Vol III, Springer Berlin, etc.
Kornhuber HH (1962) *Optisch-vestibuläre und somatisch-vestibuläre Integration an Neuronen der Großhirnrinde.* Habilitationsschrift, Freiburg i. Br.
Kornhuber HH (1972) Tastsinn und Lagesinn. pp 51 – 112 in Gauer OH, Kramer K, Jung R, eds., *Physiologie des Menschen* Vol 11. Urban und Schwarzenberg, München.
Kornhuber HH (1973a) Neural control of input into long-term memory: limbic system and amnestic syndrome in man. pp 1 - 22 in Zimppel HP, ed., *Memory and transfer of information*. Plenum, New York.
Kornhuber HH (1973b) Discussion to Kurtzberg and Vaughan pp 142 - 145 in Zikmund W, ed., *The oculomomotor system and brain functions*. Butterworth, London.
Kornhuber HH (1974a) Cerebral cortex, cerebellum and basal ganglia: an introduction to their motor functions. pp 267 – 280 in Schmitt FO, Worden FG, eds., *The Neurosciences: Third Study Program*. Cambridge, Mass: MIT Press.
Kornhuber HH (1974b) The vestibuar system and the general motor system. pp 581 – 620 in Kornhuber HH, ed., *Handbook of Sensory Physiology* Vol 6 vestibular system part 2, Springer Berlin etc.,.
Kornhuber HH (1977) Hat unsere Gesundheit noch Zukunft? Mensch, Medizin, Gesellschaft 2: 103-108
Kornhuber HH (1978a) Blickmotorik. pp 357 – 426 in Gauer, Kramer, Jung, eds., *Physiologie des Menschen* Vol 13, Urban und Schwarzenberg, München.
Kornhuber HH (1978b) Zur Situation der Familie. *Kinderarzt* 9: 1319 – 1325.
Kornhuber HH (1978c) Hilfe zur Lebensorientierung. Das Studium generale an der Universität Ulm. *Ulmer Forum* 48: 24 - 27.
Kornhuber HH (1978d) Wahrnehmung und Informationsverarbeitung. pp 783 - 798 in Stamm RA, Zeier H, eds., *Die Psychologie des 20. Jahrhunderts*. Vol VI. Kindler, Zürich.
Kornhuber HH (1978e) Geist und Freiheit als biologische Probleme. pp 1122 - 1130 in Stamm RA, Zeier H, eds., *Die Psychologie des 20. Jahrhunderts*. Vol VI. Kindler, Zürich.
Kornhuber HH (1982) Primärprävention des Zigarettenrauchens. pp 126 – 133 in *Rauchen oder Gesundheit – politische, präventive und therapeutische Aspekte*. Neuland-Verlagsgesellschaft Hamburg.
Kornhuber HH (1983) Präventive Neurologie. *Nervenarzt* 54: 57 – 68.
Kornhuber HH (1984a) Attention, readiness for action and the stages of voluntary decision – some electrophysiological correlates in man. *Experimental Brain Research* Suppl 9 (Springer, Berlin etc.,): 420 – 429.
Kornhuber HH (1984b) Von der Freiheit pp 83 -112 in Lindauer M, Schöpf A, eds., *Wie erkennt die Mensch die Welt? Grundlagen des Erkennens, Fühlens und Handelns. Geistes- und Naturwissenschaftler im Dialog*. Ernst Klett, Stuttgart.
Kornhuber HH (1984c) Bluthochdruck und Alkoholkonsum. pp 149 - 162 in Rosenthal J, ed., *Arterielle Hypertonie*. Springer Berlin etc.,
Kornhuber HH (1985) Wie Familienpolitik und der Kampf gegen die Arbeitslosigkeit zusammenhängen. *Die geistige Welt* p 17, 22. 6.
Kornhuber HH (1986a) in Weigelt K, ed., *Die Tagsordnung der Zukunft*. Bouvier/Grundmann, Bonn. pp 107 - 112.
Kornhuber HH (1986b) Bessere Gesundheit statt bloßer Kostendämpfung – Was unserer Gesundheit not tut und wie es bezahlt werden kann. *Öff Ges-Wes* 48: 167-175.
Kornhuber HH (1987) Handlungsentschluß, Aufmerksamkeit und Lernmotivation im Spiegel menschlicher Hirnpotentiale. Mit Bemerkungen zu Wille und Freiheit. pp 376 – 401 in Heckhausen H, Gollwitzer PM, Weinert FE, eds., *Jenseits des Rubikon: Der Wille in den Humanwissenschaften*. Springer, Berlin etc.,.
Kornhuber HH (1988a) Das Risiko Benzodiazepin. *Dtsch Ärzteblatt* 85: A-536.
Kornhuber HH (1988b) Mehr Forschungseffizienz durch objektivere Beurteilung von Forschungsleistungen. pp 361 – 382 in Daniel HD, Fisch R, eds., *Evaluation von Forschung*. Universitätsverlag Konstanz.
Kornhuber HH (1988c) The human brain: from dream and cognition to phantasy, will, conscience and freedom. pp 241 - 258 in Markowitsch HJ, ed., *Information processing by the brain*. Toronto, Lewiston, Bern, Stuttgart. Hans Huber.
Kornhuber HH (1989) Überwachung von Säuglingen durch Heimmonitore. *Dt Ärztebl* 86: A3589.
Kornhuber HH (1990) Vom täglichen Alkohol und der Menschwürde. *Deutsche Richterzeitung*. Feb 1990: 49 – 54.
Kornhuber HH (1992) Gehirn, Wille, Freiheit. *Revue de Metaphysique et de Morale* 2: 203 - 2223..
Kornhuber HH (1993) Prefrontal cortex and Homo sapiens: on creativity and reasoned will. *Neurol Psychiat Brain Res* 2: 1 - 6.
Kornhuber HH (1994) *Von der Forschung und vom Weg dahin*. Universitätsverlag Ulm.
Kornhuber HH (1995) Präfrontalcortex-Funktion und Homo sapiens. In: W Lang, L Deecke, HC Hopf, eds.,: *Topographische Diagnostik des Gehirns. Vh Dt Ges Neurol* 9: 40-48. Springer, Wien New York.
Kornhuber HH (1997a) Alzheimer-Demenz. pp 151–158 in Lauritzen C, ed., *Altersgynäkologie*, Thieme Stuttgart.
Kornhuber HH (1997b) On the promotion of research. *Neurol Psychiat Brain Res* 5: 157 – 162.
Kornhuber HH (1998) Cerebral microangiopathy, anaerobic infection, and the prevention of Alzheimer's disease. *Neurol Psychiat Brain Res* 5: 209 – 212.
Kornhuber HH (2001) *Alkohol – Auch der 'normale' Konsum schadet*. Urban und Vogel. München.
Kornhuber HH, Deecke L (1964) Hirnpotentialänderungen bei Willkürbewegungen, dargestellt mit Magnetbandspeicherung und Rückwärtsanalyse. *Pflügers Arch Eur J Physiol* 281: 52.
Kornhuber HH, Deecke L (1965) Hirnpotentialänderungen bei Willkürbewegungen und passiven Bewegungen des Menschen: Bereitschaftspotential und reafferente Potentiale. *Pflügers Arch* 284: 1-17 'Citation Classic'
Kornhuber HH, Widder B, Christ KJ (1980) The measurement of residual urine by means of ultrasound (sonocystography) in neurogenic bladder disturbances. *Arch Psychiat Nervenkr* 228: 1 - 6.

Kornhuber HH, Riebler R (1984) *Mit der Multiplen Sklerose leben. Die häusliche Behandlung der MS.* Schattauer Stuttgart.

Kornhuber HH, Backhaus B, Kornhuber AW, Kornhuber J (1989) Riskfactors and the prevention of stroke. pp 191 – 212 in Amery WK, Bousser MG, Rose FC, eds., *Clinical trial methodology in stroke.* Bailliere Tindall, London.

Kornhuber HH, Deecke L, Lang W, Lang M (1989) Will, volitional action, attention and cerebral potentials in man: Bereitschaftspotential, performance-related potentials, directed attention potential, EEG spectrum changes. pp 107 - 168 in Hershberger WA, ed., *Volitional action.* Elsevier Science/North Holland.

Kornhuber HH, Deecke L (1990) Readiness for movement – The Bereitschaftspotential-Story, *Current Contents Life Sciences* 33, 4 *Citation Classics* January 22: 14.

Kornhuber HH, Mauch E (1995) Immunsuppressive und symptomatische Behandlung der Multiplen Sklerose. *Klinikarzt* 3/24: 114 - 122.

Kriebel J, Grözinger B, Kornhuber HH (1990) The R-wave. Biography of a brain potential. pp 443 - 447 in Deecke L, Eccles JC, Mountcastle VB, eds., *From Neuron to Action.* Springer, Berlin etc.,.

Kristeva R, Keller E, Deecke L, Kornhuber HH (1979) Cerebral potentials preceding unilateral and bilateral simultaneous finger movements. *Electroenceph Clin Neurophysiol* 47: 229-238

Kristeva R, Deecke L (1980) Cerebral potentials preceding right and left unilateral and bilateral finger movements in sinistrals. In: HH Kornhuber, L Deecke, eds., *Motivation, motor and sensory processes of the brain: Electrical potentials, behaviour and clinical use. Progr Brain Res* Vol 54 Elsevier Amsterdam 748-754

Lang W, Lang M, Kornhuber A, Deecke L, Kornhuber HH (1983) Human cerebral potentials and visuo-motor learning. *Pflügers Arch Eur J Physiol* 399: 342 – 344.

Lang W, Lang M, Heise B, Deecke L, Kornhuber HH (1984) Brain potentials related to voluntary hand tracking, motivation and attention. *Hum Neurobiol* 3: 235 – 240.

Lang W, Lang M, Kornhuber A, Kornhuber HH (1986) Electrophysiological evidence for a right frontal lobe dominance in spatial visuomotor learning. *Arch Ital Biol* 124: 1 - 13.

Lang W, Cheyne D, Kristeva R, Beisteiner R, Lindinger G, Deecke L (1991) Three-dimensional localization of SMA activity preceding voluntary movement. A study of electric and magnetic fields in a patient with infarction of the right supplementary motor area. *Exp Brain Res* 87: 688-695

Lang W, Deecke L (1998) Psychophysiologie der Motorik. Chapter 5 in: F Rösler, ed., Volume 5 *Ergebnisse und Anwendungen der Psychophysiologie. Enzyklopädie der Psychologie,* Serie I *Biologische Psychologie.* Hogrefe Göttingen, pp 225-283

Lang W, Lalouschek W, on behalf of the Vienna Stroke Study Group (2001) The Vienna Stroke Registry – Objectives and Methodology. *Wiener Klin Wochenschr* 113: 141-147

Libet B, Gleason CA, Wright EW, Pearl DK (1983) Time of conscious intention to act in relation to onset of cerebral activity (readiness-potential). Brain 106: 623-642

Libet B (1985) Unconscious cerebral initiative and the role of conscious will in voluntary action. Behav & Brain Sci 8: 529-566

Libet B, Freeman A, Sutherland K, eds (1999) The volitional brain: towards a neuroscience of free will. Imprint Academic, Thorverton, UK

Limousin P Brown RG Jahanshahi M Asselman P Quinn N Thomas D Obeso J Rothwell JC (1999) The effects of posteroventral pallidotomy on the preparation and execution of voluntary hand and arm movements in Parkinson's disease. *Brain,* 122, 315-327

Lindauer M, Schöpf A, eds (1984) *Wie erkennt die Mensch die Welt? Grundlagen des Erkennens, Fühlens und Handelns. Geistes- und Naturwissenschaftler im Dialog.* Ernst Klett Verlag, Stuttgart.

Lindinger G, Svasek P, Lang W, Deecke L (1991) PC-supported 64-channel DC-EEG amplifier. In: Adlassnig KP, Grabner G, Bengtsson S, Hansen R, eds., *Lecture notes in medical informatics* 45. Springer Berlin etc., pp 1005-1009

Lindinger G, Baumgartner C, Burgess R, Lüders H, Deecke L (1994) Topographic analysis of epileptic spikes using spherical splines and spline laplacian (CSD). *J Neurol* 241: 276-277

Marsden CD, Obeso JA (1994) The functions of the basal ganglia and the paradox of stereotaxic surgery in Parkinson's disease. *Brain* 117: 877-897

Marsden CD, Deecke L, Freund H-J, Hallett M, Passingham RE, Shibasaki H, Tanji J, Wiesendanger M (1996) The functions of the supplementary motor area: Summary of a workshop. In HO Lüders, ed., *Advances in Neurology,* Vol. 70: *Supplementary Sensorimotor Area,* pp 477-487

Mackay MD (1984) Source density analysis of scalp potentials during evaluated action I. coronal distribution. *Exp Brain Res* 54: 73-85

Mishkin M, Ungerleider LG, Macko KA (1983) Object vision and spatial vision: the two cortical pathways. *Trends Neurosci* 6: 414-417

Nagamine T, Kkaji R, Suwazono S, Hamano T, Shibasaki H, Kimura J (1992) Current source density mapping of somatosensory evoked responses following median and tibial nerve stimulation. *Electroencephalogr Clin Neurophysiol* 84: 248-256

Noelle-Neumann E (1978) Die Wiederentdeckung der Freiheit. Mit der Demoskopie auf der Spur eines Begriffs. *Frankfurter Allgemeine Zeitung* 48: 10, 1.3.1978.

Playford ED, Jenkins IH, Passingham RE, Nutt J, Frackowiak RSJ, Brooks DJ (1992) Impaired mesial frontal and putamen activation in Parkinson's disease: a positron emission tomography study. *Ann Neurol* 32: 151-161

Penfield W, Rasmussen AI (1950) *Cerebral cortex of man. A clinical study of localization of function.* Mac Millan, NY

Rascol O, Sabatini U, Chollet F, Celis JP, Montastruc JC, Marc-Vergnes JP, Rascol A (1992) Supplementary and primary sensory motor area activity in Parkinson's disease. Regional cerebral blood flow changes during finger movements and effects of apomorphine. *Arch Neurol* 49: 144-148

Rektor I, Kubova D, Bares M (2001) Movement-related potentials in the basal ganglia: an SEEG readiness potential study. *Clin Neurophysiol* 112: 2146-2153

Renner A, Kornhuber HH (1987) An improved home-use monitor to detect apnea associated with sudden infant death. *Verh Dt Ges Neur* 4: 742 - 743.

Roth G (2001) *Fühlen, Denken, Handeln. Wie das Gehirn unser Verhalten steuert.* Suhrkamp, Frankfurt am Main.

Schreiber H, Lang M, Lang W, Kornhuber A, Heise B, Keidel M, Deecke L, Kornhuber HH (1983) Frontal hemispheric differences of the Bereitschaftspotential associated with writing and drawing. *Hum Neurobiol* 2: 197-202

Shallice T (1982) Specific impairments of planning. *Phil Trans Roy Soc Lond* 298: 199 – 209.

Shima K, Aya K, Mushiake H, Inase M, Aizawa H, Tanji J (1991) Two movement-related foci in the primate cingulate cortex observed in signal-triggered and self-paced forelimb movements. *J Neurophysiol* 65: 188-202

Simpson JA, Khuraibet AJ (1986) Readiness potential of cortical area 6 in Parkinson's disease. Evidence for a dopaminergic striatal control of postural set involving supplementary motor area. *J Neurol Neurosurg Psychiat* 49: 475

Simpson JA, Khuraibet AJ (1987) Readiness potential of cortical area 6 preceding self paced movement in Parkinson's disease. *J Neurol Neurosurg Psychiat* 50: 1184-1191

Schwäbische Zeitung (1978) Logopädie-Schule nimmt den Lehrbetrieb auf. 3.5.1978.

Talbot WH, Darian-Smith I, Kornhuber HH, Mountcastle VB (1968) The sense of flutter-vibration: comparison of the human capacity with response patterns of mechanoreceptive afferents from the monkey hand. J Neurophysiol 31: 301-334

Tanji J (1994) The supplementary motor area in the cerebral cortex. *Neurosci Res* 19: 251-268

Uhl F, Goldenberg G, Lang W, Lindinger G, Steiner M, Deecke L (1990) Cerebral correlates of imanining colours, faces and a map - II. Negative cortical DC-potentials. *Neuropsychologia* 28: 81-93

Waldvogel R (1982) Gespräch mit Hans Helmut Kornhuber, Universität Ulm: Plädoyer für das Studium generale. Weitblick kann vor Lockrufen auf Glatteis bewahren. Trotz Anfechtung bleibt der Professor beim Glauben an Wissensdurst der Studenten über das eigene Fach hinaus. *Schwäbische Zeitung* Nr. 141, 19.2.1982.

Walter WG, Cooper R, Aldridge VJ, McCallum WC, Winter AI (1964) Contingent negative variation: An electric sign of sensori-motor association and expectancy in the human brain. *Nature* 203: 380

Widder B, Kornhuber HH, Renner A (1983) Restharnmessung in der ambulanten Versorgung mit einem Klein-Ultraschall-Gerät. *Dt Med Wschr* 108: 1552 - 1555.

INDEX

321